AF830036

Monte Verità Proceedings of the Centro Stefano Franscini
Ascona

Edited by K. Osterwalder, ETH Zürich

New Approaches and Concepts in Turbulence

Edited by
Th. Dracos
A. Tsinober

1993

Birkhäuser Verlag
Basel · Boston · Berlin

Editors:

Themistocles Dracos
Institut für Hydromechanik
ETH Hönggerberg
CH-8093 Zürich
Switzerland

Arkady Tsinober
Tel Aviv University
Tel Aviv 69978
Israel

A CIP catalogue record for this book is available from the Library of Congress, Washington D.C., USA

Deutsche Bibliothek Cataloging-in-Publication Data

New approaches and concepts in turbulence / ed. by Th. Dracos ; A. Tsinober. – Basel ; Boston ; Berlin : Birkhäuser, 1993
(Monte Verità)
ISBN 3-7643-2924-6 (Basel ...)
ISBN 0-8176-2924-6 (Boston)
NE: Dracos, Themistocles [Hrsg.]

This work is subject to copyright. All rights are reserved, whether the whole or part of the material is concerned, specifically the rights of translation, reprinting, re-use of illustrations, recitation, broadcasting, reproduction on microfilms or in other ways, and storage in data banks. For any kind of use permission of the copyright owner must be obtained.

© 1993 Birkhäuser Verlag Basel, P.O. Box 133, CH-4010 Basel, Switzerland
Camera-ready copy prepared by the editors
Printed on acid-free paper produced of chlorine-free pulp
ISBN 3-7643-2924-6
ISBN 0-8176-2924-6

9 8 7 6 5 4 3 2 1

Contents

Session Four. Chairman M. Lesieur

Session Five. Chairman J. R. Herring

Session Six. Chairman J. Jiménez

Group Work

Summary of the Symposium

Foreword

The idea of organising a colloquium on turbulence emerged during the sabbatical leave of Prof. Arkady Tsinober in Zurich. New experimental observations and the insight gained through direct numerical simulations have been stimulating research in turbulence and are leading to the developments of new concepts. The organisers felt the necessity to bring together researchers who have contributed significantly to the advances in this field in a colloquium in which the current achievements and the future development in the theoretical, numerical and experimental approaches would be discussed. The main emphasis of the colloquium was put on discussions. These discussions led to an interesting and exciting exchange of ideas, but also involved its very laborious transcription onto paper. It was due to the personal efforts of Mrs. A. Vyskocil, Dr. N. Malik and Dr. X. Studerus that this work could be completed.

The colloquium was held in the relaxed atmosphere of the Centro Stefano Franscini in Monte Verità near Ascona, a locality of exceptional natural beauty, which was put at our disposal by the Swiss Federal Institute of Technology. We would like to express our gratitude for this generous financial and logistic support, which contributed considerably to the success of the colloquium.

Zurich, April 1993

Th. Dracos, A. Tsinober

Participants

Adrian, R. J.
Antonia, R. A.
Aref, H.
Betchov, R.
Bewersdorff, H.-W.
Castaing, B.
Chen, J.
Dracos, T.
Frisch, U.
Gagne, Y.
Gaster, M.
Gyr, A.
Herring, J.
Hoyer, K.
Hunt, J. C. R.
Hustad, C. W.
Jiménez, J.
Kambe, T.
Kit, E.
Landahl, M. T.
Lesieur, M.
Malik, N.
Moffatt, H. K.
Moin, P.
Mullin, T.
Novikov, E. A.
Pedrizzetti, G.
Reynolds, W. C.
Rys, F.
Shtilman, L.
Virant, M.
Wygnanski, I.
Yaglom, A. M.
Zaslavsky, G. M.

Opening Address

Themistocles Dracos

"It is not meaningful to talk of the properties of a turbulent flow independently of the physical situation in which it arises. In searching for a theory of turbulence, perhaps we are looking for a chimera."

P. G. Saffman

I would like to welcome you to Monte Verità, the Mount of Truth. At the beginning of the century this place attracted a number of painters, sculptors, performing artists, writers, but also psychologists, medical doctors, businessmen and socialites, who were seeking a new way of life. They became anarchists, dadaists, nudists, etc., and broke with all traditions and morals of society hoping to establish a new, utopian society which would make human beings free and happy. They were looking for a chimera. If Saffman is right, there is no better place to talk about turbulence!

During the first fifty years after Reynolds' formulation of the mathematical framework describing turbulent flow the emphasis of research was focused mainly on the linear stability theory and on the development of a statistical theory of turbulence. The latter is associated with the names of G. I. Taylor, G. K. Batchelor, Th. von Karman, and culminates in the work of L. Onsager, C. F. von Weizsäcker and A. M. Kolmogorov. Criticism did not wait long to be formulated. In 1949 J. v. Neumann wrote: "All these efforts may be criticized on a conceptual level... That is, there is considerable uncertainty regarding the nature of the averages that are involved." And adds: "Another aspect of these investigations, which is in conceptually unsatisfactory state, is the nature of the 'ergodicity' of 'disorder' assumptions which are inherent in all of them." His conclusion is: "The entire experience with the subject indicates that the purely analytical approach is beset with difficulties, which at the moment are still prohibitive. The reason for this is probably that our intuitive relationship to the subject is still too loose - not having succeeded at anything like deep mathematical penetration in any part of the subject - we are still quite disoriented as to the relevant factors, and as to the proper analytical machinery to be used." And then he adds: "Under these conditions there might be some hope to 'break the deadlock' by extensive, but well planned computational efforts."

I would like to extend this statement to include "the well planned *experimental* efforts". It seems to me that a hundred and two years after Reynolds published the equations bearing his name and fifty years after Kolmogorov published his work in which he introduced the concept of local isotropy it is time to ask ourselves if our intuitive relationship to turbulence has improved and if we are approaching an orientation as to the relevant factors and the proper analytical machinery to be used.

This could have been the question open for discussion in this colloquium. In view of the developments of the last decades this question must be posed in more detail.

In the same year in which v. Neumann wrote his comments on turbulence, Batchelor and Townsend found in their measurements that the small-scale movements in grid turbulence and in the turbulence wake of a cylinder were intermittent, thus questioning Kolmogorov's theory, which, as well as all other statistical theories, is based on the assumption that the random fields are continuous. Intermittency implies that vorticity and hence dissipation are mainly concentrated in spots.

It is astonishing that although Prandtl and Tietjens published as early as 1934, their nice pictures showing large flow patterns in channel flow, this observation was forgotten until Brown and Rosko and Kline and his coworkers performed their well-known experiments at the end of the sixties and the beginning of the seventies. This opened up a new area of research. The fact that in three-dimensional turbulent flows vortex filaments are convoluted and stretched and that vorticity in a given fluid volume could increase was well-known, but the appreciation of the existence of coherent "deterministic" structures in "fully developed" turbulence, which were vortical in essence, opened new perspectives. A big amount of mainly experimental work was done to elucidate the importance of coherent structures. The problem is that point measurements supply only limited information on such structures. Thus, the information on the structures depends mostly on the detection scheme used. Nevertheless, these observations strongly influenced the intuitive picture of turbulence and initiated new approaches to this problem. Research on vortex dynamics was enhanced and is considered one of the most promising ways to a better understanding of turbulence. Topological studies of the flow field were initiated and are still developing. The experimental work, however, was limited to two-dimensional observations while the dynamics of turbulent flows are closely related to its three-dimensional character.

With the development of fast computers numerical computations of turbulent flows, at least for small Reynolds numbers and simple geometries, became possible. So a new field of experimentation was opened. The missing three-dimensional information was now available, at least for the simple flows mentioned above. The information gathered by such simulations improved our picture of the structure of turbulent flows, but also provided new results, which pose new problems. One of these results, for example, which is also confirmed by measurements, is that the vorticity vector and also material lines tend to align with the direction of the intermediate principal eigenvector of strain and not with the one of maximum positive eigenvalue as expected. This observation demands an explanation. The degree of linearisation, that is to say, Beltramisation ($\omega = a\mathbf{u}$) and Eulerisation ($\nabla^2(p/\rho' + (1/2)\, u^2) = \nabla(\mathbf{u} \times \omega)$ has also been determined, which brings us to the advances made in the theoretical field. Here the most important and consequently new but also controversial approaches are the ones based on the theory of dynamical systems and the related theory of deterministic chaos. Sceptics recognize the important contribution of the latter from the point of view of the theory of cognition, but doubt that a theory of fully developed turbulence based on this concept can be developed in view of the huge number of degrees of freedom involved. However, this approach is relevant for the onset mechanism of turbulence, and reduction of the degrees of freedom in developed turbulence to the relevant ones may enable its application on such flows. Of course we are only at the beginning of a new area of research in turbulence and it is inevitable that doubts and controversies will arise and that as a result different ways of applying these theories will be found.

Some of the controversies, especially concerning fully developed turbulence, i.e. high Reynolds number flows, could be resolved if reliable experiments, either numerical or physical, could be performed under appropriate conditions. The limits imposed on numerical simulations in respect to the Reynolds number have already been mentioned. Hot-wire and Laser Doppler Anemometry, on the other hand, are well developed, allow measurements at reasonably high Reynolds numbers and have a large dynamic range. But these are point measurements and thus give limited information on the spacial structure of the flow. Also, with increasing Reynolds number the physical dimensions of the interrogation volume become large compared to the small scales to be resolved. Results concerning these scales are therefore often not reliable. In the last years new methods have been developed or are being developed. Operational at the time are Particle Image Velocimetry, which allows the measurements of instantaneous two-dimensional velocity fields with a spacial resolution comparable to the traditional methods mentioned above, and Particle Tracking Velocimetry in two- and three-dimensions. The latter allows for spatio-temporal observations of three-dimensional velocity fields in a Lagrangian frame of reference. The resolution of these methods depends on the size of the observation volume which limits their dynamic range. Other possibilities are observations of scalar concentrations at high Schmidt numbers with resolutions down to the Kolmogorov scale. In fact, this technique can be further developed to a kind of flow tomography giving insight into the three-dimensional structure of the scalar field, and into the development of material surfaces. When such observations are combined with methods which allow the determination of the flow field from the concentration field, like template matching, the information gained can be very illuminating for the turbulent processes taking place at small scales. This enumeration of methods is by far not complete, but it shows the potential for future experimental investigation, which, when well planned, can help to verify and improve theoretical concepts. This has led us to formulate the following topics to serve as a broad framework for this colloquium.

I **Prospects of physical (laboratory and field) and numerical experiments for the fundamental understanding of turbulence**

II **Future implications of chaos theory, dynamical systems and prospects of low-dimensional descriptions of turbulent flows**

III **Reduction of non-linearity in turbulent flows - further lines of research and their use in the development of the theory on turbulence**

IV **Prospects of statistical approaches to the turbulence problem - how statistical methods, e.g. the theory of random fields, should be used in view of recent developments regarding the structure of turbulence, dynamical chaos, strange attractors, fractals**

Still due to complexity the topics will remain controversial, at least in the near future. Paraphrasing Whithead I would like to say that controversies are an opportunity, not a disaster, and I hope that this will be the spirit of this colloquium.

Swiss Federal Institute of Technology, Zurich

Session One

Chairman T. Dracos

Similarity Laws for Wall Turbulent Flows: Their Limitations and Generalizations

Akiva M. Yaglom

Abstract. Similarity laws for simple turbulent wall flows are usually justified by dimensional arguments. However, the dimensional analysis is always based on idealization of the problem, which allows the number of relevant physical parameters to be severely restricted. The idealization, in turn, is based on some model for the mechanism of wall tubulence. The classical model developed by L. Prandtl, T. von Karman and G. I. Taylor in the twenties and thirties implies in particular the so-called Reynolds number similarity principles (i.e., independence of turbulent regime far from solid walls on the molecular viscosity) and the vanishing of the effect of the external length scale in the proximity of the wall. However, recent discoveries related to an important part played by the large-scale organized structures everywhere within turbulent flows make one doubt about the correctness of the formulated statements. Some suggestions and experimental data which contradict the classical similarity laws are considered. The difficulties related to the applicability of simple similarity laws for turbulence in a thermally stratified medium are noted.

1. Introduction

It is known that the nonlinearity of fluid mechanics equations makes all the dynamic equations of turbulent flows nonclosed. Therefore, these equations cannot be solved without using some speculative closure hypotheses, the selection of which always involves a great degree of arbitrariness. That is why the methods allowing to diminish the arbitrariness and to obtain some general results without any use of dynamic equations are of great importance for the theory of turbulence. The similarity and dimensional arguments are the most important methods based on general physical considerations rather than dynamic equations. These methods are widely used in all the parts of physics (see [1] to [3]) and in the mechanics of turbulence they are used specially often.

The dimensional analysis often permits one to obtain the similarity laws, indicating that the solution of the problem may depend only on some combination of variables and relevant physical parameters raised to integer powers. Moreover, it has been recently discovered by Barenblatt and Zeldovich [4], [5], that more complicated similarity solutions appear in many mechanical problems; these solutions include some arbitrary (usually non-integer) power of the main variable (most often time) divided by the scale formed by the dimensional parameters of the problem. The unknown exponent can be found in some cases from the solution of a supplementary eigenvalue problem, but most often it is determined from experimental data. These more complicated solutions are called the self-similar solutions of the second type by Barenblatt and Zeldovich; they often generalize less specific self-similar solutions of the first type implied by the dimensional analysis.

A typical example of the self-similar solution of the second type is provided by the refined Kolmogorov theory of the local structure of turbulence which takes into account fluctuations in the dissipation rate $\varepsilon(\mathbf{x},t)$. According to the original Kolmogorov theory of local structure developed by Kolmogorov and Obukhov in 1941 (see, e.g. [6], chap. 8), at large Reynolds numbers the statistical characteristics of turbulent fluctuations of the scale l, where $L \gg l \gg \eta$, (here L is the external length scale of a flow, $\eta = (\nu^3/\overline{\varepsilon})^{1/4}$ is the Kolmogorov internal scale, ν is the kinematic viscosity and the overbar symbolizes averaging) depend only on $\overline{\varepsilon}$ and ν. Therefore

$$D_n(r) = \overline{[\,u_L(\mathbf{x}+\mathbf{r}) - u_L(\mathbf{x})\,]^n} = C_n\,(\overline{\varepsilon} r)^{n/3} \qquad \text{for} \quad L \gg r \gg \eta \tag{1}$$

where $r = |\mathbf{r}|$, u_L is the velocity component in the direction of vector $\mathbf{r}$ and C_n is a universal constant. However, the dissipation rate $\varepsilon = \varepsilon(\mathbf{x},t)$ fluctuates disorderly when $Re = UL/\nu \gg 1$ (where U is the typical velocity of a flow). These fluctuations are produced by the small-scale coherent structures of a flow (which depend upon Re) and they affect the structure functions $D_n(r)$. According to the refined theory developed by Kolmogorov and Obukhov in 1962 (see [6], sect. 25), when the fluctuations in ε are taken into account, the following modification of eq. (1) can be derived:

$$D_n(r) = C_n(\overline{\varepsilon} r)^{n/3} \left(\frac{r}{L}\right)^{\zeta_n} \qquad \text{for} \quad L \gg r \gg \eta \tag{2}$$

where C_n can now be weakly dependent on Re and ζ_n is a new universal constant. Eq. (1) gives an example of the similarity law of the first type and the refined eq. (2) is the similarity law of the second type (cf. Barenblatt [5], sect. 11.5).

Generally, eq. (2) agrees well with the most recent experimental data; however, there are a number of challenging problems related to these equations which have not been solved yet. Nevertheless, these poblems will not be discussed below since the similarity laws for wall turbulent flows are the only topic treated in this paper. Many examples of such similarity laws are considered in detail in the book [6] and survey papers [7] to [9]. However, no attention is given there to the often observed deviations of data from the simple results implied by dimensional analysis, which is always based on idealization of the problem permitting the number of relevant physical parameters to be severely restricted. Below, attention will be focused on the difficulties related to the application of simple dimensional arguments and to possible generalizations of the known similarity laws allowing to describe experimental data more accurately.

2. Similarity Laws for Pipe, Channel, and Boundary-Layer Turbulent Flows

This section is devoted to steady parallel (or, at least, almost parallel) turbulent flows in a circular pipe, broad plane channel, and boundary layers over a flat plate. In the case of a boundary-layer flow, the longitudinal pressure gradient dP/dx will be considered to be negligibly small, and it will be assumed that the non-parallelism of a flow can be taken into account by the introduction of the dependence on x of the wall stress τ_w and the boundary-layer thickness L. (The x-axis will always be directed along the mean flow, and the z-axis will be perpendicular to the wall). Besides the three types of flow mentioned above, the atmospheric surface layer at neutral thermal stratification will also be considered, since it is similar to the laboratory boundary layer over a flat plate.

If the wall is smooth, then the flow at a given x is fully characterized by the fluid density ρ, the kinematic viscosity ν, the vertical length scale L (channel half-width, pipe radius or boundary-layer thickness) and the typical velocity scale U, e.g. the maximum mean velocity on a pipe or channel axis, or the boundary-layer free-stream velocity. Instead of U, the friction velocity $u_*=(\tau_w/\rho)^{1/2}$ may also be used. In the case of an atmospheric surface layer, the definition of L and U is not a simple task, but u_* can easily be measured here by a modern sonic or hot-wire anemometer.

When the scalar field T mixed by a turbulent flow (e.g. temperature in the case of forced convection in a flow with heat transfer from or to the wall) is also of interest, the following parameters must be added to the list: the molecular diffusivity χ of the scalar field T, the wall vertical flux Q of T (or the corresponding scalar scale $T_*=Q/u_*$), and the mean value of the scalar at the wall: T(0). In the case of a rough wall, the mean height h of the roughness elements and the dimensionless parameters σ_1, σ_2,..., which characterize the shapes of roughness elements and their distribution on the wall must also be added to the list. However, it is natural to assume (and this assumption agrees well with the data) that parameters h, σ_1, σ_2,... affect the flow only at heights z of the order of h or less. Therefore, these parameters will not be taken into account in this paper, i.e. the wall will be considered to be smooth, although all the results below can be applied to rough walls too, and many used data and the references in the reference list relate to rough wall turbulent flows. The scalar field T will be called below temperature, though the concentration of any passive admixture can be taken as T.

2.1 Mean Velocity and Temperature Profiles. It is easy to see that dimensional analysis implies the following general similarity law for the mean velocity profile $\overline{u}(z)$:

$$\overline{u}(z) = u_* f(\frac{zu_*}{\nu}, \frac{z}{L}) \tag{3}$$

where f is the function of two variables which can be different for pipe, channel, and boundary-layer flows. The similar law for the mean temperature profile $\overline{T}(z)$ has the form:

$$\overline{T}(z)-\overline{T}(0) = T_* f_T(\frac{zu_*}{\nu}, \frac{z}{L}, Pr) \tag{3a}$$

where $Pr=\nu/\chi$ and f_T is the function of three variables which can depend on the type of flow. However, it is difficult to deal with the function of several variables when treating the experimental data; therefore, the general laws (3) and (3a) are not very useful in practice.

It is usually assumed that the external length scale L does not affect the flow near the wall where $z << L$. If so, then the following *wall laws* (due to L. Prandtl) must be valid for $z << L$:

$$\overline{u}(z) = u_* f^{(1)}(\frac{u_* z}{\nu}), \qquad \overline{T}(z)-\overline{T}(0) = T_* f_T^{(1)}(\frac{u_* z}{\nu}, Pr) \tag{4}$$

where $f^{(1)}$ and $f_T^{(1)}$ are two universal functions. Moreover, it is also natural to assume that in the region $z >> (\nu/u_*, \chi/u_*)$, where the turbulent viscosity $K = \kappa u_* z$ ($\kappa \simeq 0.4$ is the Karman constant) is much greater than the molecular viscosity ν, and the turbulent temperature diffusity K_T is much greater than the molecular temperature diffusity χ (for simplicity, it will be always assumed that Pr is of the order of unity and therefore, χ is of the same order as ν and K_T is of the same order as K), the following *defect laws* (due to T. von Karman) are valid:

$$\overline{u}(L)-\overline{u}(z) = u_* f^{(2)}(\frac{z}{L}), \qquad \overline{T}(L)-\overline{T}(z) = T_* f_T^{(2)}(\frac{z}{L}) \tag{5}$$

The functions $f^{(2)}$ and $f_T^{(2)}$ are not universal since they can depend on the type of a flow.

If an overlap layer exists, where both the wall and defect laws (4) and (5) are simultaneously valid, then it is easy to show that in this layer the mean profile of both velocity and temperature are given by logarithmic equations:

$$\overline{u}(z) = u_*[A \ln\frac{zu_*}{\nu} + B] \;, \qquad \overline{T}(z)-\overline{T}(0) = T_*[A_T \ln\frac{zu_*}{\nu} + B_T(Pr)] \qquad (6)$$

$$\overline{u}(L)-\overline{u}(z) = u_*[-A \ln\frac{z}{L} + B^{(1)}] \;, \qquad \overline{T}(L)-\overline{T}(z) = T_*[-A_T \ln\frac{z}{L} + B_T^{(1)}] \qquad (6a)$$

where $A = 1/\kappa$, B, and $A_T = 1/\kappa_1$ are universal constants, $B_T(Pr)$ is a universal function, and $B^{(1)}$ and $B_T^{(1)}$ can depend on the type of a flow. (The overlap layers for the velocity and temperature and also the thickness of the velocity and temperature boundary layers on a flat plate must not be exactly the same, but they apparently differ only slightly and can be assumed to be indistinguishable in the first approximation). The derivation of logarithmic laws (6) from the overlap arguments was proposed by Izakson in 1937, was further developed by Millikan in 1939 and von Mises in 1941, and was included in many textbooks and monographs (see, e.g. [6], sects. 5.5 - 5.7). The ordinary defect laws (5) cannot be applied to an atmospheric surface layer since it has no clear upper edge. (However, the defect laws and the overlap arguments can be applied to a much thicker atmospheric boundary layer whose upper edge is determined by the kink in the mean temperature profile; see e.g. [6], p. 411). Nevertheless, the logarithmic equations (6) can also be derived for the atmospheric boundary layer from the assumption that the molecular coefficients ν and χ do not affect the velocity and temperature gradients if z is much larger than ν/u_* (and the mean roughness height h).

Eqs. (4) to (6a) have been repeatedly compared with the data related to all the considered types of turbulent wall flows, and the agreement has proved to be fully satisfactory. The experimental values of the coefficients in (6) and (6a) are only moderately scattered, and the values $A_{(1)}= 2.5$ (i.e. $\kappa = 1/A = 0.4$), $B = 5.0$, $B^{(1)}= 0.6$ for the pipe and channel flows, and $B^{(1)}= 2.4$ for boundary-layer flows, $A_T = 2.12$ (i.e. $\kappa_1 = 0.47$), $B_T(0.7) = 3.8$, $B_T^{(1)} = 0.3$ for the pipe and channel flows, and $B_T^{(1)}= 1.5$ for the boundary-layer flows are recommended in [8] as the mean experimental estimates. (The data for the values of $B_T(Pr)$ for $Pr = 0.7$ and B and B_T for rough-wall flows can be found in [8], [9]). However, the experimental data do not permit the upper and lower edges of a logarithmic layer (where eqs. (6) and (6a) are valid) to be estimated accurately enough. Moreover, it is even impossible to state quite definitely now that the experimental mean velocity and temperature profiles cannot be described more accurately by the general equations (3) and (3a) than by more specific equations (4) to (6a). Note that the validity of the wall laws (4) is equivalent to the existence of the finite limits $f(x,0) = f^{(1)}(x)$ and $f_T(x,0,Pr) = f_T^{(1)}(x,Pr)$ of the functions $f(x,y)$ and $f_T(x,y,Pr)$ for $y \to 0$. Similarly, the validity of the defect laws (5) is equivalent to the existence of the finite limits $f(\infty,y_1) - f(\infty,y_2) = f^{(2)}(y_1) - f^{(2)}(y_2)$ and $f_T(\infty,y,Pr) - f_T(\infty,y_2,Pr) = f_T^{(2)}(y_1,Pr) - f_T(y_2,Pr) - f^{(2)}(y_2,Pr)$ of the functions $f(x,y_1) - f(x,y_2)$ and $f_T(x,y_1,Pr) - f_T(x,y_2,Pr)$ when $x \to \infty$ and y_1 and y_2 are fixed. The validity of the assumption for f(x,y) has recently been questioned by Barenblatt [5], [10], [11] (see also [12]). This author pays attention to the fact (noted in [13], [14]) that the velocity profiles for the turbulent smooth-wall pipe flows measured by Nikuradse [15] at different Re = UL/ν can be described quite accurately by power laws (with the exponent depending on Re). Basing on this fact, Barenblatt assumes that the general similarity law (3) can be, in fact, reduced to the self-similarity law of the second type having the form

$$\overline{u}(z) = u_* C \left[\frac{u_* z}{\nu}\right]^{\alpha} \tag{7}$$

where C and α depend on the Reynolds number $Re_1 = 2LU_m/\nu$; U_m being the bulk velocity (Re_1 is the single-valued function of $Re = LU/\nu$, $U = \overline{u}(L)$, and of the modified Reynolds number $Re_* = Lu_*/\nu = (z/L)^{-1}(zu_*/\nu)$). By treating all the data collected by Nikuradse, Barenblatt [11] discovered that eq. (7) with $\alpha = 3/[2 \ln Re_1]$ and $C = [\sqrt{3} + 5\alpha]/2\alpha$ agrees better with Nikuradse's data than all the previously suggested equations for the pipe-flow velocity profiles (see Fig. 1). (Equation $\alpha = 3/2 \ln Re_1$ is related to the equation given by Castaing et al. [16] for the exponents ζ_n in (2); Barenblatt gives special attention to this fact.)

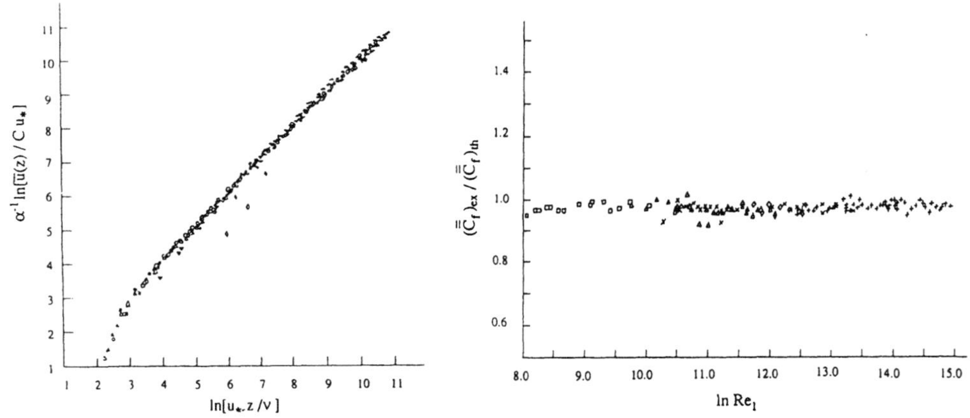

Fig. 1 Comparison of the corollaries from Barenblatt's eq.(7) with Nikuradse's data.
(a) Verification of the equation $\alpha^{-1} \ln[\overline{u}(z)/C u_*] = \ln[u_* z/\nu]$, $\alpha = 3/[2 \ln Re_1]$ and $C = [\sqrt{3} + 5\alpha]/2\alpha$. Different marks correspond to different values of Re_1.
(b) Verification of the relation $(c_f)_{ex}/(c_f)_{th} = 1$, where $(c_f)_{ex}$ and $(c_f)_{th}$ are the values measured and calculated from eq.(7) of the friction coefficient. Different marks correspond to the different values of the pipe radius.

The assumption leading to the wall laws (4) and the defect laws (5) seemed physically quite natural for a long time. Good agreement of the new equation (7) with Nikuradse's data is, of course, not enough to reject the old classical laws. However, I would like to stress that the recent development of the mechanics of turbulence related to the study of organized coherent structures makes the derivations of the wall and defect laws less convincing now than they were earlier. A number of reasons can be given at present to make these laws doubtful.

a) It has recently been discovered (e.g. [17], [18]) that many thin vortex filaments are always involved in a developed turbulent flow; the number of filaments and their topology depend strongly on Re. Therefore, it is natural to expect that the molecular viscosity may affect the flow within the filaments and on their boundaries at all Reynolds numbers; and $Re = UL/\nu$ is also important since it affects the number and the topology of vortical structures at any distance from the solid walls. This new effect clearly contradicts the Reynolds number similarity.

b) It has also been discovered that the hierarchy of attached horseshoe eddies, whose sizes are extended up to the external length scale, L, exist in the wall shear layers (see [19] to [21]). Therefore, the length L can affect the flow quite near the wall. Some data on the variances of velocity components in the wall layers, which will be discussed below, indicate the dependence of the variances in the constant-stress layer on both ν and L attributed to the attached-eddy effect. This makes the assumption of the dependence of $\overline{u}(z)$ in the same layer on ν and L less unprobable than it had seemed earlier.

c) Townsend [22] noted long ago that intensities of velocity fluctuations in constant-stress layers of developed turbulent flows differ sometimes between differnt flows of the same stress (i.e. the same u_*). To explain this he assumed that the turbulent motion in the wall region of a turbulent flow consists of (i) an "active" component, which produces the shear stress τ and satisfies the usual wall laws, and (ii) an "inactive" and practically irrotational component, which is generated by the turbulence in the outer region of a flow and depends on L and Re. Later, Bradshaw [23] showed that Townsend's hypothesis explains a number of experimental results related to turbulent spectra and fluctuation moments in the inner part of both a constant-pressure and an adverse-pressure-gradient boundary layer. Some features of the turbulent structure observed in the atmospheric surface layers and numerically simulated boundary-layer flows have also recently been attributed to "inactive" motions (see [24] to [26]).

Townsend [22] stressed that he did not know about any deviations of velocity profiles in the constant-stress region of boundary layers from the ordinary similarity laws. Then Bradshaw [23] constructed a model of the "inactive" motion which contributes significantly to the second-order moments of velocity fluctuations, but affects the mean velocity profile very little. However, in principle, the "inactive" motion can affect the mean motion and make it depend on L and Re at any distance from the wall.

The reasons indicated can explain the validity of Barenblatt's equation (7) but, of course, they do not prove it. Therefore, new highly accurate measurements of the mean velocity profiles in wall turbulent flows at moderate and especially high Re are urgently required. If new experimental data confirm the dependence of u(z) on L and ν at all distances z, then it will be necessary to study in detail the physical effects producing this dependence.

The data on the mean temperature profiles in pipe, channel and boundary-layer flows are at present much poorer than those on the velocity profiles. Therefore, the question of the accuracy of the temperature profile wall and defect laws in comparison to that of the general similarity law (3a) cannot be settled without a number of new accurate measurements.

2.2 Velocity and Temperature Variances and Spectra. Let us now consider the variances $\sigma_i^2 = \overline{u_i'^2}$, i = 1,2,3, and $\sigma_T^2 = \overline{T'^2}$ of three velocity components $u_1 = u$, $u_2 = v$, and $u_3 = w$, where prime denotes fluctuations (i.e. $u' = u - \overline{u}$). The general similarity laws for these variances have the form

$$\sigma_i = u_* \, g_i\left(\frac{u_* z}{\nu}, \frac{z}{L}\right), \qquad \sigma_T = T_* g_T\left(\frac{u_y z}{\nu}, \frac{z}{L}, Pr\right) \tag{8}$$

where g_i and g_T are the fluctions which can be different for pipe, channel, and boundary-layer flows. The laws (8) are quite similar to (3) and (3a). Assuming that the length L is not important for z << L, and the viscosity ν and the temperature diffusivity χ are not important for $z >> \nu/u_*$, they lead to the simpler wall and outer-flow laws

$$\sigma_i = u_* \, g_i^{(1)}\left(\frac{u_* z}{\nu}\right), \nu \qquad \sigma_T = T_* g_T^{(1)}\left(\frac{u_* z}{\nu}, Pr\right) \qquad \text{for } z << L \tag{9}$$

$$\sigma_i = u_* \, g_i^{(2)}\left(\frac{z}{L}\right), \qquad \sigma_T = T_* g_T^{(2)}\left(\frac{z}{L}\right) \qquad \text{for } z >> \nu/u_* \tag{10}$$

where the functions $g^{(1)}$ are universal and $g^{(2)}$ may depend on the type of a flow. It is natural to assume that the overlap layer, where the laws (9) and (10) are valid simultaneously, is close to the logarithmic layer where the laws (6) are valid; in this overlap layer eqs. (9) and (10) imply the equations

$$\sigma_i = a_i u_*, \qquad \sigma_T = a_T T_* \qquad \text{for } L >> z >> \nu/u_* \tag{11}$$

where a_i and a_T are universal constants.

Eqs. (11) have been repeatedly compared with experimental data. The laboratory velocity-fluctuation data are rather scattered, but it is usually believed that the data at high Re agree more or less satisfactorily with (11) for $a_1 \simeq 2.3$, $a_2 \simeq 1.7$, and $a_3 \simeq 1$ (see e.g. [6] to [8]). However, the assumption of the constancy of σ_1/u_* and σ_2/u_* in the layer where the velocity profile is given by (6), does not quite agree with the data, and the agreement for σ_1/u_* (cf., e.g., Fig. 26 in [6]) is especially unsatisfactory.

Townsend [22] and Bradshaw [23], [24] indicated that the "inactive" motion can produce some dependence of σ_1/u_* and σ_2/u_* (but not σ_3/u_*) on z/L in the inner part of wall flows. Högström [25] mentioned some laboratory data indicating an increase of σ_3/u_* with height in the inner layer. He attributed this to the "inactive turbulence"; this assumption obviously disagrees with Bradshaw's point of view. Smedman [27] remarked that the data referred to by Högström [25] for proving that σ_3/u_* increases with z show also that σ_1/u_* decreases with height in the wall boundary layer. Moreover, she noted that the decrease of the anisotropy index $\sigma_1/\sigma_3 = \sigma_u/\sigma_w$ with height can be explained by the diminishing influence of the wall on the pressure fluctuations since it is known that in the absence of walls the pressure fluctuations are equalizing the variances of three velocity components. (Such an explanation must of course be supplemented by the indication of the relevant physical processes making the simple similarity laws (11) invalid in the wall layer.) Quite another explanation of the decrease of σ_1/u_* and σ_2/u_* with z, which will be discussed below in this section, was proposed by Perry et al ([21], [28] to [30]).

Atmospheric measurements of the ratios σ_i/u_*, i = 1, 2, 3, usually give much more scattered results than those from the laboratory measurements. Most of the authors discussing the surface layer atmospheric turbulence came to the conclusion that the available data in near-neutral conditions in general do not contradict the assumption of the independence of the three ratios σ_i/u_*, i = 1, 2, 3, on z, and imply the mean estimates of these ratios, which do not strongly differ from the corresponding laboratory estimates (see, e.g. the books [6] and [31]). However, recently Högström [25] and Smedman [27] stated that their (and also some other) accurate measurements of wind velocity fluctuations definitely show an increase of σ_3/u_* and decrease of σ_1/u_* with height in a neutral surface layer. Högström, who considered only the measurements of the vertical velocity component, found that the values of σ_3/u_* can be represented with a reasonable accuracy by a linear function of $\ln(zf/u_*)$, where $f = 2\omega_z$ is the Coriolis parameter. Later Smedman showed that her data from five different sufficiently homogeneous sites in the Southern part of Sweden agree satisfactorily with Högström's equation for σ_3/u_* and imply a similar linear equation for σ_1/u_* (see Fig. 2). Note that u_*/f is proportional to the thickness H of the planetary boundary layer in neutral conditions; therefore, according to [25] and [27] the ratios σ_i/u_* depend on the dimensionless height z/H.

Högström attributed the increase of σ_3/u_* to the "inactive turbulence", and Smedman proposed another rough explanation of her results, which was mentioned above in this section. However, both explanations do not seem to be fully satisfactory and the interesting results shown in Fig. 2 clearly require further verification. Note in this respect that atmospheric measurements of the variances $\sigma^2{}_1 = \overline{u'^2}$ and $\sigma^2{}_2 = \overline{v'^2}$ are very often not reliable enough since the spectra of horizontal velocity components extend up to the very low frequencies. Therefore, it is difficult to determine the lower limit of the frequency range that must be taken into account. Moreover, the averaging time must be very long for attaining the adequate accuracy.

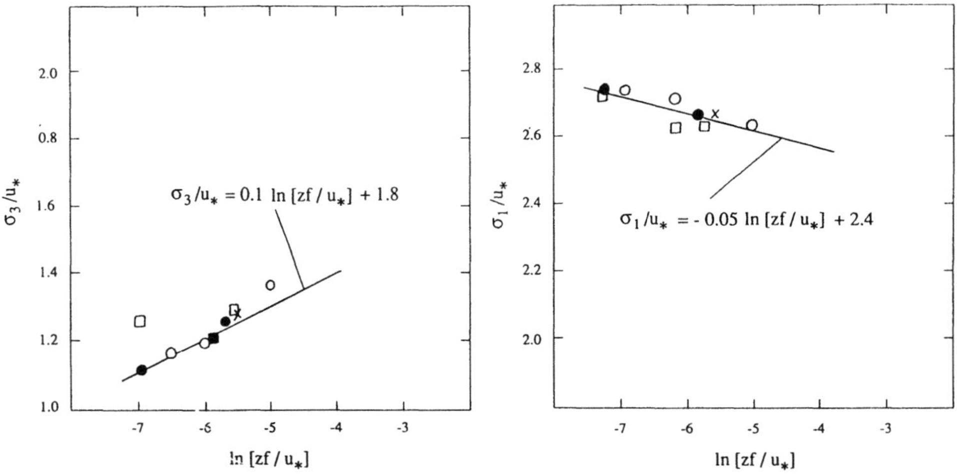

Fig. 2 Dependence of the normalized standard deviation of the (a) vertical and (b) longitudinal velocity components on the dimensionless height zf/u_* according to Smedman [27]. The height is measured from the zero-plane-displacement level. Different marks correspond to the differnt sites.

Note also that the atmospheric measurement of the vertical velocity variance implies the estimate $\sigma_3/u_* \simeq 1.3$, which is slightly higher than the corresponding laboratory estimate. Much larger discrepancy between the laboratory and atmospheric data characterizes the measurements of $\sigma_T^2 = \overline{T'^2}$. The estimate of the value of σ_T/T_* within the logarithmic layer, implied by the laboratory measurements in turbulent flows with heat transfer under the condition of forced convection, has the form $\sigma_T/T_* \simeq 1.3$ (see [7], [8], [32]). The measurement of the value of σ_T/T_* in the neutrally stratified atmospheric surface layer is a difficult task (this value must be determined as the limit of the values of σ_T/T_* for a thermally stratified surface layer as the stratification tends to disappear), but all such measurements lead to the value of σ_T/T_* in the interval between 2.5 and 3 (see [7], [8], [32]). A similar disagreement is found between the laboratory and atmospheric measurement of $\overline{u'T'}/Q = \overline{u'T'}/\overline{w'T'}$: this ratio is close to -2.5 in the laboratory flows, but its value lies between -3.5 and -4.0 in the neutral atmospheric surface layer. The differences in the values of σ_T/T_* and $\overline{u'T'}/Q$ in the laboratory flows and atmospheric surface layers are explained in [8],[32] by the influence of the thermal inhomogeneities that always exist on the Earth ground while in the laboratory measurements the wall is usually thermally homogeneous. Thermal inhomogeneities (and also the fluctuations of the ground temperature with time) produce local convective motions that increase fluctuations of the air temperature above the ground. This explanation seems plausible, but of course it requires a further experimental verification. It is clear, however, that the temperature fluctuations in the wall flows with heat transfer must be affected by some supplementary parameters (besides Q and u_*) which take different values in the laboratory and geophysical flows.

The similarity laws (11) are closely related to those for the longitudinal (in the x - direction) spatial spectra $E_{ii}(k)$, i = 1, 2, 3, and $E_{TT}(k)$ of u'_i and T', since

$$\overline{u'^2_i} = \int_0^\infty E_{ii}(k)dk , \qquad \overline{T'^2} = \int_0^\infty E_{TT}(k)dk \qquad (12)$$

Eqs. (11) are based on the assumption that turbulence characteristics depend only on u_* and T_* if $L >> z >> \nu/u_*$. This assumption also implies that

$$E_{ii}(k) = u_*^2 z\, e_{ii}(kz) , \qquad E_{TT}(k) = T_*^2 z\, e_{TT}(kz) \qquad (13)$$

Assuming that eqs. (13) are valid for all wave numbers k, one can easily obtain (11). It is clear, however, that the similarity laws (13) cannot be valid for any value of k. Very large values of k correspond to small-scale fluctuations responsible for dissipations of $\overline{u'^2}$ and $\overline{T'^2}$; hence, the spectra $E_{ii}(k)$ and $E_{TT}(k)$ at such k must depend on the molecular coefficients ν and χ. It was shown by Kolmogorov that ν is important when k is of the order of the Kolmogorov wave number $k_\eta = \eta^{-1} = (\overline{\varepsilon}/\nu^3)^{1/4}$; if $Pr \simeq 1$, then χ affects the temperature spectrum in the same wave-number range. It is natural to assume that the contributions of the dissipation ranges to the integrals in eqs. (12) is negligibly small for large Re, but this assumption needs a special verification.

If k is very small, then the corresponding wave length $\lambda = 2\pi/k$ can be of the order of L or even greater and such large disturbances must be affected by the value of L. Hence, the spectra at small k must depend on L (and also on the geometry of the flow). Also note that the statistical regime of large disturbances, whose length scales are considerably larger than the distance from the wall z, must apparantly be independent of z. If so, then the spectra $E_{ii}(k)$ and $E_{TT}(k)$ (and also the cospectra $E_{13}(k)$, $E_{1T}(k)$, and $E_{3T}(k)$) do not depend on z if kz is small enough. Perry et al. ([21], [28] to [30]) suggested that three regions of different scaling exist in the spectra of u and v (but not w) fluctuations: region 1 of the "outer-flow scaling", where the spectra can be represented as $u_*^2 L e_{ii}^{(1)}(kL)$, region 2 of the "inner-flow scaling", where eqs. (13) are valid, and region 3 of the "Kolmogorov scaling", where η and $v_\eta = (\bar{\varepsilon}\nu)^{1/4}$ are the relevant length and velocity scales and, hence, $E_{ii}(k) = v_\eta^2 \eta e_{ii}^{(2)}(k\eta)$. At large Re regions 1 and 2 and also 2 and 3 overlap; let us denote the corresponding overlap regions by the Roman numbers I and II. In the overlap region I the applicability of both the outer- and inner-flow scalings implies that

$$e_{ii}^{(1)}(kL) = \frac{G_{ii}}{kL}, \qquad e_{ii}(kz) = \frac{G_{ii}}{kz} \tag{14}$$

(i.e. "-1 power law" is valid). Similarly, in the overlap region II the applicability of the inner-flow and the Kolmogorov scalings together with the energy-budget relation $\bar{\varepsilon} = u_*^3/\kappa z$ implies the equations

$$e_{ii}(kz) = \frac{K_{ii}}{(kz)^{5/3}}, \qquad e_{ii}^{(2)}(k\eta) = \frac{K_{ii}\,\kappa^{2/3}}{(k\eta)^{5/3}} \tag{15}$$

so that the overlap region II coincides with the inertial range of the Kolmogorov theory. It is clear that eqs. (11) for σ_1/u_* and σ_2/u_* will be invalid if the contributions of the part of region 1 beyond the small-scale limit of the overlap region I to the variances σ_1^2 and σ_2^2 are not negligibly small.

Perry et al. justify their three-region model of spectra E_{11} and E_{22} by the Townsend [19] attached-eddy hypothesis, and they suggest that the region of the outer-flow scaling does not exist in the w spectra. However, there are some rough data which show that a short "-1 power range" exists in the w spectra too. Note that such a range does not necessarily belong to the outer-flow scaling region. The temperature spectra have not been considered at all by Perry and his collaborators, but their dimensional arguments can also be applied to these spectra. Here $E_{TT}(k) = T_*^2 L e^{(1)}_{TT}(kL)$ for k in region 1 and

$$e^{(1)}_{TT}(kL) = \frac{G_{TT}}{kL}, \qquad e_{TT}(kz) = \frac{G_{TT}}{kz} \tag{16}$$

within the overlap region 1. Moreover, $E_{TT}(k) = T_\eta^2 \eta e^{(2)}{}_{TT}(k\eta, Pr)$ (where $T_\eta = (\bar{N}^2 \nu/\bar{\varepsilon})^{1/4}$ is the Kolmogorov temperature scale, $\bar{N} = u_* T_*^2/\kappa_1 z$ is the mean dissipation rate for $\overline{T'^2}/2$) for k in region 3, while

$$e_{TT}(kz) = \frac{K_{TT}}{(kz)^{5/3}}, \qquad e^{(2)}{}_{TT}(k\eta, Pr) = \frac{K_{TT}\kappa_1 \kappa^{-1/3}}{(k\eta)^{5/3}} \qquad (17)$$

within the overlap region 2 (which coincides with the inertial range of temperature spectrum).

The "-1 power law" for the u spectra in the wall turbulent flows was suggested by Abell in his 1944 Ph.D. thesis (see [28]) and by Korotkov [33] on the basis of the assumption that $E_{11}(k)$ must be independent of the wall distance z. This law has also been noted many times by the experimentalists (see [29], [34], [35] containing many additional references). Kader [34] compares the "-1 power law" with the experimental data from about 30 sources related to spectra $E_{11}(k)$, $E_{22}(k)$, $E_{33}(k)$, $E_{TT}(k)$, and cospectrum $E_{13}(k)$ in pipe, channel and boundary-layer flows. The -1 power range is the shortest for w spectra, where it can be observed only for the highest Reynolds numbers reached in the analysed experiments. But the same range in the T spectra is well-defined and rather long. The data analysed by Kader enable one to obtain rough estimates of the coefficients G_{ii}, i = 1, 2, 3, G_{TT}, and G_{13} in equations of the form (14) and (16)

$$G_{11} \simeq 0.8, \quad G_{22} \simeq 0.5, \quad G_{33} \simeq 0.25, \quad G_{TT} \simeq 0.3, \quad G_{13} \simeq -0.25$$

(Here the most reliable are the estimates of G_{11} and G_{TT}). Note that these estimates agree satisfactorily with the estimate $G_{11} \simeq 0.9$ of Turan et al. [35] and the estimates $G_{11} \simeq 1.1$, $G_{22} \simeq 0.5$ of Perry and Li [21]. Kader's estimates of the coefficients K_{11}, K_{22}, K_{33}, and K_{TT} agree satisfactorily with the results of the Kolmogorov theory (which states that $K_{22} = K_{33} = 3K_{11}/4$) and the existing experimental estimates of the universal constants entering the -5/3 power laws.

The spectral results presented in [34] can be supplemented by the data related to spectra of turbulent fluctuations in the atmospheric surface layer. Extensive data on the spectra $E_{11}(k)$, $E_{33}(k)$, and $E_{TT}(k)$ in the neutrally stratified surface layer were collected during the summers of 1981 to 1984 at the Tsimlyansk Field Station of the Moscow Institute of Atmospheric Physics. These data, presented in [36], [32], (see also Fig. 3) show that for an extended wave-number range (different for E_{11}, E_{33}, and E_{TT}) the spectra can be described well enough by a two-power-range model consisting of -5/3 and -1 power subranges which adjoin each other. The spectrum $E_{22}(k)$ was not measured at the Tsimlyansk Station since the wind direction variations made such measurements unreliable. However, the cospectra $E_{13}(k)$, $E_{1T}(k)$, and $E_{3T}(k)$ were measured and they were found to contain also the -1 power subranges. The atmospheric estimates $G_{11} \simeq 0.95$ and $G_{33} \simeq 0.35$ do not differ too much from the corresponding laboratory estimates, but the estimate $G_{TT} \simeq 0.8$ is almost three times as great as the laboratory estimate of G_{TT}. This discrepancy is, of course, related to the difference in σ_T/T_* for the laboratory and atmospheric flows.

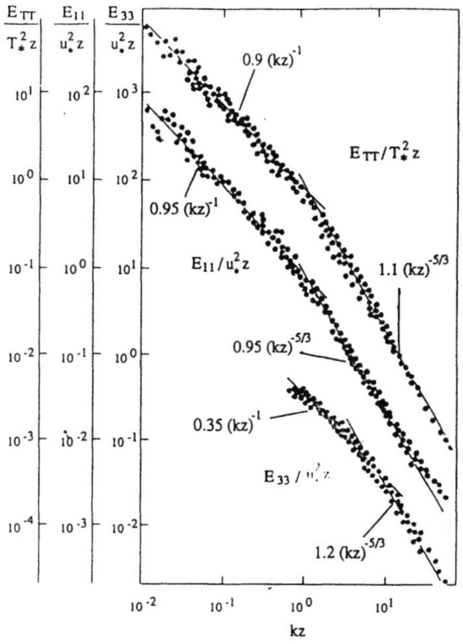

Fig. 3 The spectra $E_{11}(k)$, $E_{33}(k)$, and $E_{TT}(k)$ in the neutrally stratified atmospheric surface layer measured at the Tsimlyansk Field Station.

According to the data presented in [36, 37] the outer-flow-scaling region, where the spectra are strongly affected by the external length scale L, does not become apparent in E_{11} and E_{22} graphs for $kz \geq 10^{-2}$. However, this result disagrees with the results of Högström [25] and Smedman [27], who found that in the case of near-neutral stratification the normalized dimensionless spectra $kE_{11}(k)/u_*^2$ and $kE_{33}(k)/u_*^2$ within the surface layer depend only on kz in the high-wave-number domain, but when kz is diminishing, they start to depend also on the dinemsionless height $zf/u_* \simeq z/H$ at about kz = 0.1. Another conclusion was formulated by Kaimal [38], who also studied the spectra of surface-layer atmospheric turbulence and did not find any dependence of normalized velocity spectra on some external length L in the case of neutral stratification, but discovered that in the case of unstable stratification the normalized spectra of horizontal velocity components contain the low-wave-number region, where the thickness H of the atmospheric boundary layer definitely affects the spectral values (see sect. III below). Note that just this result of Kaimal was discussed by Bradshaw [24], who attributed it to the "inactive motion" and remarked that this motion must affect the horizontal-velocity spectra also during neutral condition. We see that the existing information about the spectra of surface-layer atmospheric turbulence is somewhat contradictory, but on the whole it supports doubts about the universal applicability of the simple law (11) to an atmospheric surface layer.

Let us now note that Perry et al. (see [21[, [28] to [30]) apply their three-scaling-region model for spectra $E_{11}(k)$ and $E_{22}(k)$ to the evaluation of ratios σ_1/u_* and σ_2/u_*. Assuming that the low wave number limit of the overlap region I is given by the equation kL = F, where the constant F may depend on the type of flow and the high wave number limit is given by kz=P where P is the universal constant, they obtained the relations

$$\sigma_i = u_*[B_i - G_{ii}\ln\frac{z}{L} + V(z_+)]^{1/2}, \qquad i = 1,2 \qquad z_+ = (\frac{u_* z}{\nu}), \qquad (18)$$

where constant B_i may be different for pipe, channel and boundary-layer flows and $V(z_+)$ is the contribution of the dissipation range $k \geq k_\eta$. (Accurate evaluation of the function $V(z_+)$ is given in [21]; apparently $V(z_+)$ is negligibly small at very high Re). Eqs. (18) are special cases of the similarity law (8) for i = 1,2; they generalize the simplest similarity law (11), but have the form that differs from that of self-similarity laws of the second type. Note that according to (18) the effect of L on σ_i/u_* increases when z decreases. This paradoxical result is the consequence of the attached-eddy model suggested in [19], [20] and it agrees satisfacotrily with the data related to σ_1/u_*.

3. Similarity Laws for Turbulence in a Thermally Stratified Medium

In the case of a thermally stratified medium, the buoyancy parameter $\beta = g\gamma$, where g is the acceleration due to gravity and γ is the coefficient of thermal expansion, which is equal to $1/T_0$ for an ideal gas, T_0 being the mean absolute temperature of a medium, must be added to the list of relevant dimensional parameters. The so-called Obukhov length scale $L_0 = u_*^2/T_*\beta$ can be formed from the parameters u_*, T_*, and β and this is the only length determined by these parameters. Hence, the dimensional analysis implies that for $z \gg \nu/u_*$, where influence of the molecular coefficients may be neglected, if the Reynolds number similarity principle is valid, all the dimensionless statistical characteristics of turbulence must be given by the single-valued universal functions of the dimensionless height $\zeta = z/L_0$; in particular,

$$\frac{z}{u_*}\frac{d\overline{u}}{dz} = \varphi_u(\zeta)\,; \quad \frac{z}{T_*}\frac{d\overline{T}}{dz} = \varphi_T(\zeta)\,; \quad \frac{\sigma_i}{u_*} = \psi_i(\zeta)\,, \quad i = 1,2,3; \quad \frac{\sigma_T}{T_*} = \psi_T(\zeta) \qquad (19)$$

The formulated statement is the core of the famous Monin-Obukhov similarity theory of turbulence in a stratified medium, which is considered in detail e.g. in chapter 4 of [6] and in [31]. This theory has been recently questioned by Barenblatt and Monin [5], [12]. They added the molecular coefficients ν and χ and the external length scale Λ (thickness of an atmospheric surface layer) to the list of relevant dimensional parameters. Then, generalizing eq. (7), they suggest that the functions φ_u and φ_T in (19) may depend also on $u_*z/\nu = z_+$ and $u_*\Lambda/\nu = Re_*$ and that at large values of z_+ and Re_* the dependence on z_+ is a power dependence:

$$\frac{z}{u_*}\frac{d\overline{u}}{dz} = C_u(\zeta)\left(\frac{u_* z}{\nu}\right)^{\lambda}, \qquad \frac{z}{T_*}\frac{d\overline{T}}{dz} = C_T(\zeta)\left(\frac{u_* z}{\nu}\right)^{\mu} \qquad (20)$$

where $C_u(\zeta)$, $C_T(\zeta)$, λ and μ may vary slowly with Re_*. (Possible dependence of φ_u and φ_T on Pr is neglected here.) It is also assumed in [5], [12] that the turbulent viscosity K and temperature diffusivity K_T must be asymptotically independent of u_* when $\zeta \to \infty$ and that $Pr_t = K/K_T$ is bound for all $u_*\Lambda/z$ and ζ. This assumption implies that $\mu = \lambda$.

It was not indicated in [5], [12] how the surface layer thickness Λ must be determined and it was stressed there that no data are available that permit eqs. (20) to be verified. Since eqs. (20) generalize (7), it is, of course, reasonable to start with the comparision of eq. (7) with the extensive data of precise laboratory measurements including, if possible, the results of newly organized special measurement. If eq. (7) is verified by such a comparison, then it will be worth-while to arrange a comparison of eqs. (20) with the appropriate data and to think of possible physical mechanism responsible for the effect of the length Λ and the molecular constants ν and χ on φ_u and φ_T. However, in principle, the possibility of such an effect cannot be excluded at present.

Let us now return to the classical Monin-Obukhov similarity theory which implies eqs. (19). In the case of unstable stratification, where $L_0 > 0$, it is natural to think that the influence of u_* on the turbulence structure will vanish as $\zeta = z/L_0 \to \infty$. The asymptotic regime corresponding to $\zeta >> 1$ is called the *free-convection regime* and it was first investigated by Prandtl [39] and Obukhov [40], [41]. However, according to the equations given in these papers, u_* does not enter the asymptotic forms of eqs. (19) for $d\overline{T}/dz$, σ_i, i = 1,2,3, and σ_T as $z/L_0 \to \infty$, but nevertheless, it enters the equations for $d\overline{u}/dz$. This controversial result was disputed by Zilitinkevich [42] and Betchov and Yaglom [43] and the conclusions given in the latter paper agree satisfactorily with some recent data analysed in [32], [37], [44]. According to Betchov's and Yaglom's arguments the equations proposed by Prandtl and Obukhov are valid in the intermediate region of moderate positive values of ζ, where there is no energy exchange between vertical and horizontal motions, and therefore it is possible to assume that horizontal and vertical lengths have different dimensions. Above this region a proper free-convection region is placed where u_* is cancelled in all the equations for mean gradients and turbulence charateristics; here equations for $d\overline{T}/dz$, $\sigma_3 = (\overline{w'^2})^{1/2}$ and $\delta_T = (\overline{T'^2})^{1/2}$ differ from the same equations in the intermediate region only by constant factors, but the equations for $d\overline{u}/dz$, σ_1 and σ_2 change much more drastically.

It has already been stated that these predictions agree satisfactorily with some recent experimental data. Nevertheless, some questions related to these predictions need further investigation. Is it true that the energy exchange between vertical and horizontal motions is negligibly small within a rather lengthy interval of moderate positive values of ζ? Does this interval coincide with that where all the corrolaries of the dimensional analysis with two different length dimensions agree satisfactorily with the data? What is the mechanism which leads to the establishment of somewhat paradoxical free-convection regime where u_* does not affect the flow but the flow regime is, nevertheless, not axially symmetric, and the gradient $d\overline{u}/dz$ does not vanish? One may hope that further development of numerical simulations for high Reynolds number flows will help to answer these questions.

Note now some experimental data which disagree with both Obukhov's older predictions [41] an more recent predictions by Zilitinkevich [42], Betchov and Yaglom [43], and Kader and Yaglom [32]. The measurements of variances of horizontal wind components σ_1^2 and σ_2^2 within an unstably stratified atmospheric surface layer are rather unreliable till now, but, nevertheless, it is worth noting that the results obtained strongly disagree with equations (19) implied by the Monin-Obukhov similartiy theory. There are many data on σ_1^2 and σ_2^2 which show that these variances are practically independent of z in the unstable surface layer for values of z/L_0 exceeding a few tenths. The analysis by Panofsky et al. [45], based on the earlier suggestion by Wyngaard and Coté [46], leads to the conclusion that

$$\sigma_i = u_* \psi_i^{(1)} \left(\frac{z_i}{L_0}\right) \qquad \text{for } i = 1,2 \text{ and } L_0 > 0 \qquad (21)$$

in an atmospheric surface layer, where z_i is the height of the capping inversion (increase of $\overline{T}(z)$ with z) above the unstable atmospheric boundary layer (i.e., z_i is the thickness of the whole boundary layer; this thickness is denoted by H in sect. II and it usually has the order of 1 km). According to Panofsky et al. [45], $\sigma_1 \sim \sigma_2 \sim (z_i/L_0)^{1/3}$ as $z_i/L_0 \to \infty$ and therefore u_* cancels in the equations (21) at strong instabilities.

The results by Panofsky et al. are also supported by those by Kaimal [38] who found that the normalized horizontal velocity spectra $kE_{11}(k)/\sigma_1^2$ and $kE_{22}(k)/\sigma_2^2$, measured at heights 4 and 32 m, depend only on kz_i in a wide range of small wave numbers making the main contribution to the variances of u' and v'. Therefore it is worth-while to think of a possible mechanism explaining the influence of z_i on the horizontal velocity components within the surface layer near the ground. The explanation which has been occasionally mentioned in the literature refers to the so-called horizontal roll vortices which are large, organized structures whose vertical scale coincides with z_i (Without mentioning the roll vortices, Bradshaw [24] expressed a similar opinion). The permanent existence of such roll vortices in an unstable atmospheric boundary layer is confirmed by many experimental studies, but the detailed description of the relation between these vortices and the validitiy of eqs. (21) is a problem to be solved in future. Note also that Bradshaw [24] explained the dependence of surface-layer turbulence on the thickness z_i of the whole atmospheric boundary layer by the influence of "inactive motions", but according to this explanation, z_i must affect the statistical regime of horizontal wind fluctuations u' and v' in the neutral conditions too. We have already noted that Panofsky et al. [45] and Kaimal [38] found no evidence for this last effect, but later Smedman [27] presented some data illustrating the dependence of the variance σ_1^2 and spectrum $E_{11}(k)$ in the neutral stratification on z/z_i. At the same time Smedman [27] and Högström [25] found that z_i strongly affects the variance of vertical velocity w in a neutral surface layer (see Fig. 3 above), though in the papers [24], [38] and [45] it was stressed that the boundary layer thickness affects only horizontal, but not vertical wind fluctuations. These contradictions in the results of different authors clearly show that the question about the factors affecting the statistical regime of wall-layer turbulence requires further careful investigation.
The number of unsolved questions related to the similarity laws for turbulent wall flows can, of course, be considerably increased. Let us only stress, in conclusion, that the recent (and also the future) discoveries concerning the organized coherent structures in such turbulent flows and their role in the mechanism of wall turbulence clearly have a direct realation to the formulation and explanation of valid similarity laws.

Acknowledgements

The author is grateful to Prof. G. I. Barenblatt for valuable discussions.

References

1. P. W. Bridgman: Dimensional Analysis, New Haven, Yale Univ. Press, 1932.
2. L.I. Sedov: Similarity and dimensional Methods in Mechanics, New York, Academic, 1959.
3. S. J. Kline: Similitude and Approximation Theory, Berlin, Springer, 1986.
4. G. I. Barenblatt, Ya. B. Zeldovich: Self-similar solutions as intermediate asymptotics, Ann. Rev. fluid Mech., **4**, 285 - 312, 1972.
5. G. I. Barenblatt: Similarity, Self-Similarity and Intermediate Asymptotics, New York, Plenum, 1981.
6. A. S. Moin, A. M. Yaglom, Statistical Fluid Mechanics, Vols. 1 and 2, Cambridge (Mass.), MIT Press, 1971 and 1975.
7. A. M. Yaglom: Similarity laws for constant-pressure and pressure-gradient turbulent wall flows, Ann. Rev. Fluid Mech., **11**, 505 - 540, 1979.
8. B. A. Kader, A. M. Yaglom: Similarity laws for turbulent wall flows. In Developments in Science and Technology, Ser. Mech. Liquid and Gas, **15**, 81 - 153, Moscow, Soviet Inst. of Sci. and Eng. Inform., 1980 (in Russian).
9. B. A. Kader, A. M. Yaglom: Roughness and pressure-gradient effects on turbulent boundary layers. In Developments in Science and Technology, Ser. Mech. Liquid and Gas, **18**, 3 - 111, Moscow, Soviet Inst. of Sci. and Eng. Inform., 1984 (in Russian).
10. G. I. Barenblatt: Dimensional Analysis, New York, Gordon and Breach, 1987.
11. G. I. Barenblatt: On the scaling laws (incomplete self-similarity with respect to Reynolds numbers) for the developed turbulent flows in tubes, C. R. Acad. Sci. Paris, **313**, Ser. II, 307 - 312, 1991.
12. G. I. Barenblatt: A. S. Moin: Similarity laws for turbulent stratified shear flows, Arch. Rat. Mech. Anal., **70**, 307 - 317, 1979.
13. H. Schlichting: Boundary Layer Theory (4th ed.), New York, McGraw-Hill, 1960.
14. J. O. Hinze: Turbulence (2nd ed.), New York, McGraw-Hill, 1975.
15. J. Nikuradse: Gesetzmässigkeiten der turbulenten Strömung in glatten Rohren, VDI-Forschungsheft Nr. 356, 1932.
16. B. Castaing, Y. Gagne, E. J. Hopfinger: Velocity probability density functions of high Reynolds number turbulence, Physica D, **46**, 177 - 200, 1990.
17. K. W. Schwarz: Evidence for organized small-scale structures in fully developed turbulence, Phys. Rev. Lett., **64**, 415 - 418, 1990.
18. A. Vincent, M. Meneguzzi: The spatial structure and statistical properties of homogeneous turbulence, J. Fluid Mech., **225**, 1 - 20, 1991.
19. A. A. Townsend: The Structure of Turbulent Shear Flow, (2nd ed.), Cambridge, Camb. Univ. Press, 1976.

20. A. E. Perry, M. S. Chong: On the mechanism of wall turbulence, J. Fluid Mech., **119**, 173 - 217, 1982.
21. A. E. Perry, J. D. Li: Experimental support for the attached-eddy hypothesis on zero-pressure-gradient turbulent boundary layers, J. Fluid Mech., **218**, 405 - 438, 1990.
22. A. A. Townsend, Equilibrium layers and wall turbulence, J. Fluid Mech., **11**, 97 - 120, 1961.
23. P. Bradshaw, "Inactive" motion and pressure fluctuations in turbulent boundary layers, J. Fluid Mech., **30**, 241 - 258, 1967.
24. P. Bradshaw, Comments on "Horizontal velocity spectra in an unstable surface layer", J. Atmos. Sci., **35**, 1768 - 1769, 1978.
25. U. Högström, Analysis of turbulence structue in the surface layer with a modified similarity formulation for near neutral conditions, J. Atmos. Sci., **47**, 1949 - 1972, 1990.
26. P. H. Spalart, Direct similation of a turbulent boundary layer up to R = 1410, J. Fluid Mech., **187**, 61 -98, 1988.
27. A.-S. Smedman, Some turbulence characteristics in stable atmospheric boundary layer flow, J. Atmos. Sci., **48**, 856 - 868, 1991.
28. A. E. Perry, C. J. Abell, Asymptotic similarity of turbulence structure in smooth- and rough-walled piper, J. Fluid Mech., **79**, 785 - 799, 1977.
29. A. E. Perry, S. M. Henbest, M. S. Chong, A theoretical and experimental study of wall turbulence, J. Fluid Mech., **165**, 163 - 199, 1986.
30. A. E. Perry, K. L. Lim, S. M. Henbest, An experimental study of the turbulence structure in smooth- and rough-wall boundary layers, J. Fluid Mech., **177**, 437 - 466, 1987.
31. H. A. Panofsky, J. A. Dutton, Atmospheric turbulence, models and methods for engineering applications, New York, J. Wiley.
32. B. A. Kader, A. M. Yaglom, Mean fields and fluctuation moments in unstably stratified turbulent boundary layers, J. Fluid Mech., **212**, 637 - 662, 1990.
33. B. N. Korotkov, Some types of local self-similarity of the velocity field of wall turbulent flows, Izv. Akad. Nauk SSSR, Ser. Mekh. Zhidk. i. Gaza (Bull. Acad. Sci. USSR, Ser. Mech. Liquid and Gas), No.6, 35 - 42, 1976.
34. B. A. Kader, Structure of anisotropic velocity and temperature fluctuations in a developed turbulent boundary layer, Izv. Akad. Nauk SSSR, Ser. Mekh. Zhidk. i Gaza, No. 4, 47 - 56, 1984.
35. O. Turan, R. S. Azad, S.Z. Kassab, Experimental and theoretical evaluation of the k spectral law, Phys. Fluids, **30**, 3463 - 3474, 1987.
36. B. A. Kader, Anisotropic fluctuations of wind velocity and temperature in neutrally stratified atmospheric surface layer. In Meteorological Researches, No. 28 (ed. by

L. R. Tsvand and B. A. Kader), 26 - 35, Moscow, Soviet Geophys. Committee Acad. Sci. USSR, 1987 (in Russian).

37. B. A. Kader, A. M. Yaglom, Spectra and correlation functions of surface layer atmospheric turbulence in unstable thermal stratification. In Turbulence and Coherent Structures, (ed. O. Métais and M. Lesieur), 387 - 412, Dordrecht, Kluwer, 1991.

38. J. C. Kaimal, Horizontal velocity spectra in an unstable surface layer, J. Atmos. Sci., **35**, 18 - 24, 1978.

39. L. Prandtl, Meteorologische Anwendung der Strömungslehre, Beitr. Phys. fr. Atmos., **19**, 188 - 202, 1932.

40. A. M. Obukhov, Turbulence in thermally inhomogeneous atmosphere. In Trudy Inst. Teor. Geofiz. Akad. Nauk. SSSR, No. 1, 95 - 115, 1946 (Transl. in Boundary-Layer Meteor., **3**, 7 - 29, 1971.)

41. A. M. Obukov, Structure of temperature and velocity fields under conditions of free convection, Izv. Akad. Nauk SSSR, Ser. Geofiz., No.9, 1392 - 1396, 1960.

42. S. S. Zilitinkevich, On the turbulence and diffusion under free convection conditions, Izv. Akad. Nauk SSSR, Ser. Fiz. Atmos. i Okeana, **7**, 1263 - 1269 (825 - 828 in the English transl. of the journal), 1971.

43. R. Betchov, A. M. Yaglom, Comments on the theory of similarity as applied to turbulence in an unstably stratified fluid, Izv. Akad. Nauk SSSR, Ser. Fiz. Atmos. i Okeana, **7**, 1270 - 1279 (829 - 832 in the English transl. of the journal), 1971.

44. B. A. Kader, Three-layer structure of an unstably stratified atmospheric surface layer, Izv. Akad. Nauk SSSR, Sr. Fiz. Atmos. i Okeana, **24**, 1235 - 1250 (907 - 918 in the English transl. of the journal), 1988.

45. H. A. Panofsky, H. Tennekes, D. H. Lenshow, J. C. Wyngaard, The characteristics of the velocity components in the surface layer under convective conditions, Boundary-layer Meteor., **11**, 355 - 361, 1977.

46. J. C. Wyngaard, C. R. Coté, The evolution of a convective planetary boundary layer a higher-order-closure model study, Boundary-layer Meteor., **7**, 289 - 308, 1974.

Address of the author

Akiva M. Yaglom
Institute of Atmospheric Physics
USSR Academy of Sciences
Moscow, USSR

A Prediction of the Multifractal Model: the Intermediate Dissipation Range.

U. FRISCH (*) and M. VERGASSOLA (*) (**)

(*) *CNRS, Observatoire de Nice - B.P. 139, 06003 Nice Cedex, France*
(**) *Dipartimento di Fisica, Università «La Sapienza»*
P.le A. Moro 2, I-00185 Roma, Italy

(received 1 November 1990; accepted in final form 12 December 1990)

PACS. 47.25C – Isotropic turbulence.

Abstract. – It is shown that the multifractal model of fully developed turbulence predicts a new form of universality for the energy spectrum $E(k)$, which can be tested experimentally. Denoting by R the Reynolds number, $\log E/\log R$ should be a universal function of $\log k/\log R$. This includes an *intermediate dissipation range* in which a continuous range of multifractal scaling exponents are successively turned off by viscosity.

1. Introduction.

The multifractal model was introduced in 1983 by Parisi and Frisch (PF) [1] as a possible interpretation of the inertial range scaling properties of structure functions of variable order [2, 3]. The model involves a function $D(h)$ assumed to be universal, but not given by phenomenological theory. In so far as the inertial range scaling properties of structure functions are concerned, the multifractal model is indistinguishable from Mandelbrot's «weighted curdling» model [4]. Here, we show that the multifractal model predicts a behaviour of the energy spectrum in the dissipation range which can in principle be tested experimentally. For this, we make use of the existence of local scaling exponents in the multifractal model.

First, we recall some key features of the multifractal model of turbulence. We assume that the turbulence is statistically homogeneous and isotropic. For simplicity we shall here assume that the scale l_0 and the velocity v_0, characteristic of large energy-containing eddies, are equal to unity (1).

In the limit of vanishing viscosity, the Navier-Stokes equations are scale-invariant: the space variable may be rescaled by an arbitrary positive factor λ, the velocity by λ^h and the time by λ^{1-h}. The scaling exponent h is arbitrary. In Kolmogorov's 1941 (K41) theory, *global* scale invariance (in a statistical sense) is implicitly assumed and the exponent $h = 1/3$ is

(1) Fluctuations of v_0 are important in their effect on the probability distribution function of velocity gradients [5].

singled out by an energy argument. In the PF multifractal model *local* scale-invariance is assumed: there is a continuous range of scaling exponents. For any h in this range, there is a set $\mathcal{S}(h)$ in $\boldsymbol{R}^3$ of Hausdorff dimension $D(h)$ such that, when $r \in \mathcal{S}(h)$, the velocity increment $v_l(r) = |v(r+l) - v(r)|$ behaves for small l as

$$v_l(r) \sim l^h. \tag{1}$$

Note that in the present phenomenological approach the vector character of quantities such as r, v and l is not stressed. The argument r of $v_l(r)$ will be dropped whenever no ambiguity arises.

The multifractal model implies simple scaling properties for structure functions. The structure function of order p is defined as $S_p(l) = \langle v_l^p \rangle$. The contribution of the set with scaling exponent h to the structure function of order p is proportional to $l^{ph}\, l^{3-D(h)}$. The second factor is the probability of encountering the set $\mathcal{S}(h)$ within a ball of radius l (controlled by the co-dimension $3 - D(h)$). A smooth weight $\mathrm{d}\mu(h)$ (not given by the theory) is then attributed to the various scaling exponents. Hence

$$S_p(l) \sim \int \mathrm{d}\mu(h)\, l^{ph+3-D(h)}. \tag{2}$$

In the limit of small l's, the power law with the smallest exponent dominates. Thus

$$S_p(l) \propto l^{\zeta_p} \quad \text{with } \zeta_p = \min_h [ph + 3 - D(h)]. \tag{3}$$

The scaling exponent ζ_p of the structure functions is thus the Legendre transform of the co-dimension.

In this formulation, the multifractal model is capable of interpreting an arbitrary function ζ_p (subject to a convexity constraint). However, it does so using another unknown function, thus leaving no room for experimentally verifiable predictions. This situation will now change by consideration of viscous effects.

2. Viscous cut-offs.

The effect of a finite amount of kinematic viscosity ν can be estimated as follows. For a given scaling exponent h, the viscous cut-off $\eta(h)$ is obtained simply by equating the turnover time $t_l = l/v_l$ to the viscous diffusion time l^2/ν. Hence, an h-dependent viscous cut-off is obtained:

$$\eta(h) \sim \nu^{1/(1+h)}. \tag{4}$$

When $h = 1/3$, the Kolmogorov dissipation scale is recovered $\nu^{3/4}$. Equation (4) has been previously derived by Paladin and Vulpiani [6]. Mandelbrot had already observed in his cascade models that there can be more than one dissipation scale [7].

A confirmation that the h-dependent viscous cut-off given by (4) is appropriate is provided by the following argument. The mean energy dissipation is given by $\varepsilon \sim \nu \langle (\nabla v)^2 \rangle$. Associated with the exponent h, there is a cut-off $\eta(h)$ and a velocity gradient $\nabla v \sim v_{\eta(h)}/\eta(h)$. The contribution to the energy dissipation coming from the scaling exponent h is thus proportional to $\nu [v_{\eta(h)}/\eta(h)]^2\, \eta(h)^{3-D(h)}$. The second factor comes from the probability of intercepting a set of dimension $D(h)$ in a ball of radius $\eta(h)$. Using (1) and (4), we find that for small ν the dominant contribution to the mean energy dissipation is $\varepsilon \sim \nu^{\rho}$, where

$$\rho = 1 + \min_h \left[\frac{2(h-1) + 3 - D(h)}{h+1} \right]. \tag{5}$$

In all intermittency models it is assumed that $\zeta_3 = 1$, so as to ensure that the third-order longitudinal structure function (in the zero viscosity limit) is directly proportional to l, a relation deduced from the Navier-Stokes equations by Kolmogorov (see, *e.g.*, ref. [8]). In view of (3), this relation implies $\min_h [3h + 3 - D(h)] = 1$. Simple algebra shows then that the energy dissipation exponent $\rho = 0$, thereby ensuring a finite-energy dissipation in the inviscid limit (2).

For later use, we shall need an estimate of the minimum scaling exponent $h_{\min}$. The experimental data of Anselmet *et al.* [3] for the exponents ζ_p up to order 18 suggest that $h_{\min} \leqslant 0.08$. Preliminary direct measurements of the scaling exponents by wavelet techniques indicate that $h_{\min}$ may be smaller, possibly negative [10]. The log-normal model of Obukhov and Kolmogorov (cf. ref. [8]) has arbitrary negative scaling exponents. A negative inertial range exponent h implies by (1) an unbounded velocity, after the limit $\nu \to 0$ and $l \to 0$ have been taken (in that order). If the velocity becomes greater than the speed of sound, the basic assumption of incompressibility breaks down and the whole Navier-Stokes framework becomes questionable [11]. We shall not, here, open this Pandora's box. As for exponents $h < -1$, they would correspond to singularities so strong that viscosity cannot damp them. We shall thus assume that $h_{\min} > -1$.

The fact that exponents h significantly smaller than 1/3 have viscous cut-offs much smaller than the Kolmogorov dissipation scale, can be a severe constraint on experimental techniques where multifractality is measured from the statistics of the local dissipation. Such measurements can be spurious for those exponents which have cut-offs less than the probe size [12].

3. The intermediate dissipation range.

The presence of an h-dependent viscous cut-off, discussed in the previous section, has an immediate implication for the behaviour of the structure functions at those scales where viscosity begins to be felt. For simplicity, we limit ourselves to the second-order structure function, generalizations being obvious. It now suffices to restrict the integration in (2) to those scaling exponents h such that the viscous cut-off $\eta(h)$ is less than the scale l under consideration. Hence, for the second-order structure function, we obtain

$$S_2(l) = \langle v_l^2 \rangle \sim \int_{\eta(h) < l} d\mu(h)\, l^{2h+3-D(h)}. \tag{6}$$

Instead of a sharp cut-off in the integration, we could use a smooth cut-off function of $l/\eta(h)$; if the Fourier transform of this function decreases fast enough (say, faster than algebraically), no difference with a sharp cut-off is obtained to leading order in the subsequent analysis. To study the behaviour predicted by (6), let us define h_K to be the scaling exponent contributing mostly to the second-order structure function in the inertial

(2) The above argument can clearly be extended to evaluate the scaling properties (as a function of viscosity) of moments of arbitrary order of ∇v. This is done in a recent work of Nelkin [9], received after completion of the present work.

range. Since deviations from K41's $-5/3$ law are very minute, $h_K \simeq 1/3$ and the corresponding viscous cut-off $\eta_K \simeq \nu^{3/4}$. Let us also denote by $\eta_{\min}$ the minimum viscous cut-off determined by (4).

In the inertial range, for $1 \gg l > \eta_K$, we obtain the usual result, that $S_2 \propto l^{\zeta_2}$ (with $\zeta_2 = 2h_K + 3 - D(h_K) \simeq 2/3$). Novel scaling is obtained for $\eta_K > l > \eta_{\min}$, in the intermediate dissipation range ([3]). As we reach scales less than η_K, the contribution from h_K is suppressed by viscous dissipation in favour of lesser scaling exponents not yet influenced by viscosity. Specifically, assuming the viscosity to be small, we find that the dominant contribution to the integral (6) comes from the scaling exponent $h(l)$ such that $\eta(h) = l$. To leading order the result is given by

$$S_2(l) \propto l^{2h(l)+3-D(h(l))}, \qquad \nu^{1/(1+h(l))} = l, \qquad \text{for } \eta_K > l > \eta_{\min}. \tag{7}$$

Let us consider the energy spectrum $E(k)$, related to the second-order structure function by Fourier transformation. We observe that a power law l^{α} contributes a term proportional to $k^{-1-\alpha}$ to its Fourier transform. Here, the exponent of l is actually logarithmically dependent on l, but this simple formula remains true to leading order. We thus obtain the following results. In the inertial range, to leading order in ν:

$$E(k) \propto k^{-1-\zeta_2} \simeq k^{-5/3}, \qquad \text{for } 1 \gg k > \eta_K. \tag{8}$$

In the *intermediate dissipation range*:

$$\begin{cases} E(k) \propto k^{-1-2h(k)-(3-D(h(k)))} \\ h(k) = -1 - \dfrac{\log \nu}{\log k}, \end{cases} \qquad \text{for } \eta_K^{-1} < k < \eta_{\min}^{-1}. \tag{9}$$

We see that the predicted energy spectrum in the intermediate dissipation range is a *pseudo-algebraic* law, *i.e.* a power law with a slowly (logarithmically) varying exponent. At the top of the intermediate dissipation range, for $k = \eta_K^{-1}$, it is easily checked that the exponent recovers its inertial range value $-1 - \zeta_2 \simeq -5/3$. Beyond the bottom of the intermediate dissipation range, for $k > \eta_{\min}^{-1}$, much faster decrease of $E(k)$ is expected, possibly of exponential type, as suggested by an analyticity argument [13].

We mention that the question of the dissipation range spectrum for (log-normal) intermittent turbulence has also been addressed by Ellison and Keller and Yaglom (see references in [8], Sect. 25.4). Their result (an exponential of minus a squared logarithm of the wavenumber, extending to infinite wavenumbers) is different from (9). Their derivation assumes that the cascade is stopped uniformly at the Kolmogorov scale.

4. Universality of multifractal type and consequences.

The main result which we now want to stress is that (9) actually constitutes a *prediction* of the multifractal model. The input of our analysis is an arbitrary function $D(h)$ of one variable, but the output is a family of spectra for varying viscosity, thus a function of *two* variables. Because the functional expression (9) involves more than a single dissipation

---

([3]) If there is a single scaling exponent, as in the «unifractal» β-model, the intermediate dissipation range obviously disappears.

wavenumber, it is not possible to collapse onto a single curve the various energy spectra obtained for different Reynolds numbers, simply by log-log plotting and shifting, as done for example in fig. 75 of [8]. In other words, multifractality invalidates Kolmogorov's first similarity hypothesis. We now observe that the viscosity can be eliminated and a function of a single variable can be constructed from (9) by the following procedure. Define $F = \log E(k)/\log(1/\nu)$ and $\theta = \log k/\log(1/\nu)$. From (8) and (9), we find that

$$F(\theta) = \begin{cases} (-1-\zeta_2)\,\theta, & \text{for } \theta < \dfrac{1}{1+h_K}, \\[2ex] -2-2\theta+\theta D\left(-1+\dfrac{1}{\theta}\right), & \text{for } \dfrac{1}{1+h_K} < \theta < \dfrac{1}{1+h_{\min}}. \end{cases} \tag{10}$$

Since the function $F(\theta)$ depends solely on the universal function $D(h)$, it is itself universal ([4]). The rescaling procedure which brings all the spectra on a universal curve is actually identical to a reduction procedure used by Wu *et al.*, in analysing temporal spectra of temperature in high Rayleigh number convection [14]. They refer to it as «multifractal», although it is not introduced by the kind of theory presented here. It is likely that our analysis can be carried over to thermal convection.

Among the other consequences of the novel scaling predicted, we shall mention only two.

It has been proposed that polymers in very dilute concentrations can suppress the Kolmogorov turbulence cascade below a scale r^{**} obtained by balancing kinetic and elastic energies [15]. When multifractal scaling is assumed, there will be a different «elastic» cut-off for each scaling exponent. Hence, a behaviour similar to the one predicted here for the intermediate dissipation range may already be present in the inertial range.

It has been noted that multifractality may change Landau's estimate of the number of degrees of freedom of high Reynolds number turbulence from $O(R^{9/4})$ to $O(R^{3/(1+h_{\min})})$, thereby modifying the scaling law for the computational work required to simulate increasing Reynolds number flows [6]. In practice, it may not be necessary to resolve all scales down to $\eta_{\min}$. However, we wish to point out that if the computational mesh is between η_K and $\eta_{\min}$, spectral methods will achieve only pseudo-algebraic rather than exponential accuracy. Still, spectral simulations are presently limited to resolution from 240^3 to 864^3, depending on the symmetries of the flow. At such resolutions, intermediate dissipation range scaling may barely begin to manifest itself.

5. Experimental validation.

The prediction of universal scaling of multifractal type for the energy spectrum can be tested experimentally.

Recently, Gagne and Castaing [16] have analysed a wide sample of turbulence data with Reynolds numbers ranging from (roughly) 10^3 to 10^7. Remarkable agreement with scaling of multifractal type is obtained over the entire range of available scales. Traditional turbulence measurements involve probes with a length of one millimeter or more which can resolve only the beginning of the intermediate dissipation range. Multifractality predicts that there is

([4]) In the work of Nelkin mentioned in footnote ([2]), it is suggested that some universal scaling of multifractal type could hold for turbulence. No attempt however is made to derive such scaling from the multifractal hypothesis.

considerable life in turbulence well below one-millimeter scales. Smaller probes or high-resolution nonintrusive techniques ([5]) should be developed to meet the challenge.

* * *

This work was directly stimulated by a question of V. YAKHOT concerning the possibility of existence of intermediate asymptotics with power law behaviour in the dissipation range. We have benefitted from extensive discussions with R. BENZI, Y. GAGNE and L. KADANOFF. The support from the EEC (SC1-0212-C) and from DRET (90/1444) is acknowledged.

([5]) One promising techniques is the Raman excitation plus laser-induced electronic fluorescence (see ref. [17]).

REFERENCES

[1] PARISI G. and FRISCH U., in *Turbulence and Predictability in Geophysical Fluid Dynamics and Climate Dynamics, Proceedings of the International School of Physics «E. Fermi», Varenna, Italy*, edited by M. GHIL, R. BENZI and G. PARISI, Course LXXXVIII (North-Holland, Amsterdam) 1985, p. 84.
[2] VAN ATTA C. W. and PARK J., in *Statistical Models and Turbulence*, edited by M. ROSENBLATT and C. VAN ATTA (1972), 402.
[3] ANSELMET F., GAGNE Y., HOPFINGER E. J. and ANTONIA R. A., *J. Fluid Mech.*. **140** (1984) 63.
[4] MANDELBROT B., *J. Fluid Mech.*. **62** (1974) 331.
[5] FRISCH U. and SHE Z. S., *On the probability distribution function of velocity gradients in fully developed turbulence*, submitted to *Fluid Dynamics Res.* (1990).
[6] PALADIN G. and VULPIANI A., *Phys. Rev. A*, **35** (1987) 1971.
[7] MANDELBROT B., in *Turbulence and Navier-Stokes Equation*, edited by R. TEMAN, *Lecture Notes in Math.*, Vol. **565** (Springer, Berlin) 1976, p. 121.
[8] MONIN A. S. and YAGLOM A. M., *Statistical Fluid Mechanics*, Vol. **2**, edited by J. LUMLEY (MIT Press, Cambridge, Mass.) 1975.
[9] NELKIN M., *Phys. Rev. A*, in press (1990).
[10] BACRY E., ARNÉODO A., FRISCH U., GAGNE Y. and HOPFINGER E. J., *Turbulence and Coherent Structures*, edited by O. MÉTAIS and M. LESIEUR (Kluwer Academic Publishers) 1990, p. 203.
[11] FRISCH U., in *A. N. Kolmogorov*, special issue of *Proc. R. Soc. London, Ser. A* (1991).
[12] AURELL E., FRISCH U., LUTSKO J. and VERGASSOLA M., unpublished (1990).
[13] FRISCH U. and MORF R., *Phys. Rev. A*. **23** (1981) 2673.
[14] WU X. Z., KADANOFF L., LIBCHABER A. and SANO M., *Phys. Rev. Lett.*. **64** (1990) 2140.
[15] TABOR M. and DE GENNES P. G., *Europhys. Lett.*, **2** (1986) 519.
[16] GAGNE Y. and CASTAING B., *A universal nonglobally self-similar representation of the energy spectra in fully developed turbulence*, submitted to *C.R. Acad. Sci. Paris* (1990).
[17] MILES R. B., CONNORS J. J., MARKOVITZ E. C., HOWARD P. J. and ROTH G. J., *Exp. Fluids*, 8 (1989) 17.

A New View of Developed Turbulence

B. Castaing
and
Y. Gagne, E.J. Hopfinger

Abstract. In this paper, we present a detailed statistical analysis of selected experimental data which leads to a new model of the energy cascade process. At fixed inertial scale r, experiments are not inconsistent with a Gaussian distribution of $\ln\varepsilon(r)$ where $\varepsilon(r)$ is the local transfert rate. The same experiments suggest, however, that the variance $\langle\delta(\ln\varepsilon(r))^2\rangle$ has a power law behaviour $r^{-\beta}$ and not a logarithmic one as assumed in the log-normal model. From this point of view $\beta\ln r$ appears as a universal variable. This is confirmed by energy spectra E(k), measured during the past thirty years, which merge on a single curve when plotted as $\ln(E(k))/\ln Re$ versus $\ln(k)/\ln Re$, suggesting that β is inversely proportional to lnRe. An extremum principle approach gives all these results and suggests that β would be the co-dimension of the dissipative structures.

1. Experiments

1.1 Flow configuration and instrumentation

The measurements were made in two different flows : in a very weakly heated axisymmetric jet and in the very large wind tunnel S1 of the O.N.E.R.A. in Modane. The experimental conditions of the axisymmetric jet were the same as those already reported in part 2 of the paper by Anselmet *et al.* (1984). The wind tunnel S1 in Modane has a transonic section of 8 m diameter. The subsonic back flow occurs through a very large axisymmetric pipe at atmospheric pressure. The pipe section in which the measurements were made has a length of 153 m, and is composed of three parts. An up-stream cylindrical one (42 m in length, 15 m in diameter) holds two vertical rectangular entries of atmospheric air. A middle part (65 m in length) is a diffuser with a half angle of 3°25' and has a transverse wire netting which makes a kind of a grid. The third part is a cylindrical pipe, 24 m in diameter, downstream of this diffuser. The velocity probe was on the axis. This flow has the advantage that the turbulent intensity is low ($\lesssim 7$ %) which favours the use of Taylor's hypothesis used in all measurements reported here. Because of the large size the turbulent Taylor microscale Reynolds number is $R_\lambda \simeq 2720$. Table 1 contains flow characteristics of the wind tunnel S1 and also of the axisymmetric jet.

Table 1 Flow characteristics of the wind tunnel SI of the ONERA and of the axisymmetric jet.

	U (ms^{-1})	$\frac{(<v^2>)^{1/2}}{U}$	λ(mm) Taylor Scale	η (mm) Kolmogorov Scale	l (m) Integral Scale	R_λ
Jet	6.32	26.7	7.5	0.13	0.24	852
S1	20.80	7.08	36	0.35	~12	2720

For the velocity measurements a DISA 55M10 constant temperature system was used together with a standard gold plated, 5 μm diameter, 1.1 mm long wire in the wind tunnel S1 and a 3 μm diameter, 0.35 mm long Wollaston wire in the jet. The Wollaston wire with $l_w/d_w = 120$ gave results consistent with previous measurements made with a wire of $l_w/d_w > 200$ (Gagne 1987). The output voltage was first conditioned by subtracting the d.c. component and by low pass filtering (at 48 db/octave) at 50 kHz for the jet and 10 kHz for the wind tunnel S1, before it was amplitued.

In the jet case, velocity signals were transmitted on-line to a Preston A/D converter of a Nord 100 computer, giving a dynamic range of the whole acquisition system equal to 3×10^4. For the wind tunnel S1, velocity and temperature fluctuations were recorded on an analog magnetic tape Schlumberger MP5522 in a FM mode, with a signal to noise ratio of 54 db and a frequency range of 0-20 kHz. The analog data tape was then converted in the laboratory, using the same Preston A/D as in the case of the jet.

For the jet, measurements showed that, because of the high turbulence level ($\gtrsim 25$ %), linearization had a drastic effect on the probability density functions (cf. Anselmet *et al.* 1984). It was carried out on the computer by fitting a power law to the calibration curve. No linearization was necessary in the case of wind tunnel S1 measurements. In this flow, the turbulence level is low ($\lesssim 7$ %) and the mean velocity ($\gtrsim 20$ m/s) is sufficiently high, in order for the velocity fluctuations to be always in the nearly linear part of the calibration curve of the hot wire.

1.2 Probability density functions

Measurements of two points velocity fluctuation differences $u_r = v(\vec{x}+\vec{r}) - v(\vec{x})$, give mainly the velocity of eddies of size r. So, we examine the scale hierarchy of the energy transfer when the distance r is varied over the whole inertial range. For the jet of $R_\lambda=852$ the inertial range lies within $r/\eta=164$ and 20; for the wind tunnel S1 with $R_\lambda=2720$ this range is within $r/\eta=2500$ and 30, where η is the Kolmogorov scale. The p.d.f. of the velocity differences presented here are normalized, i.e. the random variable α on the abscissa is equal to $u_r/(<u_r^2>)^{1/2}$, and the ordinate value P_α is such that the area under the p.d.f. curve is equal to one. A typical p.d.f. of u_r of the jet is shown in semilogarithmic coordinates (figure 1). A small but crucial negative asymmetry is observed when fluctuations are large ($|\alpha| \gtrsim 3$). This indicates that rare (less than 2 %) but large fluctuations of the velocity give the negative sign of the assymmetry coefficient of the p.d.f.'s, which is due to the stretching of vortex lines in fully developed turbulence. When the distance r is comparable with the smaller scales of the inertial range, the maximum value of P_α increases slowly up to 0.5. Note that for distances r close to the dissipative scale ($r/\eta \gtrsim 10$) this maximum value reaches 0.65.

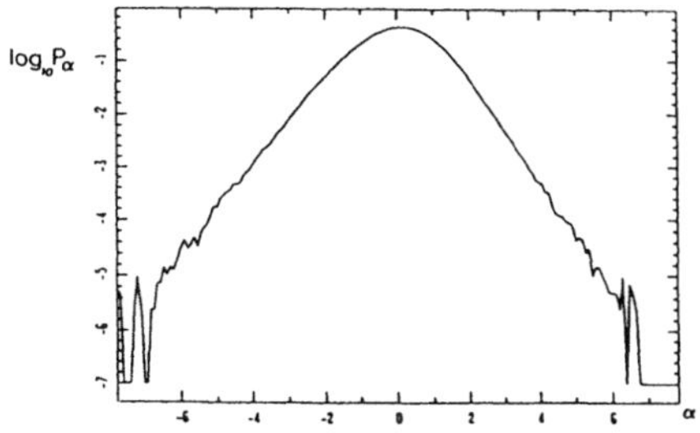

Fig. 1 Pdf of u_r in the tunnel case, in the semilogarithmic coordinates (r/η) = 2971.

The main property of the pdf is the apparent linear aspect of the "tails" of the pdf curves in the semilogarithmic coordinates (figure 1). It means that for large fluctuations of α ($|\alpha| > 3$), the pdf can be fitted roughly by an exponential law such as $P(\alpha) \sim \exp(-b^+.\alpha)$ for $\alpha > 0$ and by $P(\alpha) \sim \exp(b^-.\alpha)$ for $\alpha < 0$.

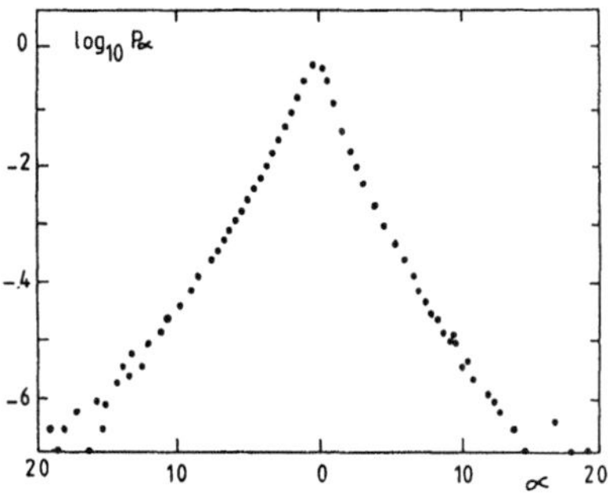

Fig. 2 Pdf of velocity in the dissipation range (jet, r/η =.3.3).

However, when the separation r is in the dissipative range, the tails of the pdf of u_r have not more an exponential behaviour, and in semilogarithmic coordinates these tails are weakly curved (for example, see figure 2). We have checked that the bending of the pdf tails is not due to an insufficient convergence of u_r, which can cause problems at very small r.

2. The empirical distribution

In this section, we develop an empirical model for the pdf of velocity differences u_r for separations r lying in the inertial range. This model is based on the distinction between two levels of fluctuations :

i) fluctuations of u_r for a given transfer rate ε ,

ii) and fluctuations of the transfer rate itself.

As soon as we distinguish, as Kraichnan (1974) did, the energy tranfer rate ε (from here on we drop the indice r) from the dissipation rate ε^*, a clear definition of ε is not obvious. Indeed, the instantaneous transfer rate can be negative, and the averaged value must be equal to the dissipation. One could take for ε the average on an intermediate time τ_t much smaller than the time necessary for the energy transfer down to the dissipative range. However, in this paper, ε is defined by the properties we expect for it: it must be positive, and plays the role it had in the Kolmogorov 41 theory.

In any model the intermittency only comes from the large distribution of ε. For a fixed value of ε it is thus simpler to assume that any quantity has a gaussian distribution (we shall come back to the skewness later). Such a distribution is completely defined by its variance σ. The distribution of u for fixed ε is therefore given by:

$$P_\varepsilon(u) = \frac{1}{\sigma\sqrt{2\pi}} \exp - \frac{u^2}{2\sigma^2}$$

Introducing now the intermittency through the fluctuations of σ,we postulate that σ has a log normal distribution :

$$Q_\lambda(\sigma)\, d\sigma = \frac{1}{\lambda\sqrt{2\pi}} \exp(- \frac{\ln^2(\sigma/\sigma_0)}{2\lambda^2})\, d\ln\sigma$$

where σ_0 is the most probable variance of u, and λ is the variance of $\ln\sigma$. Such an hypothesis on σ is natural if we consider that σ is entirely determined by ε and r. Dimensional arguments, as in Kolmogorov (1941), yield :

$$\sigma \propto (\varepsilon r)^{1/3}$$

Combining equations, we obtain the distribution of u :

$$\Pi_\lambda(u) = \frac{1}{2\pi\lambda}\int_0^{+\infty} \exp(-\frac{u^2}{2\sigma^2})\,\exp(-\frac{\ln^2\sigma/\sigma_0}{2\lambda^2})\,\frac{d\sigma}{\sigma^2}$$

This distribution is symmetric. As remarked in section 1, the velocity distributions are not symmetric.

In order to take into an account the skewness, we introduce a positive number a_s. Physical arguments discussed in details in Castaing *et al.* (1990) lead to the probability density for u :

$$\Pi_{a_s,\lambda}(u) = \frac{A(a_s)}{2\pi\lambda}\int \exp\left[-\frac{u^2}{2\sigma^2}\left(1+a_s\frac{u/\sigma}{(1+\frac{u^2}{\sigma^2})^{1/2}}\right)\right]\exp\left(-\frac{\ln^2(\sigma/\sigma_0)}{2\lambda^2}\right)\frac{d\sigma}{\sigma^2}$$

where $A(a_s)$ is a normalisation constant. From a simple dimensional argument, as no number can be made with only ε and r, a_s might be "universal", independent of ε and r in the whole inertial range.

3. Comparison with experiments

Here we show how the functions defined previously fit the pdf's presented in section 1. In order to cancel the spurious non zero mean value of u a shift is applied to the expression. The skewness parameter a_s is adjusted to experiments but is the same for all of them. We have checked that the whole results are consistent with a constant value for a_s :

$$a_s = 0.18$$

Figure 3 shows the experimental distributions for the small scale of the inertial range, and the corresponding fits with a_s equal to 0.18 and 0.26 . It gives an idea both of the quality of the fit and of its sensivity to the value of a_s .

3.1 Dependence of λ^2 on r

We now examine the variation of $<\ln^2\varepsilon>$ with distance r, as determined from fitting the experimental data. This fitting gives the value of $<\ln^2\sigma> = \lambda^2$ where σ is the variance of the velocity difference at fixed ε :

$$\sigma^2 = <(u_r^2)>_\varepsilon$$

Following Kolmogorov 41, λ is related to $<\ln^2\varepsilon>$ by :

$$<\ln^2\varepsilon> = 9 <\ln^2\sigma> = 9\lambda^2$$

We can thus *verify* the third hypothesis of Kolmogorov (1962) and Obukhov (1962), that $<\ln^2\varepsilon>$ depends linearly on lnr. Figure 4 shows $9\lambda^2$ plotted versus lnr for the tunnel experiment (circles), and for the jet experiment (triangles). Clearly, the data cannot be fitted in the whole range by a formula :

$$9\lambda^2 = \Lambda_o^2 - \mu \ln r$$

with a universal value of μ. The discrepancy could be solved defining a restricted "inertial range" (filled symbols) where the smallest distance is several times the Kolmogorov range (30 times).

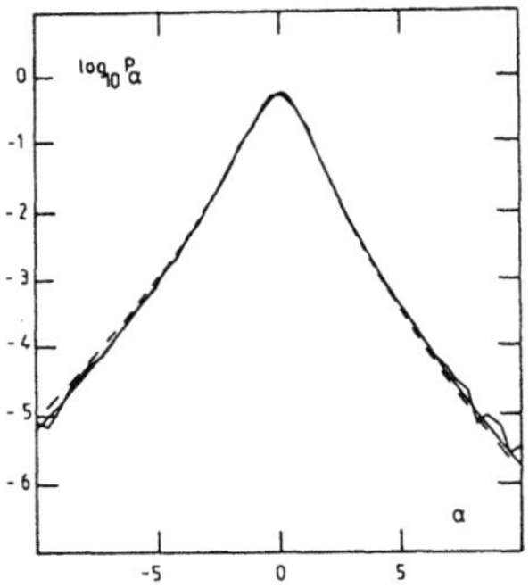

Fig. 3 Comparison of empirical model with experimental data (tunnel experiment). The experimental curve has been copied through a H.P. digit table.
continuous line : $r/\eta = 11.6$; $a_s = 0.18$; $\lambda = 0.428$
dashed line : $r/\eta = 11.6$; $a_s = 0.26$; $\lambda = 0.437$

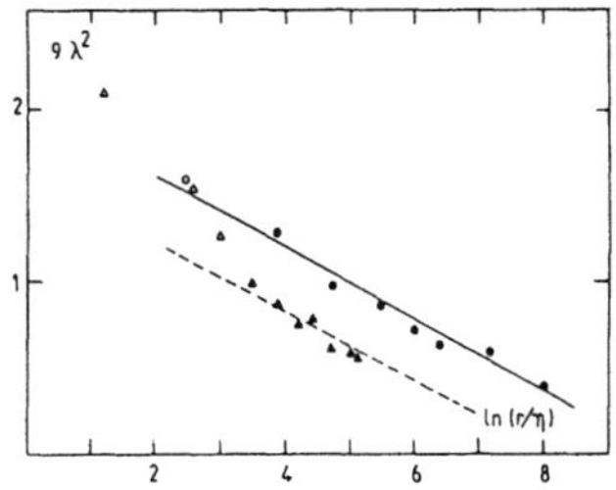

Fig. 4 Verification of the H2 hypothesis of Kolmogorov and Obukhov. $9\lambda^2$ represents the variance squared of $\ln\varepsilon_r$. The filled symbols correspond to the range where $<u_r^3>/r$ is independent on r ("inertial range" in ref. 1) : jet experiment : Δ tunnel experiment O.

It is more interesting however to plot $\ln\lambda^2$ versus lnr which is shown on Figure 5. A straight line approximates the data over the whole range of measurement, down to only 4 times the Kolmogorov scale η, suggesting that a power law

$$9\lambda^2 = (r/r_o)^{-\beta}$$

is physically more meaningfull than the logarithmic one. β is not universal : $\beta = 0.37$ for

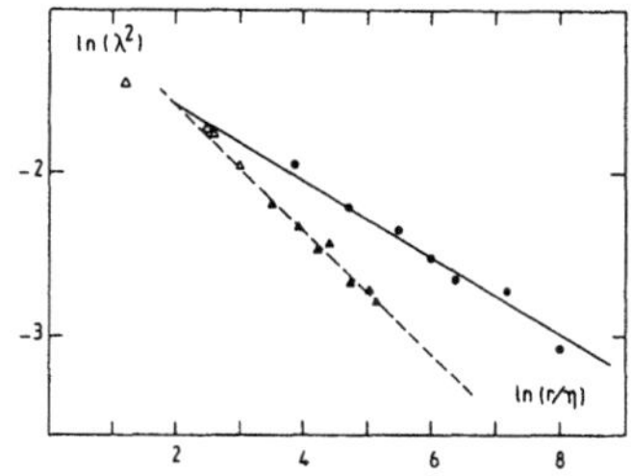

Fig. 5 Same symbols as in figure 4. The linearity of $\ln\lambda^2$ versus $\ln r$ reveals a power law behaviour. Only the smallest distance in the jet experiment shows some saturation of the behaviour. The slopes give $\beta = 0.37$ for the jet and $\beta = 0.24$ for the tunnel experiment.

the jet experiment, and $\beta = 0.24$ for the tunnel one ; it could thus appear as a new adjustable parameter. We shall see however that $\beta = \beta_o/\ln(R_\lambda/R^*)$ where R^* is independently given by comparing various spectra. β_o is thus the only adjustable parameter as μ did.

3.2 Interpretation of β

The fact that β is smaller for the highest Reynolds number is qualitatively in agreement with a recently published theory. Every detail can be found in Castaing (1989). We shall simply recall its foundations and the main results. At the origin is the well known remark that the energy transfer rate ε is constant only on average. It can punctually depend on the scale k. The starting point of the theory is then to assume that the probability to observe a given dependence $\varepsilon(k)$ is such that its logarithm is proportional to :

$$Q = V \int q(\varepsilon, \frac{d\varepsilon}{dk}, k)\ dk$$

where V is the volume. Q $\{\varepsilon(k)\}$ is thus a functional depending on the random function $\varepsilon(k)$. The arguments on which such a form is based are the universally accepted ones for turbulence, homogeneity, independence of scales .... In the same spirit, as the viscosity ν cannot appear explicitely in the inertial range, the only relevant variable is :

$$\psi = -\frac{k}{\varepsilon}\frac{d\varepsilon}{dk}$$

and $$q = k^{d-1}\, f(\psi)$$

where d is the space dimension.

The following results are then be derived :
- The statistics of ψ is gaussian and ε has log normal fluctuations.
- The variance of $\ln\varepsilon$: $\wedge^2 = <(\delta\ln\varepsilon)^2>$ is proportional to the most probable ψ : ψ_m.
- ψ_m, and thus $\wedge^2$ has a power law dependence with k. Namely :

$$\psi_m = \left(\frac{k}{k_o}\right)^\beta ; \wedge^2 = \frac{2}{\beta}\left(\frac{k}{k_o}\right)^\beta$$

- The exponent β has the following asymptotic behavior : $\beta\ln R_e \to$ cste when $R_e \to \infty$.

Beside of these derived results, this formalism suggests geometrical interpretation for β. The preceeding results imply that q is proportional to $k^{\beta-1}$. The quantity Q, logarithm of a probability, can be seen also as a number of degrees of freedom. The fact that its density q behaves as $k^{\beta-1}$ suggests that β acts as an effective dimension for the flow.

Now if the gradient of velocity is concentrated on some structure, the velocity will have significant variations only in the direction perpendicular to this structure. Then β is also the co-dimension of this "gradient structure". This is the geometrical interpretation we propose.

4. Relation with the universal representation of energy spectra

In this section we show that experimental data of energy spectrum are in good agreement with the asymptotic behaviour discussed above ($\beta\ln R_e \to$ cste). The Kolmogorov (1941) similarity hypothesis implies that spatial spectra of fully developed turbulence are proportional if plotted versus $k\eta$ where k is the wave number and η the Kolmogorov length. This is experimentally verified only for very high Reynolds numbers (Monin & Yaglom 1975). In the usual way, the kinetic energy per unit mass is written

$$\xi = \frac{\nu^2}{\eta^2}\int E(k\eta)dk\eta$$

ν is the kinematic viscosity $\eta = (\nu^3/\underline{\varepsilon}))^{1/4}$ where $\underline{\varepsilon})$ is the mean energy flux per unit time and mass from the large to small scales in the Kolmogorov-Richardson cascade : η is the Kolmogorov length, E(K) is the k-spectrum and $K = k\eta$ the wave number in non-dimensional units.

Figure 6 (from Gagne & Castaing 1991) shows spectra corresponding to different turbulent flows whose Taylor scale based Reynolds numbers R_λ extend over 2 decades (130 to 13000). We see that all these spectra merge onto a single curve f(x) if we define:

$$x = \ln(K/K_o)/\ln(R_\lambda/R^*)$$
$$f(x) = \ln(E(K)/E_o)/\ln(R_\lambda/R^*)$$

This figure clearly shows that all spectra down to rather low Reynolds number merge within a translation and homothecy in logarithmic coordinates.The best scaling is obtained for $R^* = 75$ which is nicely consistent with the measured values $\beta(2720) = 0.24$ and $\beta(852) = 0.37$.

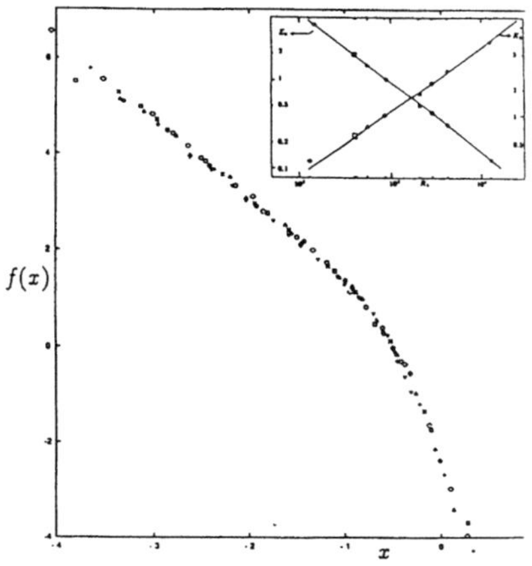

Fig. 6 Unidimensional energy spectrum f(x) with $R_\lambda = 852$ as reference spectrum. Inserted figure : Power scaling of the E_o and K_o coefficients versus R_λ.

On the inserted figure, we observe that the normalization factors E_o and K_o corresponding to translations have a power law dependence on R_λ. Here in this figure, the jet spectrum at $R_\lambda = 852$ has been chosen as the "reference spectrum" (i.e. $K_o(852) = E_o(852) = 1$). With this choice, the insert figure leads to :

$$K_o(R_\lambda) = \left(\frac{R_\lambda}{852}\right)^{\rho} \text{ and } E_o(R_\lambda) = \left(\frac{R_\lambda}{852}\right)^{\sigma}$$

with $\rho = 0.695\pm0.02$ and $\sigma = -0.77\pm0.02$. It can be shown however (Gagne & Castaing 1991) that the exponents ρ and σ depend on the Reynolds number value of the "reference spectrum". The dependence is $\rho(r) = \rho_o/\ln(R_\lambda/R^*)$ and $\sigma(r) = \sigma_o/\ln(R_\lambda/R^*)$. We obtain from our fittings :

$$\rho_o = 1.69\pm0.05 \; ; \; \sigma_o = -1.87\pm0.05$$

In conclusion, the turbulent spectra show no similarity in the sense of Kolmogorov but anyway allow a universal representation. What is important is that this representation is coherent both with the extremum principle approach, and the multifractal model (Frisch & Vergassola 1991). Indeed, the main ingredient responsible of such a scaling is the behaviour of the width of the distribution of ε, close to the dissipation length. This yields a distribution of viscous cut-offs. In all the models of intermittency (K62, multifractal, extremum principle) the variance Λ of the distribution is proportional to $\ln R_e$. In the extremum principle approach it comes from the behavior of β, inversely proportional to $\ln R_e$. The resulting growth ot the cutt-off distribution yields the spreading of the spectrum and this universal representation.

References

Anselmet, F., Gagne, Y., Hopfinger, E.J., and Antonia, R.A. (1984), J. Fluid Mech. 140, 63-89.

Castaing, B. (1989), J. Phys. (Paris) 50, 147-156.

Castaing, B., Gagne, Y. and Hopfinger, E.J. (1990), Physica D 46, 177-200.

Champagne, F.H., Harris, V.G. and Corrsin, S. (1970), J. Fluid Mech. 41, 81-139.

Champagne, F.H. (1978), J. Fluid Mech. 86, 67-108.

Frisch, U. and Vergassola, M. (1990), Europhys. Lett. 14, 439-444.

Gagne, Y. (1987), Thesis (Institut National Polytechnique de Grenoble, unpublished).

Gagne, Y. and Castaing, B. (1991), C.R.A.S. (Paris) Série II 312, 441-445.

Hannart, B., Gagne, Y. and Hopfinger, E.J. (1985), Série II 301, 669-674.

Grant, H.L., Stewart, R.W. and Mollier, A. (1962), J. Fluid Mech. 12, 241-268.

Kistler, A.L. and Vrebalovich, T. (1966), J. Fluid Mech. 26, 37-47.

Kolmogorov, A.N. (1941), Dokl. Acad. Nauk. 30, 301-305.

Kolmogorov, A.N. (1962), J. Fluid Mech. 13, 82-85.

Kraichnan, R.H. (1974), J. Fluid Mech. 62, 305-330.

Monin, A.S. and Yaglom, A.M. (1975), Statistical Fluid Mechanics 2 (M.I.T. Press).

Obukhov, A.M. (1962), J. Fluid Mech. 13, 77-81.
Sandborn, V.A. and Marshall, R.D. (1965), Colorado State University C.E.R. 65.
Williams, R.M. and Paulson C.A. (1977), J. Fluid Mech. 83, 547-567.

B. Castaing
Centre de Recherches sur les Très Basses Températures
C.N.R.S.
BP 166
38042 Grenoble-Cédex
France

Y. Gagne and E.J. Hopfinger
Institut de Mécanique de Grenoble
U.M.R.101
BP 53 X
38041 Grenoble-Cédex
France

Discussion

Zaslavsky: The idea that there is non-self similarity seems to be an experimental fact. It would be good to find an explanation for this. There are two approaches: one is phenomenological, i.e. to propose something and this will immediately give us what we would like to have. The second way is to get the non-self similarity from first principles. Multi-fractality relates to the non-homogeneity of real space (or some other spaces of parameters or variables) and it may be a good way to get corrections of the Kolmogorov law. The problem is to do this in a consistent way with experiments, which does not merely fit the theoretical curve with an experimental one, but is in full correspondence with the real experimental physical processes.

Castaing: I totally agree with you. The problem is to define new quantities, different from the quantities measured up to now. That was the only goal we had about this distribution. To find a way to measure the quantity which was not measured up to now: the parameter Λ^2, i.e. the variance of the fluctuations of $\ln \varepsilon$. Our only goal was to measure a physical quantity with which we can test the different theories.

Novikov: Of course the best way to get all these results is somehow to get it from the Navier-Stokes equation, but nobody can do this right now so we have to compromise and use various models and I think maybe there are two criteria for using the model. One is pure aesthetics. If the model is beautiful, if you like it, okay. Another, maybe of more specific character, is how many free parameters you have in the model. If you have the minimum of parameters, this model will probably be preferable, at least when we are dealing with phenemenological theories. So my question is how many free parameters have you got in your model in comparison with other models?

Castaing: It depends on what you look at, on how many experimental facts you want to fit, and that is the way to rephrase what you say. It would be interesting to verify that the same fitting parameter works for two different measurements. For example, we are very happy to see that the R^* we have from the measurement of our β is the same as the R^* we obtain from the merging of all the spectra together. An interesting point is the one that Frisch pointed out, i.e. that the universality of the spectra provides an opportunity for comparing the D(h) obtained from measurement in the inertial range to the D(h) obtained from the universal spectrum. I hope to use this as a test of the theory. I think this is another way to answer your question, to obtain the same adjustable quantity, which explains the different set of data.

Novikov: That is correct, but how many parameters do you have? That is my question.

Castaing: We use the same number as others. For the variation of Λ^2 we have β, which is adjustable with each Reynolds number R_λ. You think we have more adjustable parameters than, for example, the Kolmogorov-Obukov law, which has only the μ, namely $\lambda^2 = -\mu \log(r/L)$, but as we know the behaviour of β versus $\log R_\lambda$, we also have only one adjustable parameter, $\beta_0 = \beta \log(R/R^*)$. But again, it is important to find new experimental results, new quantities which have not been found up to now. Then it will be interesting to see which theory introduces new adjustable parameters and which theory does not.

Jimenez: But what happens when, for instance, you start to rotate turbulence or when you sheer it strongly or when you put on a strong pressure gradient. That would presumably be different for different models, and that's something that can be tested and there are a lot of experiments on that.

Castaing: Certainly for out of-equilibrium experiments you need an extension of theories for a completely steady state situation. We cannot say anything about the out-of-equilibrium situation. That's a very interesting point, certainly, but it hasn't been solved up to now.

Herring: So far you dealt with the velocity differences. Is it possible to extract from your formalism the distribution for the velocity itself, which is probably Gaussian or sub-Gaussian, somehow eliminating one of the points in your formalism. Can you connect this in some way to a simple distribution function for the velocity field at a single point?

Castaing: That's hard to answer. In general velocity at a single point is not considered as universal.

Herring: Just one point. There are some measurements by Townsend, I believe in Batchelor's monograph, which show distribution functions which are sub-Gaussian (narrower than Gaussian). Such sub-Gaussian velocity PdF's have been consistently found in direct numerical simulations, for example that of Kimura and Kraichnan. (This issue is discussed in Kimura's paper in Topological Aspects of the Dynamics of Fluids and Plasmas, Proceedings of the Program of the Institute for Theoretical Physics, University of California at Santa Barbara (eds. Moffatt, Zaslavsky, Tabor & Comte, Kluwer Academic Publishers, Dordrecht, Netherlands).

Yaglom: Maybe I'll continue the question raised by the Russian participants. One small remark is related to the question raised by Zaslavsky. You said that it is necessary to explain the deviation from self-similarity. From my point of view it is in all cases necessary to explain the reason for self-similarity. It is clear that any self-similarity assumption represents some mathematical idealisation, which must be clearly formulated and justified. The multifractal model is a rather reasonable generalisation of the Kolmogorov self-similarity model.
Then another question related to your spectral data. Of course your result must agree - you said that your result agrees - with Kolmogorov's inertial range result, but my feeling is that the spectral data obtained in the wind tunnel by Sandborn and Marshall and also the data of sea measurement by Grant et. al. and of atmospheric measurements by Pond et. al. agree rather well not only in the inertial range, but also in a part of the dissipation range at least. (Monin and Yaglom 1975, Statistical Fluid Mechanics, MIT Press, **2**, Fig. 7b, 489). This agreement seems to disagree with your statement that the data in the dissipation range contradict the ordinary concept of self-similarity. Is this explained by the fact that all three experiments mentioned above correspond to very high Reynolds numbers?
My next remark is related to your use of the data by Kistler and Vrebalovich (J. Fluid Mech., 1966, **26**, 37 - 48). I think that these data are incorrect. They disagree with most of the other data (in particular, they lead to too great a value of the Kolmogorov constant and contradict local isotropy of the velocity fluctuations. (See, e.g., Schedvin, Stegen and Gibson, 1974, J. Fluid Mech., Vol. **65**, 561 - 579, and Yaglom, 1981, Izv. Akad. Nauk, SSSR, Ser. Phys. Atmosph. and Ocean, Vol. **17**, 1235 - 1957) and were obtained at the facilities that were dismantled shortly after the measurements which made impossible any further checks. It is a pity that only these dubious data were used to illustrate the Kolmogorov theory in the well-known books by Hinze (Turbulence, 2nd. ed., Mc Graw-Hill, 1975) and Townsend (The Structure of Turbulent Shear Flow, 2nd ed., Cambridge Univ. Press, 1976). Therefore I don't think that it is appropriate to use these data for verification of your theory.

Moffatt: This is perhaps rather a question for Uriel Frisch although it is related to what you said also. In the multi-fractal approach, one of the main physical effects that comes out is the spread in the cut-off length scale that is implied. We are accustomed to thinking of the turbulence spectrum as being the Kolmogorov $k^{-5/3}$ with a rather well-defined cut-off at the inner Kolmogorov scale and an exponential cut-off beyond that. Do we have to revise this view? It's not clear that this traditional picture of turbulence needs to be revised at all, but if it does have to be revised, then what is the modified form of the spectrum beyond the traditional spectrum cut-off where the multi-fractal effects have some effect?

Frisch: I don't believe that for Navier-Stokes turbulence the data are sufficiently good at the moment to decide in a clean way if Kolmogorov type universality should be replaced by multi-fractal universality, I mean, I just hope that this question will be reasonably settled within the next, say, couple of years. It certainly has motivated a number of new experiments, which are on their way. According to the multifractal model, the energy spectrum in the 'intermediate dissipation range' has a pseudo-algebraic form: a power-law which steepens slowly (the exponent grows logarithmically). The exact functional form depends on the function D(h) which is not (so far) given by theory. At the moment, unless somebody very clever comes up with a theoretical idea to calculate this function, you have to measure it. The interesting thing is that you can measure it in different ways. You can measure it by looking at structure functions, their exponents - and that's a purely inertial range measurement - or you can take the Fourier transform of the usual energy spectrum, and if you are able to make careful measurements, down to scales considerably smaller than the Kolmogorov cut-off, you can also get the function D(h) independently of that. Actually, you can do that also from Fourier transformed high order structure functions, so you can get, in principle, the same function from very different measurements and then see how well they agree. That is a necessary condition for this kind of description to be right. Of course one can try to think of theories to produce this function, but I personally am not so eager to do that at the moment because I'm not yet sure all this is right. You see it doesn't take much energy to produce the kind of thing I presented to you today. It was born in about 15 to 20 minutes after a vivid discussion with Victor Yahot. In contrast, it takes a lot of energy to make good experiments or good simulations, and that's, I think, where the effort has to go as a priority.

Jimenez: Yes, I would like to put a question to all of you: Is there any physical counterpart to this multifractal model, something in terms of a structure that I can think of?

Frisch: There are examples in related fields where one can answer this question in a specific way. For example, in the problem of the growth of a magnetic field by dynamo action. If one looks at it from a deterministic point of view, the folding and stretching of the magnetic field leads to a distribution of magnetic field, which can be characterized as multifractal, and one can then look at the geometry of these tubes and ribbons and all these things. One might be tempted to apply similar considerations to the vorticity, but I'm

always a little bit hesitant to treat the vorticity as if it were a passive vector. That isn't quite true, but that is the closest thing I can think of at the moment.
At the moment the answer is just whatever you do on the velocity, you can transpose to passive scalar, of course with the usual precaution that the Prandtl or Schmidt number or whatever is equal to unity. You have to take the usual precautions, but you can transpose it very easily. It may be equally wrong, but it is easy to transpose.

Lesieur: This is slightly puzzling because you may have a different behaviour of the temperature with respect to the velocity - so I wonder whether you interpret that from the point of view of small-scale intermittency.

Frisch: I am talking about exponents and not about numbers of course. Exponents are hard to extract from simulations.

Aref: One of the things that was mentioned in the first talk this morning was that direct numerical simulations are restricted to relatively low Reynolds numbers. And I was wondering - essentially the Reynolds number of the simulation is a measure of the ratio between the smallest and the largest scale. Now the other talks suggested that there is really a lot of structure at the small scales and that different things are cut off at different small scales. That would suggest that if you are doing a numerical simulation, you really need to capture that entire range of small scales in order to get good predictions for the scaling and self-similarity. And that would suggest that you must be limited to even smaller Reynolds numbers because you would have an even smaller range of scales to go up. I was wondering if any of the people who thought about these theories have thought about the restrictions that this puts on numerical simulations.

Frisch: Actually, Paladin and Vulpiani (1987, Phys. Rev. **A35**, p. 1971) were the first to comment on this when they noticed that there is a range of cut-offs depending on the scaling exponent. They pointed out briefly that this has some consequences for direct numerical simulation of turbulence, that the smallest scale may be even smaller than you thought, and this is the question which we also briefly addressed in the Frisch-Vergassola Europhysics letter. At the Reynolds number which can be presently simulated on the largest super-computer or parallel machine, the question of one scaling versus another one is a marginal one. For example, I could show you some simulations done by Ishyi and Sanada at resolution 256 cube and various $R\lambda$'s where, if you look at the spectra carefully, you see that the higher $R\lambda$'s have a fall-off in the dissipation range that is a little bit less curved, more open. But this is just marginal evidence. As the computers will grow, and they definitely are growing, if anything like the multifractal scaling holds, then you would predict the growth of the maximum Reynolds number that can be achieved as a function of the power of the computer. With multifractal scaling, achievable Reynolds numbers would be growing more slowly than with Kolmogorov scaling. This would be bad news. But I believe that this is not really going to be felt before maybe another 20 years or so. So it's not an urgent matter.

Castaing: If the scaling occurs, it means that the Reynolds number by the Taylor scale of 130 is on the same asymptotic regime as for a Reynolds number of 13000. Obviously, we have to understand in which way we are in the same asymptotic regime. I guess that Reynolds numbers obtained by numerical simulations are already in the universal regime, and that could be interesting.

Antonia: I don't understand this universal regime. If the spectra are re-plotted in a manner which weights the high wavenumber end of the spectrum (e.g. Champagne 1978, J. Fluid Mech., **86**, 67-108), then you might expect to observe a distinct dependence on the Reynolds number. Champagne's spectra, which were weighted by k_1^4, showed a clear Reynolds number dependence. In this sense, I don't understand the claim of universality.

Frisch: That has nothing to do with the question of whether you have a universal shape or not.

Antonia: Yes, it does.

Wygnanski: You simply don't seem to think of this type of plot because of the very small logarithmic scale. But if you replot it, then the little differences that seem to be just a bit of a scatter now become very prominent. I think that's what Antonia is probably trying to say.

Frisch: This is another issue: multifractal scaling implies that spectra for different Reynolds numbers can be collapsed onto a single curve, provided (i) they are plotted in log-log coordinate, (ii) both coordinates are renormalized by a one over log-Reynolds factor, (iii) suitable shifts are performed. In contrast, only step (i) and (iii) are needed for Kolmogorov scaling.

Novikov: I would like to comment on the situation raised by Lesieur about the difference between scalar properties like temperature and the velocity. The velocity is determined by a nonlinear differential equation. It creates multifractal structure itself as a consequence of the nonlinear differential equation. But the passive scalar is something which can only reflect this. Apart from dissipation of energy you have dissipation of the scalar quantity, and you have to assume how the probability distributions of these two quantities are related. Are they independent? If they are dependent, it makes the theory much more complicated, it involves additional assumptions, so it is not just a trivial transition from the standard multifractal description.

Aref: I agree with this because we know that you can get very highly intermittent passive scalar fields even from very regular velocity fields. So we don't always know if a passive scalar exactly mimics the fluctuations in the velocity field.

Novikov: I'm not talking about chaos. I'm talking about turbulence. It is a different thing. The problem is that you have a linear equation for temperature. You can assume how the rate of supply of temperature is related to the rate of supply of energy in order to construct multifractal models. It's an additional complication.

Frisch: The fact that the equation for a passive scalar is *linear* is an illusory simplification. *Linear* does not imply *simple*. We deal with a linear partial differential equation and if, for example, you neglect diffusivity, the passive scalar follows a fluid particle trajectory. Finding fluid particle trajectories is an exercise in Hamiltonian dynamics, which is highly non-trivial and which can lead to structures of arbitrary complexity. We don't even fully understand what happens in the prescribed velocity field although some progress has been made thanks to Hénon, Aref and others. By the way, what is known as "chaotic advection" is closely related to the old problem of "Lagrangian versus Eulerian" statistics.

Herring: I have a simple question related to practical issues in your talk, and also in Uriel's. What is the feedback from what you know or what you suppose happens at very small scales to the large scales? I believe people dealing with large scale effects usually put in only the number of degrees of freedom they need for these large scales (together with some form of eddy viscosity) and then the feedback from small scales in the real system eventually caused things to go wrong on the large scales. Does the multifractal perspective enable you to see what small scale effects feed back to the large scales?

Aref: I was wondering, maybe there is a way to address this question of things cutting off at different lengths by numerical experiments. I believe that what happens, for example, if you run a large spectral simulation and you set your viscosity too low compared to numerical resolution, is that at the smaller scale your energy spectrum starts to curve up. Now, maybe you should look at the spectrum of different quantities at different resolutions; you could actually see them curve up at different points, and that would tell you that different things were being cut off. So, essentially, by running somewhat under-resolved simulations, you might be able to check this idea.

Frisch: That's an interesting suggestion. It connects with something that van de Water et al. (van de Water, van der Vorst and van de Wetering 1991, Europhys. Lett. **16**, 443), attempted recently. Another prediction of multifractral scaling, which I didn't have time to mention is that structure functions of different orders have different cut-offs, just because they are influenced by different scaling exponents. Van de Water et al. have reported some experimental evidence (from wind tunnel data) that the cut-off depends indeed on the order. When I looked at it, it's not entirely clear: The Reynolds number is not very high; it's just a little bit higher than what numerical simulations can achieve, but I think that's an interesting question to look at.

Zaslavsky: This seems to be a very serious question and it deserves more discussion. Experiments would be planned in such a way that one could obtain more details for the problem of large- and small-scales connection. We have to consider large scale behaviour and see what will happen with small scale; we have to consider small-scale and control it to understand what will happen to large scales and their interaction; and that is perhaps a good reason to discuss how such kinds of experiments, real experiments or conputer simulations, can be provided. It is not enough to consider just spectrum or its deviation from the Kolmogorov-Obukhov law. It would be desirable to look at the real process, which could explain the deviations in the spectrum and which could put something essential into the word "cascade".

Yaglom: I want to stress that the experimental confirmation of the old Kolmogorov theory of 1941 is much less convincing for the scalar field than for the velocity field. The available data clearly show that the small-scale temperature fluctuations T' are not statistically isotropic, e.g., $\overline{(\partial T' / \partial x)^3} \neq 0$ (the overbar symbolizes averaging) and the temperature structure functions of odd orders do not vanish. Moreover, even the value of the Kolmogorov constant (more correctly, of the Obukhov-Corrsin constant - the dimensionless coefficient of the -5/3 power law for the temperature spectrum) is possibly not universal - in any case the scatter in the experimental values of this constant is much greater than the scatter of experimental values of the Kolmogorov constant for the velocity field. So probably the deviations from the Kolmogorov theory in the case of a scalar field are much greater than in the case of the velocity field.

Castaing: Intermittency is much larger for temperature. Don't you think that one is the consequence of the other? That the deviation from Kolmogorov (1941) is due to the very large intermittency compared to the intermittency of velocity?

Yaglom: Maybe, but it is not so clear how the intermittency can produce the local anisotropy of scalar fields. Maybe the strong intermittency necessarily leads to a considerable increase of the averaging time to obtain the reliable result, and therefore many of the available experimental estimates of the statistical characteristics of scalar fields are unreliable. In any case, at present we must state that the agreement of the measurements of scalar field fluctuations with the theoretical universal laws is still unsatisfactory.

Frisch: I would like to come back to the talk given by Akiva Yaglom , the problem of the Kármán law, the logarithmic law and these issues. I think that in a sense there is not enough attention paid to such matters. It is strange that before the war this was one of the very first nontrivial predictions made in turbulence and everyone was very excited. Then this thing was taken for granted. Certainly people who do practical turbulence, such as engineers, pay a lot of attention to the logarithmic law, but people who are more interested in fundamental questions seem not to have put too much energy in trying to really understand what is going on there. It seems to me that it is well worthwhile to carry out clean experiments. For example, some years ago there was a kind of pannel which suggested something like "a super pipe" (See my review paper with Orszag in the January 1990 issue of Physics Today). Whether it is super in size or in low viscosity, there are various options, where a Reynolds number of maybe 10^8 so can be achieved, together with a very, very large aspect ratio. I mean people are still fighting about the law in the 'buffer region', assuming there is indeed a buffer region. I see very few theoretical attempts; one noticeable exception being the work of Barenblatt (presented at the "First European Fluid Dynamics Conference" in Cambridge, September 1991). Think of all the efforts to make new models for homogeneous turbulence; there is hardly any counterpart for well-bounded flows although there are clean and well-defined problems. Mostly, I don't see such a massive effort to make good experiments. I wonder if someone has something to say about that.

Wygnanski: I would like to make a remark about the wall jet for which the classical formulation of the "law of the wall" does not seem to work. This was first observed by Bradshaw and Gee (1960, Aero. Res. Council R & M 3252). We have recently made measurements in a wall jet (Wygnanski, I., Katz, Y. & Horev, E., 1991, J. Fluid Mech. **234**, 669 - 690) trying to understand this problem in some detail. Unfortunately, our Reynolds number is fairly low, though it is comparable to other experiments in this flow in which universality in terms of the classical formulation of the law of the wall was claimed. In the case of the wall jet one cannot get the velocity distribution or its rate of change to be independent of viscosity at $y^+ = 30$. One may fit all kinds of semi-logarithmic laws to the velocity profile, but this is not universal.

Frisch: That is a rather complex flow we are talking about.

Wygnanski: Yes, but it might be significant because surprisingly enough the "defect law" works extremely well all the way towards the wall, provided $y^+ > 30$. I think this is something worth exploring.

Novikov: I would like to make a comment about the non-similarity generally. Don't forget that we are given this Navier-Stokes equation and we have to trust it, and there are some exact consequences of the Navier-Stokes equation, and there are some speculations based on these exact consequences, and this is the reason why there is no similarity. The first exact consequence of Navier-Stokes equations is that the momentum flux in a boundary layer is constant, and this momentum flux is proportional to the friction velocity squared. Then people make assumptions that this friction velocity is the main parameter and from dimensional argument they get the logarithmic profile. But this logarithmic profile is proportional not to the flux, but to the root of the flux, and because of this it is not universal. Because the exact result from Navier-Stokes equation says the flux is constant and is not the root of flux. The same thing holds for the Kolmogorov similarity law.
The same thing holds for the local structure of turbulent flows. From the Navier-Stokes equation you get exactly the result obtained by Kolmogorov, that the third order structure function of velocity is proportional to ε multiplied by the universal constant. Then, there is an assumption that maybe ε is the main parameter and the full theory of Kolmogorov and others is based on that. But all other characteristics contain ε in various forms and they are not universal. So non-universality of logarithmic profile and non-universality of higher order moments are the same type of non-universality because they are not exact consequences of the Navier-Stokes equation. We have to remember that.

Frisch: I think I would have to explain in detail my revisionist view of Landau's argument and - I think that Landau's argument can be applied also to the non-universality of the Kármán constant, and I wouldn't make any comment about the Navier-Stokes equation; that is not the issue in so far as the Navier-Stokes equation is not used directly.

Novikov: It's used directly because how else do you get the flux to be conserved, it's an exact consequence of Navier-Stokes. The same thing holds for the moments of flux of energy and many other quantities.

Frisch: Yes, but the crucial thing is that if you have fluctuations on large scales, very large scales, where the flux that wants to go down to the wall is itself modulated by something that varies very slowly in space or time, then the square root of the average flux is not the same as the average of the square root. Just as in Landau's argument.

Zaslavsky: It depends on how one does the averaging. There are some good examples that the averaging can be done in such a way that the universality is lost.

Frisch: But people could argue that large scales don't determine the flux. In the Kolmogorov homogeneous type of turbulence one usually argues that the amount of energy dissipation is essentially determinded by what the large scales want to send the smaller scales. I am not saying that I necessarily accept this; that's a traditional view. And in the theory of the boundary layer, the turbulent boundary layer, I am not so sure that people would accept the view point that the flux to the wall is something that is prescribed at large scales. If it is, one can apply a Landau argument; otherwise not.

Session Two

Chairman H. K. Moffatt

On Vortices And Cascades In Turbulence

Marcel Lesieur, Olivier Métais and Pierre Comte

Abstract. Using direct or large-eddy simulations, we study coherent vortices in free-shear or wall-bounded flows. In free-shear flows, we show that the Kelvin-Helmholtz instability is responsible for the condensation of vorticity into large tubes corresponding to strong depressions. These tubes may be quasi two-dimensional, or strongly distorted due to helical pairings. Smaller-scale hairpin vortices may also be stretched between the large vortices. In wall-bounded flows, hairpin vortices are also stretched by the basic shear. Kelvin-Helmholtz and hairpin-type eddies seem then to be the basic coherent ingredients of turbulent-shear flows. In 3D decaying isotropic turbulence, we show again the existence of large coherent vortices corresponding to strong depressions, which stretch intense passive temperature fluctuations inbetween. The temperature spectrum has an anomalous k^{-1} range in the large scales. The temperature pdf is exponential in the wings and behaves like the vorticity and the low pressures. An attempt is made for explaining these observations in terms of superposition of a quasi 2D slow turbulence and a Kolmogorov rapid turbulence.

1. Introduction

Among the important fundamental issues posed by the extremely difficult (and still largely unsolved) problem of turbulence, the questions related to coherent vortices and their interaction with smaller scale turbulence are of paramount importance. Related to this is the relevance of standart statistical theories (such as the Kolmogorov cascade for instance) if the large-scale intermittency due to these vortices is taken into account. In this paper, we will address such questions using numerical experimentation as a tool: an ensemble of original results concerning the generation and dynamics of coherent vortices in shear flows or isotropic turbulence will be presented. These results have been obtained with the aid of direct or large-eddy simulations of the 3D Navier-Stokes equations (with constant density), developed in Grenoble by pseudo-spectral or finite-volume methods. With the precious help provided by efficient 3D graphics which allow to capture in details the topology of vorticity, pressure and temperature fields for instance, these simulations complement in a valuable manner the numerous laboratory experiments performed on turbulent flows for many decades. Due to the experimental (laboratory and computations) knowledge thus acquired, and the informations provided by the theory (stability and statistical theories), it seems that we are at a point where a unified understanding of turbulence and coherent structures is at hand within a near future. This paper will try to be a modest contribution to this unification.

2. Coherent vortices in free-shear flows

2.1 Two-dimensional case. The two-dimensional dynamics of mixing layers developing between two parallel flows presenting velocity differences is well understood, due to numerous 2D numerical simulations. This is true both in the temporal or spatially-growing case (see Lesieur, 1990, for a review). These calculations show in particular how coherent eddies form by instability and roll up of the initial vortex sheet, after the development of Kelvin-Helmholtz instability, and undergo successive pairings. The computation permits also to check that the coherent vortices do correspond to strong pressure lows, in good agreement with simple models of recirculating eddies, where the centrifugal force is balanced by the pressure gradient (cyclostrophic balance). Simulations of 2D wakes or jets display also Karman streets of alternate eddies of opposite sign, still corresponding to pressure troughs. These eddies may either pair (if they are of same sign), or form couples (if they are of opposite sign) which travel under their mutual induced velocity.

Even in 2D isotropic turbulence are these vortices found (see e.g. McWilliams, 1991): in this case, they originate from local Kelvin-Helmholtz instabilities, and grow through various pairings. For instance, it was shown by McWilliams (1991), in a calculation starting with a gaussian initial field, that the relative area occupied by fluid with $|\omega|$ smaller than 10% of the extremum grows from a very low value to, say about 0.5, in 60 initial turnover times. This condensation of vorticity in spots indicates a large-scale intermittency, where $|\omega|$ will take either low or high (close to the initial maxima) values. As checked by McWilliams (1989), this corresponds to a pdf $P(\omega)$ which is strongly nongaussian, with a peak at $\omega = 0$ (since large regions of the flow are irrotational), and exponential wings. Since the vortices correspond to marked pressure troughs, one expects the pressure pdf in the lows to be also exponential. To our knowledge, the pressure pdf has not been determined in 2D isotropic turbulence calculations, and it would be of interest to check this point.

2.2 Three-dimensional stability of coherent structures. In fact, two-dimensionality is a fiction, permitted only by the computer (or requiring external constraints such as a rapid rotation for instance). Laboratory experiments show important 3D effects, for instance under the form of hairpin vortex filaments stretched between Kelvin-Helmholtz billows in the mixing layers (see, e.g., Breidenthal, 1981; Bernal and Roshko, 1986; Lasheras and Meiburg, 1991; Comte et al., 1991). Intense 3D perturbations develop, leading to the appearance of Kolmogorov $k^{-5/3}$ kinetic energy cascades. The three-dimensionalization of an initially quasi-2D shear flow is, again, very easy to study numerically. We have done that in the case of a temporal mixing layer, using Orszag's pseudo-spectral methods, with a resolution of 128^3 wavevectors. The basic flow is a hyperbolic-tangent velocity profile $U\text{th}\ 2y/\delta_i$, to which are superposed two small gaussian random perturbations of wide spectrum peaking below the most unstable 2D mode. One of the perturbations is 2D, of kinetic energy $\epsilon_{2D}\ U^2$, and the other is 3D, of kinetic energy $\epsilon_{3D}\ U^2$: we have studied the *three-dimensionally forced transition* ($\epsilon_{2D} = 0, \epsilon_{3D} = 10^{-4}$) and the *quasi two-dimensionally forced transition* ($\epsilon_{2D} = 10^{-4}, \epsilon_{3D} = 10^{-5}$) cases.

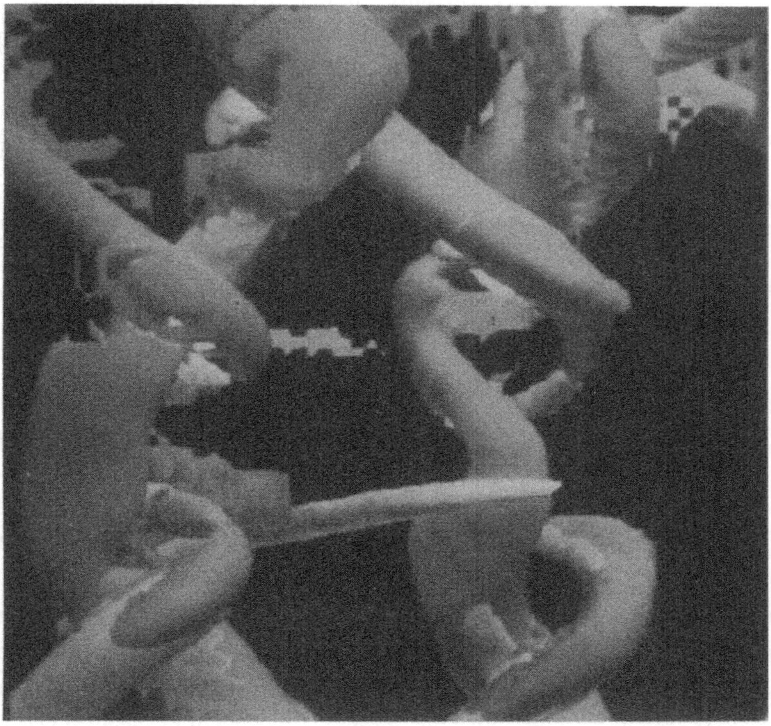

Figure 1: direct-numerical simulation of the vorticity field during a helical pairing in a three-dimensionally forced mixing layer.

Figure 1 shows a top view of the vorticity field in a box of size 4 fundamental wavelengths, for the three-dimensionally forced transition calculation. This picture is taken at a time $t = 30\ \delta_i/U$, corresponding in the 2D case to the pairing. Here, one can see a *helical pairing* between primary vortices. One notices also the presence of thin longitudinal hairpin vortices stretched in the stagnation zone between the large vortices. They carry longitudinal vorticity of $\approx |\omega_0| = 2\ U/\delta_i$, the modulus of the maximum initial spanwise vorticity. The negative leg of this hairpin vortex is weaker than the positive one, due to a scattering of the vortex filaments in this zone, as shown when examining the vortex filaments (see Comte et al., 1991, 1992). Figure 2 shows the low pressure field in the same calculation as in Figure 1. It is clear that pressure troughs are very good indicators of the large vortices. Here pressure is also affected by the positive leg of the hairpin vortex. This shows that low pressure reacts to intense vorticity. A detailed examination of the vortex-line time evolution, as well as the maximum and minimum vorticity components (Comte et al., 1992) shows that the zones of helical pairing between primary vortices do correspond to regions of maximum vortex stretching: at $t = 20$, vortex filaments in these regions start being convoluted, which implies in particular the growth of spanwise vorticity of sign opposite to the primary vorticity: it reaches a maximum value of $\approx 3|\omega_0|$ at $t = 40$, where the three vorticity components all reach this intensity (with both positive and negative values). This indicates a sort of isotropization of turbulence, and also a very spectacular production of enstrophy by vortex stretching. Within this context, the growth of *anomalous anti-spanwise vorticity*[1] is quite remarkable, and might be a powerful indicator of the ultra-violet kinetic energy cascade.

Figure 2: same calculation as in Figure 1, low pressure field.

In the forced-transition case, we see (Figure 3, see Comte et al., 1991, 1992) at the same time quasi-2D large vortices, between which longitudinal hairpin vortex

[1] Anomalous, in the sense that such a growth is not permitted in two dimensions, where the very low initial vorticity of this type due to the random perturbation will rapidly decay.

filaments are stretched. The latter are weaker than in the former case, of vorticity modulus $O.1 \approx 0.2\ |\omega_0|$ of the initial maximum spanwise vorticity. In this case, we have checked that the big spanwise vortices correspond to strong depressions, but the longitudinal vorticity is too weak to be visible on the pressure.

Longitudinal kinetic energy spectra in the 2D mixing layer have a slope comprised between k^{-3} and k^{-4} (Lesieur et al., 1988), and could be interpreted as an enstrophy cascade, where vorticity braids in the stagnation region between coherent vortices are strained, and eventually wrapped into the cores of the vortices. On the other hand, many laboratory measurements display well formed $k^{-5/3}$ Kolmogorov spectra above the wavenumber characteristic of the coherent vortices. In the 3D calculations presented above, the longitudinal kinetic energy spectra do show in the large scales a tendency to a $k^{-5/3}$ inertial range in the three-dimensionally forced case, whereas they are very close to the $k^{-3} \approx k^{-4}$ spectra of the 2D calculations in the quasi two-dimensionally forced case (Comte et al., 1992). One might be tempted to believe that the strong stretching needed by a 3D Kolmogorov cascade requires local helical pairings. The latter have, in fact, been observed in the laboratory by Browand and Troutt (1980), as recalled by Pierrehumbert and Widnall (1982). However, it is also possible that more intense hairpin vortices may be obtained in the quasi two-dimensionally forced case, if a higher level of three-dimensionality is put initially. In this case, the crushing of these vortices between the billows during pairing will also yield an ultra-violet Kolmogorov cascade. Let us mention also Moffatt's theory (1991), where a $k^{-5/3}$ cascade may be obtained by 3D randomly distributed spiralling KH vortex sheets.

We have mentioned earlier the difficulty to maintain two-dimensional turbulence: more specifically, a 3D numerical stability study of a 2D evolved isotropic turbulence has been carried out by Bartello (1991), on the basis of a large-eddy simulation: this calculation shows that the three-dimensionalization occurs with hairpin vortex filaments strained between the 2D vortices. When a rotation $\vec{\Omega}$ (of axis parallel to the vortex axes) is present, a high rotation rate[2] stabilizes the 2D turbulence, in good agreement with Taylor-Proudman theorem. On the other hand, a Rossby number of the order of 1 leads to a 2D stabilization of cyclonic vortices, and a disruption of the anticyclonic vortices which are rapidly strained and mixed by the ambient shear. This is in good agreement with a theory based on the straining of absolute vorticity proposed by Lesieur et al. (1991). The result is the reorganization of the flow, composed initially of an equal number of cyclonic and anticyclonic 2D vortices[3], into a set of only cyclonic vortices: this is an interesting example of "how order may

[2] Defined by a low Rossby number. Here, the Rossby number $\omega_0/2\Omega$ is built with the initial 2D rms vorticity ω_0.

[3] since, in a 2D periodic box, the circulation of the flow around the domain boundary is zero

Figure 3: direct-numerical simulation of the vorticity field in a forced transitioning mixing layer.

arise from chaos". The same result holds for a wake (see Yanase et al., 1991), and even with an initially 3D isotropic turbulence (Bartello, 1991): for a wide range of Rossby numbers aroud 1, rotation still reorganizes the flow into quasi-2D cyclonic vortices parallel to the axis of rotation. This gives an explanation for the laboratory experimental results of Hopfinger et al. (1982), where only cyclonic vortices are found. This phenomenon might also play a role in the generation of tornadoes, where the anticyclonic branches are nearly always disrupted.

2.3 Backward-facing step. The separated flow in a channel behind a backward-facing step is another example of mixing layer, with further complexities brought by the boundaries and the recirculation. With the aid of a finite-volume simulation code developed by Grand et al (1988), and using a subgrid-scale model called the structure-function model (see Métais and Lesieur, 1991), we have simulated two configurations, respectively a high step and a low step (with respect to the upstream channel width, see Silveira, 1991, and Silveira et al., 1991 for details). In the high-step case, the calculation shows a vortical structure very similar to the mixing layer in the case of quasi two-dimensionally forced transition, with Kelvin-Helmholtz vortices shed behind the step, and a secondary longitudinal hairpin structure stretched by the flow (Figure 4). The corresponding maximum longitudinal vorticity thus generated is here 40% of the maximum spanwise primary vorticity. In the low-step case, the structure of the flow is much more intricate, due to the influence of the lower side wall. However, the low-pressure plots indicate the presence of intense coherent billows, highly distorted in the spanwise direction (Figure 5).

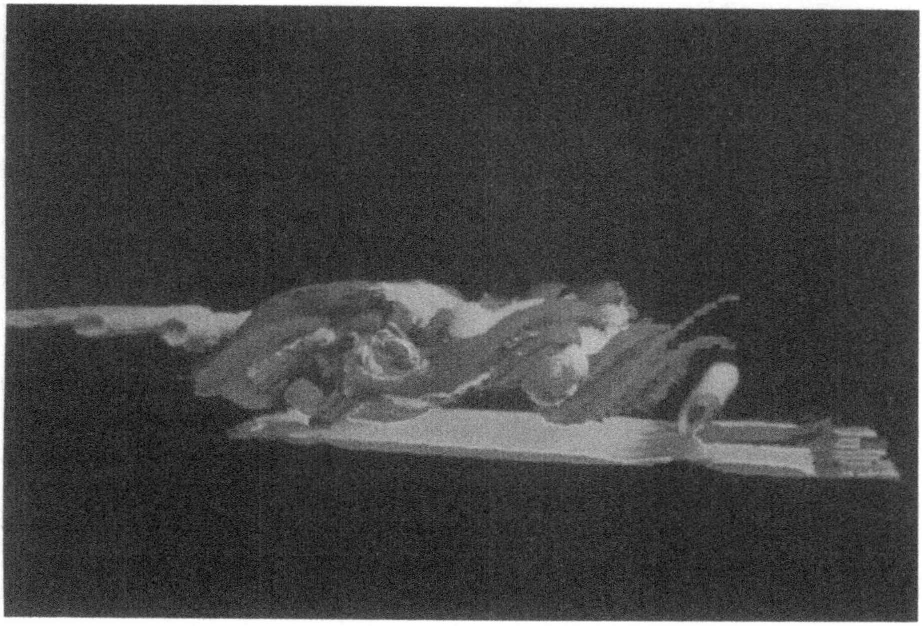

Figure 4: spanwise and longitudinal vorticity in a large-eddy simulation of the flow behind a backward-facing step (from Silveira, 1991).

3. Boundary layer

We briefly recall some results of 3D direct-numerical simulations in the case of the temporal boundary-layer over a flat plate (see Normand and Lesieur, 1991). The calculation corresponds to compressible Navier-Stokes equations, but at an external Mach number of 0.5, so that the boundary-layer is not far from incompressibility. We start initially with a set of 2D Tollmien-Schlichting waves, perturbed by 3D white noise of small amplitude: the situation is thus of the quasi two-dimensionally forced

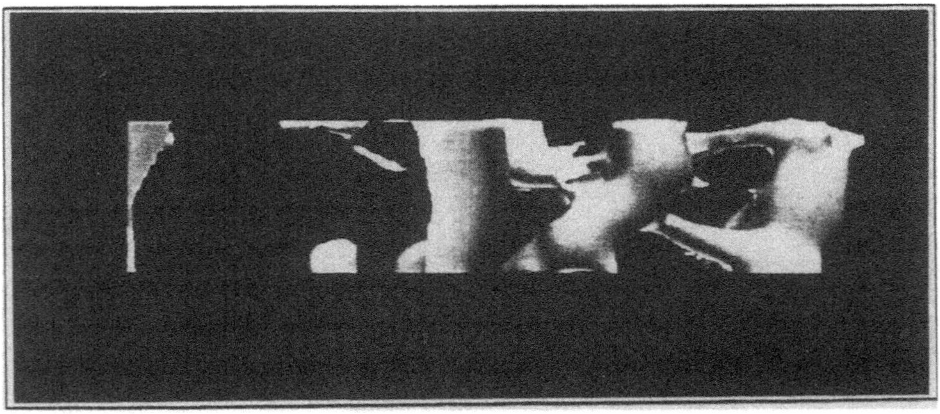

Figure 5: low-pressure field in the low-step case (from Silveira, 1991).

transition type. It is observed (Figure 6) that vortex filaments of weaker vorticity (in the crests of the TS waves) are stretched by the basic shear into hairpin vortices, corresponding to the peak-valley topology of Klebanoff et al. (1962). We have checked (see Normand and Lesieur, 1991) that the velocity profile becomes locally inflexional in a peak plane, with shedding of small Kelvin-Helmholtz vortices (spikes) which will abruptly degenerate into developed turbulence. After the transition, the calculation still shows hairpin vortices (as observed in the laboratory by Kline et al., 1967). They are no more in phase in the spanwise direction, as in Klebanoff's mode, but staggered, as predicted by Herbert (1988) on the basis of a secondary instability analysis in the case of natural transition[4]. Remark that these mechanisms of transition to developed

[4] We have also found the staggered mode developing in a forced-transition calculation of the highly-supersonic boundary-layer (Mach 5), which confirms that compressibility enhances three-dimensionality (Normand and Lesieur, 1991).

turbulence in a boundary layer are highly three-dimensional. In our calculations in particular, no roll up of TS waves could be observed in the 2D case: it is only by the mediation of 3D hairpin vortices that the TS wave can break up into turbulence. This is a point where the boundary layer differs from the channel flow: in the latter case, it can be shown in two dimensions that vortices of alternate sign develop on the boundaries (see e.g. Jimenez, 1987).

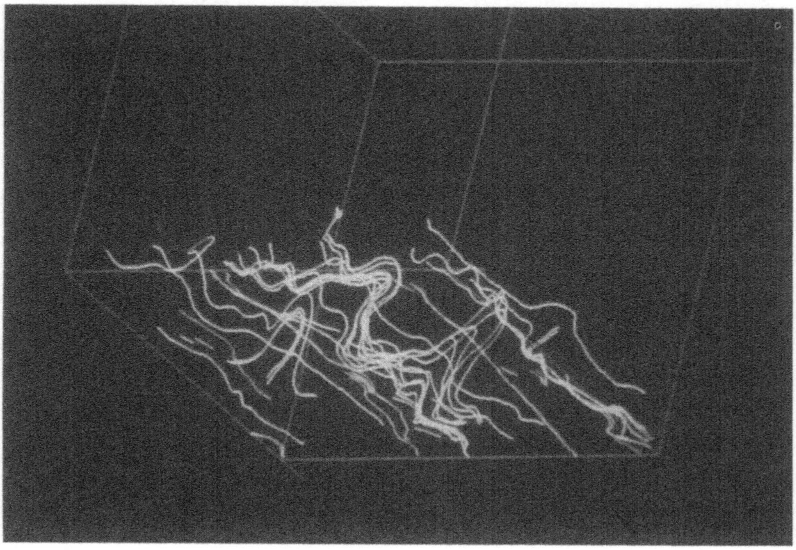

Figure 6: direct-numerical simulation of hairpin vortices stretched in the boundary-layer during the transition (from Normand and Lesieur, 1991).

4. 3D isotropic turbulence

4.1 Coherent structures and probability density functions (pdf). For a long time, people have speculated about the nature of the organized vortices which might exist in isotropic turbulence. For instance, Tennekes (1968) proposed a tube form. Forced calculations by Siggia and Patterson (1978) and Siggia (1981) did show some tube-like vortex structures in direct-numerical simulations. Calculations at higher resolution performed by Vincent and Meneguzzi (1991) show also the existence of vortex tubes. Simultaneously, various velocity and velocity derivatives pdf where determined in such calculations (see e.g. Herring and Métais, 1989, Métais and Lesieur, 1989, Vincent and Meneguzzi, 1991). The results were that the velocity components were gaussian, while the spatial velocity derivatives and the vorticity components departed strongly from a gaussian distribution, with a tendency to an exponential (or even shallower) distribution in the wings. Another strange result (which we will try to interpret later) was that the passive temperature had an exponential pdf (Métais and Lesieur, 1989, 1992).

Figure 7: direct-numerical simulation of isotropic decaying 3D turbulence, low-pressure (blue) and passive temperature (green) contours.

This velocity and velocity-derivative behaviour obtained numerically was attributed to a manifestation of *small-scale intermittency*, as proposed by Anselmet et al. (1984) to interpret similar findings in laboratory experiments. We will propose here another explanation based upon a concept of *large-scale intermittency*, very close to what happens in 2D turbulence.

4.2 Pressure and passive temperature. We have done high-resolution (128^3) pseudo-spectral direct-numerical simulations of isotropic turbulence in the unforced case. The initial velocity field is gaussian, with a spectrum peaking at k_i and exponentially decreasing at high k. A passive temperature advection equation is solved simultaneously, with a molecular Prandtl number of 1. Here we present (Figure 7), at a time of about 17 initial turnover times[5], the low-pressure (blue) and high temperature fluctuations (green). We have checked (Métais and Lesieur, 1992) that the high

[5] Which is higher than the time necessary for the initial large-scale perturbations to

vorticity regions are well correlated with the pressure troughs. Which is striking is the fact that temperature is strained in the "braid" region between the large billows exactly like in 2D turbulence. The pressure pdf in this calculation is shown on Figure 8: it is exponential in the lows and gaussian in the highs. Therefore, pressure lows are well correlated with high vorticity, and this could be explained by a large scale intermittency analogous to the one due to the condensation of vorticity into coherent vortices in 2D turbulence: here (in 3D turbulence), it is feasible to envisage that local initial vortex sheets will be unstable, and will roll up into locally quasi-2D spiralling large coherent vortices of vorticity of the order of the initial maximum vorticity. This mechanism has been proposed by Moffatt (1991). Hence, as in 2D turbulence, this large-scale intermittency will be responsible for the highly nongaussian pdf of the vorticity and of the low pressure[6]. The passive temperature and hairpin-type vortex filaments will be strained between the big rollers, as in an enstrophy cascade in 2D turbulence. This is confirmed by the presence of anomalous k^{-1} passive temperature spectra at large scales found in these decaying calculations (Lesieur and Rogallo, 1989; Lesieur et al., 1989). This is reminiscent of the inertial-convective range in the enstrophy cascade of 2D turbulence, where the kinetic energy spectrum is $\propto k^{-3}$, while the temperature is proportional to the enstrophy spectrum and therefore $\propto k^{-1}$ (see Lesieur and Herring, 1985). Since high concentrations of temperature are inbetween the vortices, it is natural to expect that the temperature pdf for maximum amplitudes will follow the vorticity[7].

4.3 Slow and rapid turbulence. In the above description, the slow turbulence corresponds to the coherent structures, and to the vorticity strained in a quasi-2D fashion in the stagnation regions between these structures. The latter might correspond to hairpin vortices of the genre discussed in the free-shear flow calculations. Even a quasi 2D roll up of these vortices into the core of the large eddies is not excluded. This quasi-2D turbulence[8] could have a 3D kinetic energy spectrum $E^S(k)$ of slope k^{-3}, corresponding to a generalized enstrophy cascade, and obtained in the

---

reach dissipative scales (see Lesieur, 1990).

[6] Notice, however, that a pressure distribution calculated (with the aid of the usual Poisson equation) from a fictitious gaussian velocity field having the same spectrum as the actual velocity, yields also a skewed pdf, but with a less intense exponential branch in the lows (Métais and Lesieur, 1992).

[7] In Lesieur and Rogallo (1989), the large-scale k^{-1} range was explained assuming that the scalar flux ϵ_T was controlled by the rate ϵ/u^2 imposed by the shear in the energetic scales: $\epsilon_T \sim k\ E_T(k)\ \epsilon/u^2$. In fact, this is exactly a 2D inertial-convective argument in the enstrophy cascade, where the shearing rate is constant and equal to $\beta^{1/3}$ (β is the enstrophy dissipation rate, see Lesieur, 1990).

[8] Which is statistically 3D isotropic.

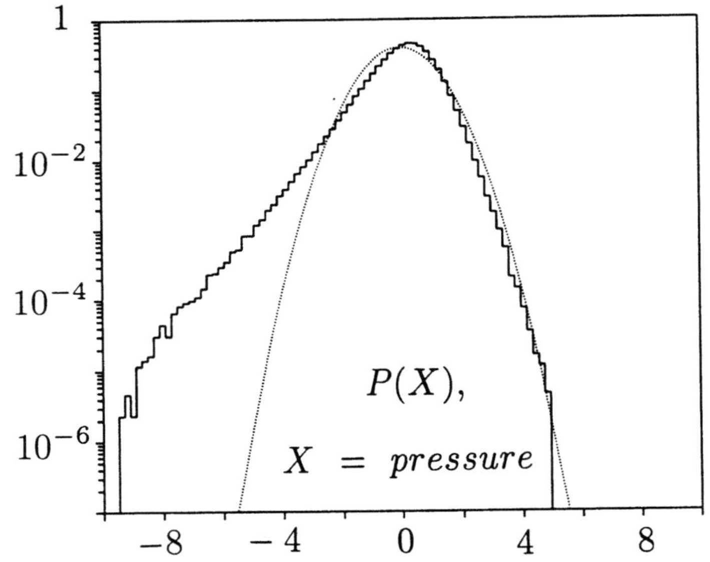

Figure 8: same calculation as in Figure 7, pressure pdf.

following way: $E^S(k) \sim \beta^{2/3}k^{-3}$, with $\beta^{1/3} \sim \epsilon/u^2$. This yields for the slow kinetic energy and temperature spectra:

$$E^S(k) \sim (\frac{\epsilon}{u^2})^2 \; k^{-3} \quad , \quad E_T^S(k) \sim \epsilon_T \; \frac{u^2}{\epsilon} \; k^{-1} \quad . \tag{1}$$

The rapid turbulence could be created for instance in the cores of the large vortices, by mechanisms discussed above in the mixing layer case, and where pairings (two-dimensional or helical) should play a major role. A small-scale $k^{-5/3}$ Kolmogorov cascade would then be rapidly created. More specifically, the rapid kinetic energy and scalar spectra are of the form

$$E^R(k) \sim \epsilon^{2/3} \; k^{-5/3} \quad , \quad E_T^R(k) \sim \epsilon_T \epsilon^{-1/3} \; k^{-5/3} \quad . \tag{2}$$

The slow and rapid kinetic energy (resp. temperature) spectra are equal for $k = k_i \sim l^{-1}$, where l is the integral scale of turbulence[9]. As schematically shown on Figure 9, $E^R(k)$ and $E_T^R(k)$ (if the Prandtl number is of the order of 1) drop at the Kolmogorov wavenumber k_D. On the other hand, $E^S(k)$ and $E_T^S(k)$ drop at $k_\lambda \approx \lambda^{-1}$, where λ, the Taylor microscale, may easily be shown to correspond to the two-dimensional enstrophy-dissipative scale. It turns out (see Figure 9) that the the slow k^{-3} kinetic energy component will be masked under the Kolmogorov $k^{-5/3}$ rapid component.

[9] The temperature integral scale l_T is here assumed to be $\approx l$.

As for the passive temperature spectrum, the k^{-1} slow component will dominate the rapid Corrsin-Oboukhov $k^{-5/3}$ one, at least at scales larger than the Taylor microscale λ. Within the present interpretation, the k^{-1} temperature spectrum we observe in the simulations is the "emerged part of the quasi-2D turbulence iceberg" hidden in the ocean of Kolmogorov turbulence.

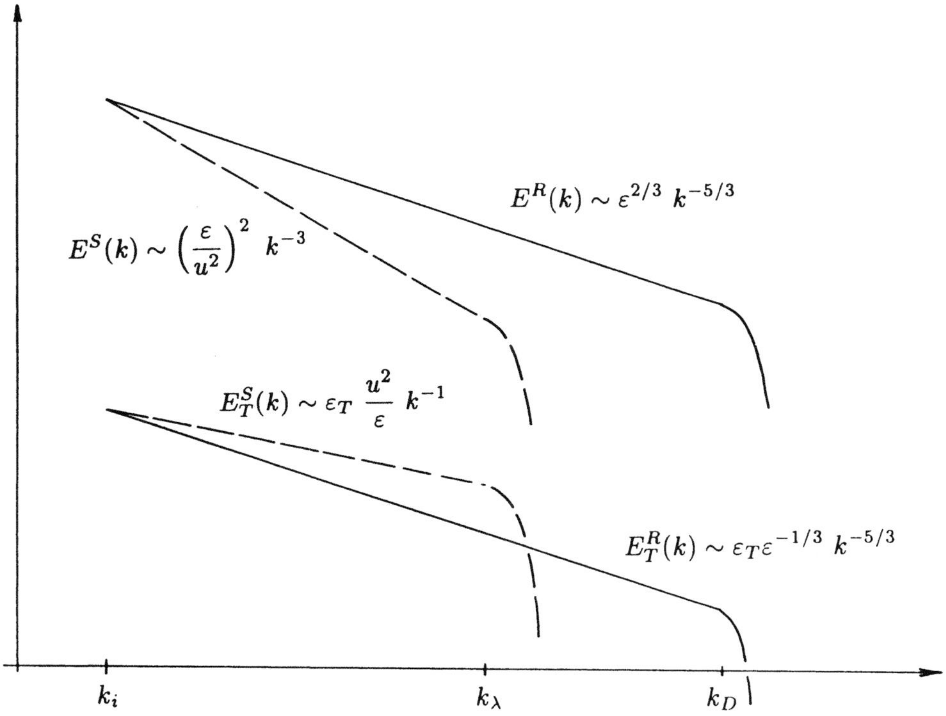

Figure 9: schematic picture of rapid and slow kinetic energy and passive temperature spectra.

The next step of the study will be to write evolution equations for the slow quasi-2D turbulence. A good starting point could be to consider space-filtered Navier-Stokes equations, within the context of large-eddy simulations. The structure-function model, where an eddy-viscosity is based on a local kinetic energy spectrum in physical space, seems to provide a good description of the large-scale coherent field and of the statistics of the flow, both in shear flows (Silveira, 1991; Normand and Lesieur, 1991) and in isotropic turbulence (Métais and Lesieur, 1992). It gives in particular very neat pressure contours (see the backward-facing step above) in the large scales, and good Kolmogorov spectra in the small scales. Notice that, in this "infinite-Reynolds number calculation", the kinetic energy (in the isotropic case) decays like $t^{-1.38}$, in good agreement with the statistical two-point closure predictions for this type of initial conditions (see Lesieur, 1990). The passive temperature decay is not far from t^{-2}, which is much steeper than the closure predictions ($t^{-1.48}$), but not far from the classical phenomenology of 2D turbulence, where it should decay like the enstrophy (Lesieur and Herring, 1985).

5. Conclusion

We have provided examples of the immense potentialities of 3D direct or large-eddy simulations in fluid mechanics, in particular for the investigation of coherent structures. In mixing layers between flows of different velocity, we have confirmed the fact that hairpin vortices are stretched between the big vortices, even in the case of helical pairing. We have also found that this latter type of pairing is very efficient for vortex stretching, since it produces isotropically three times more vorticity than the basic two-dimensional vorticity. Similar vortices have been found behind a backward-facing step. In all these cases, intense vortices[10] are very well marked by low-pressure contours. We have also investigated the role played by hairpin vortices in the transition above a flat plate.

Afterwards, we have developed high-resolution pseudo-spectral direct and large-eddy simulations of 3D decaying isotropic turbulence. We show the formation of large quasi-2D low-pressure tubes, which might result of the roll up of vortex sheets existing initially. We observe that intense passive temperature fluctuations are stretched between the large billows, and that the vorticity, low pressure and temperature fluctuations have pdf departing strongly from gaussianity. This, conjugated with the evidence of k^{-1} large-scale temperature spectra, leads us to propose the existence of a *slow quasi-2D turbulence* hidden under a *rapid Kolmogorov turbulence*. The slow turbulence has 3D kinetic-energy spectra and passive temperature spectra behaving respectively like k^{-3} and k^{-1}, corresponding to a 3D enstrophy cascade, and is subject to a large-scale intermittency which determines its various pdf. The rapid turbulence is originally created in the small scales by roll up and 3D isotropization within the cores of the big vortices of hairpin secondary longitudinal vortices stretched between the large vortices. Pairings (in particular helical pairings) should play a major role in this context. We insist that this scenario is, up to now, only a conjecture about what might happen in this still mysterious and intricate world of turbulence.

Acknowledgements

E. Lamballais and A. Silveira-Neto were very helpful for determining the pressure contours in the free-shear flow section, and Mrs A. Mokhtari for handling the colour pictures. We had stimulating discussions with P. Bartello, Y. Gagne, J.R. Her-

[10] Intense, that is of vorticity of the order of the vorticity resulting from primary instabilities.

ring, H.K. Moffatt, M. Morkovin, J. Riley, A. Roshko and J. Smagorinsky. We are pleased to acknowledge the support of CEA, CNRS (GDR *Mécanique des Fluides Numérique*), CCVR, DRET and INSU.

References

Anselmet, F., Gagne, Y., Hopfinger, E.J. and Antonia, R.A. (1984), "High-order velocity structure functions in turbulent shear flows", J. Fluid Mech., **140** , pp 63-89.

Bartello, P. (1991), "Quasi-2D rotating turbulence", Conference given at the CNRS-INSU GFD Summer School, Roscoff, 29 July-10 August 1991.

Bernal, L.P. et Roshko, A., 1986, "Streamwise vortex structure in plane mixing layers", J. Fluid Mech., **170**, pp 499-525.

Breidenthal, R. (1981), "Structure in turbulent mixing layers and wakes using a chemical reaction", J. Fluid Mech., **109**, pp 1-24.

Browand, F.K. and Troutt, T.R. (1980), "A note on spanwise structure in the two-dimensional mixing layer" J. Fluid Mech., **93**, pp 325-336.

Comte, P., Fouillet, Y., Gonze, M.A., Lesieur, M. et Normand, X. (1991), "Generation of coherent structures in free shear layers", in *Turbulence and coherent structures*, O. Métais et M. Lesieur ed., Kluwer Publishers, pp 45-73.

Comte, P., Lesieur, M. and Lamballais, E. (1992), "Large and small-scale stirring of vorticity and a passive scalar in a temporal mixing layer", submitted to Phys. Fluids.

Grand, D., Villand, M. and Coulon, C. (1988), Proc. 3rd Int. Symp. on Refined Flow Modeling and Turbulence Measurements, Nippon Toshi Center Tokyo, pp 427-434.

Herbert, T. (1988), "Secondary instability of boundary layers", Ann. Rev. Fluid Mech., **20**, pp 487-526.

Herring, J.R. and Métais, O. (1989), "Numerical experiments in forced stably stratified turbulence", J. Fluid Mech., **202**, pp 97-115.

Hopfinger, E.J., Browand, F.K. and Gagne, Y. (1982), "Turbulence and waves in a rotating tank", J. Fluid Mech., **125**, pp 505-534.

Jimenez, J. (1987), "Bifurcations and bursting in two-dimensional Poiseuille flow", Phys. Fluids, **30**, pp 3644-3646.

Klebanoff, P.S., Tidstrom, K.D. and Sargent, L.M. (1962), "The three-dimensional nature of boundary layer instability", J. Fluid Mech., **12** , pp 1-34.

Kline, S.J., Reynolds, W.C., Schraub, F.A. and Runstadler, P.W. (1967), "The structure of turbulent-boundary layers", J. Fluid Mech., **30**, pp 741-773.

Lasheras, J.C. and Meiburg, E. (1991), "On the three-dimensional dynamics of coherent vortical structures forming in free, shear flows", in *Turbulence and coherent structures*, O. Métais et M. Lesieur ed., Kluwer Publishers, pp 91-111.

Lesieur, M. (1990), *Turbulence in Fluids* (second revised edition), Kluwer Publishers.

Lesieur, M. and Herring, J.R. (1985), "Diffusion of a passive scalar in two-dimensional turbulence", J. Fluid Mech., **161** , pp 77-95.

Lesieur, M., Staquet, C., LeRoy, P. and Comte, P. (1988), "The mixing layer and its coherence examined from the point of view of two-dimensional turbulence", J. Fluid Mech., **192**, pp 511-534.

Lesieur, M. and Rogallo, R. (1989), "Large-eddy simulation of passive scalar diffusion in isotropic turbulence", Phys. Fluids A, **1**, pp 718-722.

Lesieur, M., Métais, O. and Rogallo, R. (1989), "Etude de la diffusion turbulente par simulation des grandes échelles", C.R. Acad. Sci. Paris, Ser. II, **308**, pp 1395-1400.

Lesieur, M., Yanase, S. and Métais, O. (1991), "Stabilizing and destabilizing effects of a solid body rotation on quasi-2D shear layers", Phys. Fluids A, **3**, pp 403-407.

McWilliams, J. (1989), "Statistical properties of decaying geostrophic turbulence", J. Fluid Mech., **198**, pp 199-230.

McWilliams, J. (1991) "The coherent vortices of two-dimensional and geostrophic turbulence", in *Turbulence and coherent structures*, O. Métais et M. Lesieur ed., Kluwer Publishers, pp 323-342.

Métais, O. and Lesieur, M. (1989), "Large-eddy simulation of isotropic and stably-stratified turbulence", in *Advances in Turbulence 2*, H.H. Fernholz and H.E. Fiedler eds, Springer-Verlag, pp 371-376.

Métais, O. and Lesieur, M. (1992), "Spectral large-eddy simulations of isotropic and stably-stratified turbulence", J. Fluid Mech., in press.

Moffatt, H.K. (1991), same proceedings.

Normand, X. and Lesieur, M. (1991), "Numerical experiments on transition in the compressible boundary layer over an insulated flat plate", Theor. and Comp. Fluid Dynamics, in press.

Pierrehumbert, R.T. and Widnall, S.E. (1982). "The two- and three-dimensional instabilities of a spatially periodic shear layer", J. Fluid Mech., **114**, pp 59-82.

Siggia, E. (1981), "Numerical study of small-scale intermittency in three-dimensional turbulence", J. Fluid Mech., **107**, pp 375-406.

Siggia, E. and Patterson, G.S. (1978), "Intermittency effects in a numerical simulation of stationary three-dimensional turbulence", J. Fluid Mech., **86**, pp 567-592.

Silveira-Neto, A. (1991), PhD Thesis, Institut National Polytechnique de Grenoble.

Silveira-Neto, A., Grand, D., Métais, O. and Lesieur, M. (1991), "Large-eddy simulation of the turbulent flow in the downstream region of a backward-facing step", Phys. Rev. Letters, **66**, pp 2320-2323.

Tennekes, H. (1968), "Simple model for the small-scale structure of turbulence", Phys. Fluids, **11** , pp 669-671.

Vincent, P. and Meneguzzi, M. (1991), "The spatial structure of homogeneous turbulence", in *Turbulence and coherent structures*, O. Métais et M. Lesieur ed., Kluwer Publishers, pp 191.

Yanase, S., Flores, C., Métais, O. and Lesieur, M. (1992) "Stabilizing and destabilizing effects of solid-body rotation on shear flows", Turbulent Shear Flows VIII, Munich, 9-11 Septembre 1991.

Address of the authors

Pr. Marcel Lesieur, Drs. Olivier Métais and Pierre Comte
Institut de Mécanique de Grenoble
BP 53 X - 38041 Grenoble-Cedex, France.
lesieur@frgren81.bitnet

A Dynamical Theory of Cascade in Turbulence and Non-Gaussian Statistics

T. KAMBE

Department of Physics, University of Tokyo, Bunkyo-ku, Tokyo 113, Japan

1. Introduction

A dynamical mechanism is considered which connects the cascade with non-gaussian statistics of velocity gradients. Turbulence is characterized by the continuous excitation of all scales, but in the Fourier space of the velocity field, the excited amplitude decreases rapidly with increasing wave numbers so that contribution to the total kinetic energy from the small scale components is negligibly small. Roughly speaking by the central limit theorem, the sum of a large number of Fourier modes is distributed normally when the Fourier amplitudes of different wave numbers are independent in the energy-containing eddies (Batchelor 1953).

However it is well-known that non-gaussian statistics are observed at small scales. Two simplest measures of non-gaussianity are the skewness and flatness.
(*a*) Experimental observations concerning the small-scale motion show that the flatness of the probability distribution of the various velocity derivatives increases steeply with the order of the derivative. For example, the flatness factor of the n-th order longitudinal derivative $g_n = \partial^n u/\partial x^n$ for a coordinate x and the corresponding velocity u, defined as $F_n = \overline{g_n^4}/(\overline{g_n^2})^2$, shows the value of about 3.9, 4.9 and 5.9 (for n =1, 2 and 3 respectively: Batchelor 1953; Monin & Yaglom 1975) which are greater than the gaussian value 3, where the over-bar denotes the ensemble average. A large flatness factor of a distribution implies that the probability density function has a higher central peak and broader skirts than the gaussian function of the same standard deviation. These properties are considered to be intrinsic to the micro-structures of turbulence in general.
(*b*) The measurements of the skewness factor $\overline{g_1^3}/(\overline{g_1^2})^{3/2}$ suggest that the value is around $-0.3 \sim -0.5$. The probability density of g_1 is characterized by the properties that small positive values of g_1^3 are more probable than small negative values, but this positive contribution to $\overline{g_1^3}$ is less than balanced by the larger contribution of negative large values of g_1^3 as compared with positive ones.

In the study of numerical simulation of incompressible Navier-Stokes turbulence at moderately large Reynolds number, She *et al.* (1988) isolated the contributions of different flow scales by considering Fourier-band filtering of the velocity field, and investigated the non-gaussian behaviors. In particular the probability distribution functions are found to have near-exponential forms for the velocity and its derivatives. Exponential behaviors of the distribution functions are observed in various context of experiments and numerical simulations (*e.g.* Van Atta & Chen 1970; Sheih *et al.* 1971; Yamamoto & Hosokawa 1988; Kida & Murakami 1989; Gagne 1990). These properties are examined also by the statistical analysis (Yamamoto & Kambe 1991) about the contributions from different Fourier subspaces of a decaying turbulence obtained numerically.

Kraichnan (1990) considered a heuristic model for evolution of the probability distribution of velocity gradient and an exponential distribution. The idea is that an initial gaussian field is mapped dynamically into a non-gaussian field. The initial statistics of the fluctuations are preserved during the dynamical cascade. The dynamics is governed by a nonlinear ordinary differential equation that models competition between the inertial straining and viscous damping and produces small scales. Similar nonlinear evolution equation has been considered by She (1991). His model equation is distilled from the vorticity equation by taking account of a random background field and a dynamical exponent α for a local self-stretching mechanism with a scaling argument.

Study of the vorticity dynamics based on the vorticity equation for a viscous incompressible fluid was performed previously for arbitrary flat shear layers (Kambe 1983, 1986) or an arbitrary axisymmetric shear layer (Kambe 1984). In these analyses, motions of rotational (shear) layers superimposed on irrotational straining field are investigated, in which the dynamical evolution of the vorticity proceeds under the influence of three elemental processes: convection, stretching and viscous diffusion. With using local expressions of the background straining velocity field which are linear with respect to the space coordinates, it is found to be able to represent the evolution of the rotational layers exactly in terms of the initial vorticity distribution. This formulation can describe exponential growth of the vorticity when the viscous term is neglected. However this is contrasted with the algebraically explosive behaviors which are known in the solution of the quadratically nonlinear model equation of the vorticity (Rose & Sulem 1978) or the equation for the enstrophy in the statistical theory of turbulence with the quasi-normal approximation (Proudman & Reid 1954). The present study is an endeavor to improve the previous study by incorporating the property of scale invariance of turbulence field described below.[1]

[1]The idea has been inspired by the presentation of She (1991 a, b) at the international workshop held in Japan, October 1990. His model equation includes the effect of scale invariance.

We investigate a dynamical mechanism that connects the cascade in turbulence with the near-exponential tail in the distribution of velocity and their derivative fields. It will be shown below that the difference in behavior between the lateral and longitudinal derivatives may lead to different probability distributions: slower-than or steeper-than exponential decay.

One of the remarkable symmetries of the Euler equation is the scale invariance which is considered by Frisch & Parisi (1985) in relation to a multifractal model of turbulence that they introduced. This property is as follows. The Euler equation is invariant if we simultaneously scale the distance by λ, the velocity by λ^h and the time by λ^{1-h} (the pressure by λ^{2h}), where h is an arbitrary local scaling exponent. Recent wavelet analysis of an experimental data suggests $-0.5 < h < 1.0$ (Bacry *et al.* 1990). The Kolmogorov's self-similar cascade corresponds to the value $h = \frac{1}{3}$ which describes scale invariance of the rate of energy transfer between eddies of different scales.

2. Straining of a test field by large-scale fields

2.1 Local representation

Taking account of the features of turbulence described in the introduction, we try to construct a model of cascade in turbulence. We consider the vorticity equation,

$$\omega_t + (\mathbf{v}\cdot\nabla)\omega - (\omega\cdot\nabla)\mathbf{v} = \nu\nabla^2\omega\ , \tag{1}$$

where $\mathbf{v}$ is the velocity and $\omega = \nabla\times\mathbf{v}$ is the vorticity. Suppose that the velocity field $(u,\ v,\ w)$ is locally represented in the cartesian coordinate system $(x,\ y,\ z)$ as

$$\begin{aligned} u &= Ax - \Psi_z(x,z) - \psi_y(x,y,t)\ , \\ v &= By \qquad\qquad\quad + \psi_x(x,y,t)\ , \\ w &= Cz + \Psi_x(x,z) \end{aligned} \tag{2}$$

near the origin $(0,0,0)$, where A, B, C are constants with the constraint relation $A+B+C=0$. The function $\Psi(x,z)$ satisfies the equation, $\Psi_{xx}+\Psi_{zz}=0$, hence $\Psi(x,z)$ being a streamfunction describing a steady incompressible irrotational flow in x,z plane, while $\psi(x,y,t)$ is a time-dependent streamfunction describing a rotational flow in x,y plane, hence $\psi_{xx}+\psi_{yy}\neq 0$, and considered to be a rotational perturbation to the steady irrotational field $\Psi(x,z)$. Obviously the velocity field (2) satisfies the solenoidal condition, $\operatorname{div}\mathbf{v}=0$. The velocity consists of three components: a regular irrotational field $\mathbf{v}^{(r)} = (Ax, By, Cz)$, a singular irrotational field $\mathbf{v}^{(s)} = (-\Psi_z, 0, \Psi_x)$

whose functional form is given in the next subsection, and an unsteady test field $\mathbf{v}^{(t)} = (-\psi_y, \psi_x, 0)$. It is readily shown that the vorticity has only the z component associated with ψ :

$$\boldsymbol{\omega} = (0, 0, \omega) , \quad \omega(x,y,t) = \psi_{xx} + \psi_{yy} \quad .$$

We consider the evolution of the test field $\psi(x, y, t)$ strained by the background singular field $\mathbf{v}^{(s)}$ and the regular field $\mathbf{v}^{(r)}$, which are supposed to be of the scale of inertial range and larger one, respectively. It is important that the singular component $\mathbf{v}^{(s)}$ is introduced here to represent the turbulent field having the scale invariance mentioned in the end of the previous section. In turbulence, the field $\mathbf{v}^{(s)}$ would be quasi-steady in comparison with the rapid cascading process of fluctuations.

The x and y components of the vorticity equation (1) vanish identically. Using the explicit expression of the velocity (2), we have

$$\frac{\partial \omega}{\partial t} + Ax\frac{\partial \omega}{\partial x} + By\frac{\partial \omega}{\partial y} - C\omega - \frac{\partial}{\partial x}(\Psi_z \omega) + \frac{\partial(\psi, \omega)}{\partial(x,y)} = \nu \nabla^2 \omega \quad , \tag{3}$$

for the z component $\omega = \psi_{xx} + \psi_{yy}$.

2.2 Singular irrotational field

Based on the view that straining of the test field by only the regular component $\mathbf{v}^{(r)}$ is insufficient to describe the cascade in turbulence, we introduce a singular component $\mathbf{v}^{(s)}$ having the scale-invariant property. Assume that

$$\Psi(x, z) = \sum_{n=0}^{\infty} a_n x^{h-2n} z^{2n+1} = a x^h z \left(1 + \zeta_1 \frac{z^2}{x^2} + \zeta_2 \frac{z^4}{x^4} + \cdots\right) \tag{4}$$

where h is a fractional parameter in the range $-0.5 < h < 1.0$, $a = a_0$ and $\zeta_n = a_n/a_0$. The irrotationality condition $\Psi_{xx} + \Psi_{zz} = 0$ leads to

$$\zeta_n \equiv \frac{a_n}{a_0} = (-1)^n \frac{\{2n-1-h\}}{(2n+1)!} \quad ,$$

where $\{n-h\} \equiv (n-h)(n-1-h)\cdots(1-h)(-h)$ for an integer n. This series solution converges for $| z/x | < 1$, but has a singular behavior at $x = 0$ due to the terms of the form x^{h-2n} with a fractional power $h - 2n$. Here we restrict our consideration to the part of positive x: $0 < | z | < x$. Clearly the solution (4) satisfies the viscous vorticity equation identically in the region of definition because of the vanishing vorticity.

The velocity fields, *i.e.* x and z components, represented by the streamfunction $\Psi(x, z)$ are

$$-\Psi_z \quad = \quad -a x^h \left(1 + 3\zeta_1 \frac{z^2}{x^2} + 5\zeta_2 \frac{z^4}{x^4} + \cdots\right) \quad ,$$

$$\Psi_x = ax^{h-1}z\,(h + (h-2)\zeta_1\frac{z^2}{x^2} + (h-4)\zeta_2\frac{z^4}{x^4} + \cdots)$$

respectively. It is evident that the above velocities are scaled by λ^h for the simultaneous scaling of x and z by λ.

2.3 Test field $\psi(x,y,t)$

Consider a test field whose vorticity ω is localized and characterized by a length scale $\lambda(t)$. The scale $\lambda(t)$ is supposed to decrease by the straining of the background fields. First we assume that the fluctuation $\psi(x,y,t)$ is represented in the form

$$\psi(x,y,t) = F(x,t) \tag{5}$$

for simplicity. (Later we will consider the case, $\psi(x,y,t) = yF(x,t)$ which shows an additional property of the test field). Further we assume that the vorticity $\omega(x,t) = F_{xx}$ is given by the form,

$$F_{xx}(x,t) = s_0\left(\frac{\lambda_0}{\lambda(t)}\right)^{\gamma} f(\xi) \ , \tag{6}$$

where

$$\xi = \frac{x - \lambda\,\xi_0}{\lambda} \ , \qquad \text{or} \quad x = \lambda(\xi + \xi_0) \ , \tag{7}$$

ξ_0 being a constant. The coefficient s_0 denotes a typical initial value of the test field vorticity. Note that the assumption (6) is given a special form to describe the cascade process in which the scale parameter $\lambda(t)$ is supposed to decrease. This is, in a sense, a *phenomenological* setup on which the subsequent consideration is based.

2.4 Velocity fields

Now the velocity components are given by

$$u = Ax - \Psi_z \ , \quad v = By + F_x \ , \quad w = Cz + \Psi_x \tag{8}$$

where Ψ is approximated by ax^hz with a positive parameter a. From this we obtain the longitudinal derivatives,

$$\frac{\partial u}{\partial x} = A - \Psi_{xz} \ , \quad \frac{\partial v}{\partial y} = B, \quad \frac{\partial w}{\partial z} = C + \Psi_{xz} \tag{9}$$

where $\Psi_{xz} = ahx^{h-1}$. The lateral derivatives are

$$\frac{\partial u}{\partial z} = -\Psi_{zz} \sim -x^{h-2}z, \quad \frac{\partial v}{\partial x} = F_{xx} \ , \quad \frac{\partial w}{\partial x} = \Psi_{xx} \sim x^{h-2}z \ . \tag{10}$$

The other lateral derivatives vanish. Since the vorticity has only the z component, the rate of vortex stretching is given by the longitudinal derivative $\partial w/\partial z = C + \Psi_{xz}$.

Consider a favorable case for the vortex stretching, that is $C > 0$ and $A, B < 0$. Cubic sum of the longitudinal derivatives is

$$S_3 = (A - \Psi_{xz})^3 + B^3 + (C + \Psi_{xz})^3 = 3ABC + 3(C^2 - A^2)\Psi_{xz} - 3B\Psi_{xz}^2 \qquad (11)$$

where $A + B + C = 0$ is used. Although each derivative of (9) takes both negative or positive value, the sum S_3 of the three cubic terms is found to be positive since each term on the right hand side is positive due to the assumed properties. Consideration of the relation to the turbulence statistics is given in the section 3.2 .

Let us consider a fluid particle advected passively by the flow, whose motion is denoted as $(X(t), Y(t), Z(t))$. The x, y positions must satisfy the equation,

$$\dot{X} = AX - \Psi_z(X, Z) \ , \quad \dot{Y} = BY + F_x(X, t) \ ,$$

where the over dot means the time derivative with respect to the fixed particle in motion and Ψ_z is approximated by ax^h. A small x-separation $\Delta_X(t) = X_1(t) - X_2(t)$ between two nearby particles marked as 1 and 2 will be governed by

$$\dot{\Delta}_X = (A - \Psi_{xz}(\bar{X}, Z))\, \Delta_X \ , \qquad (12)$$

where $\bar{X} = (X_1 + X_2)/2$.

2.5 A model of cascade

Intermittency in turbulence is considered to be an outcome of the dynamical cascade of the turbulent fluctuations. So far, we have described analytically a local representation of the turbulence field. We now try to derive a model equation of cascade of the fluctuation given above. Substituting (6) to ω in (3) and using (4) and (5), we obtain

$$-\, g(\xi)\dot{\lambda} = p(\xi)\,\lambda + G(\xi)\,\lambda^h + \nu f''(\xi)\frac{1}{\lambda} \qquad (13)$$

by dropping the common factor $s_0\lambda_0^\gamma \lambda^{-\gamma-1}$, where

$$g(\xi) = \gamma f(\xi) + (\xi + \xi_0) f'(\xi), \qquad G(\xi) = \frac{d}{d\xi}[\Psi_z(\frac{x}{\lambda}, \frac{z}{\lambda})\, f(\xi)],$$

$$p(\xi) = C\, f(\xi) - A\,(\xi + \xi_0)\, f'(\xi) \ ,$$

where $\Psi_z \approx ax^h$. Let us now introduce the normalized variable

$$\kappa(t) = \lambda(0)/\lambda(t) \qquad (14)$$

signifying a typical wave number of the test field with unit initial value, $\kappa(0) = 1$. The equation for $\kappa(t)$ is readily obtained from (13) by multiplying λ_0/λ^2 as

$$\frac{d}{dt}\kappa = \frac{p(\xi)}{g(\xi)}\kappa + \lambda_0^{h-1}\frac{G(\xi)}{g(\xi)}\kappa^{2-h} + \nu\frac{1}{\lambda_0^2}\frac{f''(\xi)}{g(\xi)}\kappa^3 \tag{15}$$

From now we examine only the general feature of the test field, *i.e.* the scaling parameter $\kappa(t)$. So that we approximate the function $f(\xi)$ by $f(0)\exp(-\frac{1}{2}\xi^2/b^2)$ and restrict our consideration to the behavior at $\xi = 0$ (and $z = 0$). Since $f''(0) = -b^{-2}f(0)$, the parameter b^{-2} signifies the magnitude of $f''(\xi)$. In addition, we have $f'(0) = 0$, $g(0) = \gamma f(0)$, $G(0) \approx ah\xi_0^{h-1}f(0)$, $p(0) = Cf(0)$. Thus we obtain from (15)

$$\frac{d}{d\tau}\kappa(\tau) = C\kappa + L_0\,\kappa^{2-h} - \nu k_0^2\,\kappa^3\ , \tag{16}$$

where $\tau = t/\gamma$, $L_0 = \Psi_{xz}(x_0, 0) \approx ah\,x_0^{h-1} = ah(\lambda_0\xi_0)^{h-1}$ and $k_0 = 1/b\lambda_0$. Since $-0.5 < h < 1.0$, the exponent of the second term is $1 < 2-h < 2.5$, falling in between the first and third terms.

It is interesting to find that the form of the equation (16) for κ is equivalent to that of She (1991) except for the coefficients. Especially noted that the same exponent $2 - h$ appears at the second term. Kraichnan's model equation takes the form , $\dot{J} = | s_0 | J^2 - \nu k_d^2 J^3$, where $J(s_0, t) = s(t)/s_0$ for a transverse shear s, and k_d is a characteristic dissipation wave number for s_0. One of the major differences lies in that the fractional exponent h is not included in this model, instead a quadratic term is taken into account to express the nonlinear straining. In the present analysis (and She's model, too), the fractional exponent is adopted to represent the fact that the turbulent straining has a singular nature not expressible by an integer power.

The idea of Kraichnan (1990) is as follows. The field variables like the velocity gradients develop under the nonlinear dynamical equation representing the cascade and viscous damping, during which the probability measure of the distribution is kept unchanged, probably due to the rapid process of the cascade. So that the distribution is distorted when expressed in terms of the dynamically evolving variable, due to a nonlinear dependence on the initial value. This is a dynamical mapping of the distribution function (the mapping closure, due to Kraichnan). Let s denote a field variable, and suppose that s has an initial variable s_0 with the initial distribution given by the gaussian function, $P_0(s_0) \sim \exp[-s_0^2]$. The probability distribution function will be given by

$$P(s) = P_0(s_0)\frac{\partial s_0}{\partial s}\ .$$

Kraichnan represented as $s = s_0 J(s_0, t)$ and sought the solution $J(s_0, t)$ by his nonlinear equation.

3. An improvement of the model dynamics

3.1 Improved representation

In the previous form of the test field (5) which has only the y component of velocity $v^{(t)} = F_x(x,t)$, the longitudinal velocity derivative $\partial v^{(t)}/\partial y$ vanishes identically. In order to examine more general properties of the cascade including the longitudinal derivative as well as self-interaction, we consider the test field in the form,

$$\psi(x,y,t) = y\,F(x,t) \quad . \tag{17}$$

Then the vorticity is given by

$$\omega = y\,F_{xx}(x,t)\ , \quad = s_0 \frac{y}{\lambda_0}\left(\frac{\lambda_0}{\lambda(t)}\right)^{\gamma} f(\xi)\ , \tag{18}$$

where the function $F_{xx}(x,t)$ is assumed to be of the same form as (6) before. The velocity fields in this case take the form,

$$u = Ax - \Psi_z - F(x,t)\ , \quad v = By + yF_x(x,t)\ , \quad w = Cz + \Psi_x\ . \tag{19}$$

Therefore the longitudinal derivatives are

$$\frac{\partial u}{\partial x} = A - \Psi_{xz} - F_x, \quad \frac{\partial v}{\partial y} = B + F_x, \quad \frac{\partial w}{\partial z} = C + \Psi_{xz} \tag{20}$$

and the lateral derivatives are

$$\frac{\partial u}{\partial z} = -\Psi_{zz} \sim -x^{h-2}z, \quad \frac{\partial v}{\partial x} = yF_{xx} = \omega, \quad \frac{\partial w}{\partial x} = \Psi_{xx} \sim x^{h-2}z\ . \tag{21}$$

Substituting (18) in (3) and using (17), we obtain

$$-\,g(\xi)\dot{\lambda} = p(\xi)\,\lambda + G(\xi)\,\lambda^{h} + q(\xi)\,\lambda^{-\gamma+2} + \nu\, f''(\xi)\frac{1}{\lambda} \tag{22}$$

where the third term $q\lambda^{-\gamma+2}$ newly appears to represent the self-interaction, and

$$\begin{aligned} p(\xi) &= (C-B)f(\xi) - A(\xi+\xi_0)f'(\xi)\ , \\ q(\xi) &= s_0[f'(\xi)\phi(\xi) - f(\xi)\phi'(\xi)]\ , \end{aligned} \tag{23}$$

with the definition $\phi'' \equiv f(\xi)$. The coefficients $g(\xi)$ and $G(\xi)$ are the same as before.

The equation of $\kappa(\tau) = \lambda(0)/\lambda(\tau)$ corresponding to (16) is now obtained as

$$\frac{d}{d\tau}\kappa(\tau) = (C-B)\kappa + L_0\,\kappa^{2-h} + q_0\,\kappa^{\gamma} - \nu\,k_0^2\,\kappa^3\ , \tag{24}$$

where $q_0 = s_0 f(0) = s_0$, by taking as $f(0) = 1$, and γ is a free parameter unspecified so far. The coefficient $q_0 = s_0$ denotes the initial magnitude of the test field vorticity, that is the transverse shear.

3.2 Remarks

(a) Scaling exponent γ

The equations for the vorticity ω and its derivative ω_x are

$$\dot{\omega} = (C + \Psi_{xz})\omega \ , \qquad \dot{\omega}_x = 2(C + \Psi_{xz})\omega_x + B\omega_x + \Psi_{xxz}\omega \ , \tag{25}$$

from (3) for the inviscid motion ($\nu = 0$), where the over-dot denotes the convective material derivative. Hence we have

$$-2\frac{\dot{\omega}}{\omega} + \frac{\dot{\omega}_x}{\omega_x} = B + \Psi_{xxz}\frac{\omega}{\omega_x} \ . \tag{26}$$

Assuming that $\omega \sim \lambda^{-\gamma}(t)$, $\omega_x \sim \lambda^{-\gamma-1}(t)$ and $\Psi_{xxz} \sim ah(h-1)(\lambda\xi_0)^{h-2}$ and neglecting B, we obtain

$$(\gamma - 1)\frac{\dot{\lambda}}{\lambda} \approx ah(h-1)(\lambda\xi_0)^{h-2}\lambda \ .$$

The equation (12) implies $\dot{\lambda}/\lambda \sim -ah(\lambda\xi_0)^{h-1}$ (neglecting A). Hence we have $\gamma \approx 1 + c(1-h)$ where $c = 1/\xi_0 > 0$. Since $h < 1$, $\gamma > 1$. It is likely that $2 - h \geq \gamma(> 1)$, since the vortex stretching by the singular component will be more effective than the self-stretching for the cascade. Here we take $c = 1$, therefore

$$\gamma = 2 - h \ . \tag{27}$$

This means that the vorticity (18) has the same behavior as the lateral derivative of $\mathbf{v}^{(s)}$ varying like x^{h-2} as x decreases (see (21)). For the value of (27), the second and third terms on the right hand side of (24) are put together with the common factor κ^{2-h} and the coefficient given by $(L_0 + s_0)$ which is taken to be positive. The two terms are considered to be of the same order : $L_0 \sim s_0$.

(b) Skewness

The cubic sum of the longitudinal derivatives is

$$\begin{aligned}\Sigma_3 &= (A - \Psi_{xz} - F_x)^3 + (B + F_x)^3 + (C + \Psi_{xz})^3 \\ &= S_3 + 3(-(A - \Psi_{xz})^2 + B^2)F_x - 3(C + \Psi_{xz})F_x^2 \ ,\end{aligned}$$

where S_3 is given by (11). In the case $C > 0$ and $F_x > 0$ (given above), the third term on the right hand side is negative, while the second term becomes negative for $A \leq B < 0$. In particular it is expected that at the peak position of the test field, the new negative contributions exceed in magnitude the positive S_3, although each of the three terms of (20) takes positive or negative values. Thus we find that our field is simulating qualitatively the property (*b*) described in the introduction for the probability density function. Over most points (in the neighborhood of the origin and $| z | < x$) the cubic sum will be given approximately by S_3 which is positive, but not very large. However at points in a region near the peak of the test field which could be intensified by the cascade, the cubic sum Σ_3 becomes negative. This infers qualitatively the relation of the skewness factor and the distribution function mentioned in §1, by supposing that the time average of the longitudinal derivative at a fixed point in the experiment corresponds to the spatial average in the present model.

4. Distribution functions of velocity derivatives

4.1 Near-exponential distributions of lateral and longitudinal derivatives

(a) Growth of the scale parameter κ

As far as the condition for the initial growth $\dot{\kappa}(0) > 0$ is satisfied, $\kappa(t)$ tends to a non-zero stationary value $\kappa_* = \kappa(\infty)$ at which the right hand side of (24) vanishes. For large values of $L_0 + s_0$, the scale parameter $\kappa(t)$ will grow sufficiently so that the stationary value κ_* is given by the balance of the last three terms of (24), that is we have $[(L_0 + s_0)\kappa_*^{2-h} - \nu k_0^2 \kappa_*^3] \approx 0$. Thus we obtain

$$\kappa_* \approx \left(\frac{L_0 + s_0}{\nu k_0^2} \right)^{1/(1+h)} . \tag{28}$$

(b) Lateral derivative

As shown in (21), the lateral shear $s(t)$ is given by $\omega = y F_{xx}$. Therefore from the expression (18) we find the scaling relation (for $\xi = 0$):

$$\frac{s(t)}{s_0} \approx \kappa^{2-h}(t) \tag{29}$$

for y of the order of λ_0, using (27). For the stationary state, we obtain

$$\frac{s_*}{s_0} \approx \kappa_*^{2-h} \sim s_0^{(2-h)/(1+h)} ,$$

where $s_* = s(\infty)$, assuming $L_0 \sim s_0$ in (36) (except for the proportionality constant). Hence

$$s_0 \sim s_*^{\alpha(h)/2}, \qquad \text{with} \quad \alpha(h) = \frac{2}{3}(1+h) . \tag{30}$$

Thus the distribution of s_* will be

$$P(s_*) = P_0(s_0)\frac{\partial s_0}{\partial s_*} \sim s_*^{\mu_\alpha} \exp[-s_*^\alpha] \tag{31}$$

with $\mu_\alpha = -(2-h)/3$. For $-\frac{1}{2} < h < 1$, we obtain $\frac{1}{3} < \alpha(h) < \frac{4}{3}$. The Kolmogorov scaling $h = 1/3$ gives $\alpha = 8/9$. This leads to a slower-than exponential (*i.e.* $\alpha < 1$) distribution with respect to s_* and predicts an upward flare of the skirt of the distribution observed in the linear-log plot from the simulations of turbulence (She *et al.* 1988; Hosokawa & Yamamoto 1989; Kraichnan 1990; Vincent & Meneguzzi 1991).

(c) *Longitudinal derivative*

The longitudinal derivative g concerned with the cascade is given by $\partial u/\partial x$ or $\partial v/\partial y$ of (20), which include F_x. The component of intermittency $\sigma(t)$ is

$$\sigma(t) = F_x = s_0\left(\frac{\lambda_0}{\lambda(t)}\right)^{1-h} \phi'(\xi), \qquad \phi'(\xi) = \int_{-\infty}^{\xi} f(\xi')\, d\xi' \tag{32}$$

from (27). Note that $\partial u/\partial x$ includes $-F_x$, while $\partial v/\partial y$ includes $+F_x$. An argument similar to that for the lateral derivative suggests the relation

$$\frac{\sigma_*}{\sigma_0} \approx \kappa_*^{1-h} \sim \left(\frac{L_0 + s_0}{\nu}\right)^{(1-h)/(1+h)}, \tag{33}$$

where $\sigma_0 = s_0\phi'(0) > 0$ and $\sigma_* = \sigma(\infty)$. Note that, for a large σ_0, $\partial v/\partial y$ will be positive, if $\partial u/\partial x$ is negative.

For the derivative $g = \partial u/\partial x$, the initial value will be negative with the value $g_0 = A - \Psi_{xz}(x_0, 0) - \sigma_0 \approx -\sigma_0(1 + c_-)$ where $c_- = \mid A - \Psi_{xz} \mid /\sigma_0$. The final value will be

$$g_* \approx -\sigma_* \approx -\sigma_0\kappa_*^{1-h} \tag{34}$$

from (33). Since $L_0 = \Psi_{xz}(x_0, 0)$, we have $g_0 \approx -L_0 - s_0\phi'(0) + A \approx -L_0 - s_0$, assuming $\phi'(0) = O(1)$ and neglecting A. The equation (28) suggests $\kappa_* \sim\mid g_0 \mid^{1/(1+h)}$. Thus we have

$$\mid g_* \mid \sim \sigma_0 \mid g_0 \mid^{\varepsilon(h)} \approx (1 + c_-)^{\varepsilon(h)}\sigma_0^{2/\beta(h)} \tag{35}$$

from (42), where

$$\beta(h) = 1 + h\ , \qquad \varepsilon(h) = \frac{1-h}{1+h}\ . \tag{36}$$

This corresponds to the negative side of the g_* distribution. On the other hand, the initial value of $g = \partial v/\partial y$ will be positive with $g_0 = B + \sigma_0 \approx \sigma_0(1 - c_+) > 0$ for a large fluctuation σ_0, where $c_+ = | B | / \sigma_0$. It is assumed that $1 > c_+ > 0$ and $c_+ < c_-$. The final value will be

$$g_* \approx +\sigma_* \sim \sigma_0 g_0^{\epsilon(h)} \approx (1-c_+)^{\epsilon(h)} \sigma_0^{2/\beta(h)} \,, \tag{37}$$

corresponding to the positive side of g_*.

Suppose that σ_0 is a gaussian random variable with the probability $Q_0(\sigma_0) \sim \exp[-\sigma_0^2]$. Then the distribution of g_* will be

$$Q(g_*) = Q_0(\sigma_0)\frac{\partial \sigma_0}{\partial g_*} \sim \begin{cases} g_*^{\mu_\beta} \exp[-k_- g_*^\beta] & (g_* < 0) \\ g_*^{\mu_\beta} \exp[-k_+ g_*^\beta] & (g_* > 0) \end{cases} \tag{38}$$

where $\mu_\beta = -(1-h)/2$,

$$k_- = \left(\frac{1}{1+c_-}\right)^{1-h} \,, \qquad k_+ = \left(\frac{1}{1-c_+}\right)^{1-h} \,, \tag{39}$$

k_- being smaller than k_+. This kind of asymmetry is observed in both experiments (*e.g.* Gagne 1990) and numerical simulations (*e.g.* Hosokawa & Yamamoto 1989; Yamamoto & Kambe 1991). For $-\frac{1}{2} < h < 1$, we obtain $\frac{1}{2} < \beta(h) < 2$. In the Kolmogorov scaling $h = 1/3$, we have $\beta = 4/3$. This leads to a steeper-than exponential (*i.e.* $\beta > 1$) distribution with respect to the longitudinal derivative σ_*. This behavior too is observed in the numerical simulations as well as in the experiment (Makita 1991).

4.2 Flatness factors

Once the probability distribution functions (pdfs, in short) are found by the method described above, it is possible to estimate the flatness factors of the velocity derivatives. Now we consider the n-th order longitudinal derivative $g_n(t) \equiv g(n,t) = \partial^n u/\partial x^n$. The (normalized) pdf is

$$Q_0(\sigma_0)\, d\sigma_0 = \frac{1}{\sqrt{\pi}} e^{-\sigma_*^\beta} \frac{\beta}{2} \sigma_*^{\frac{1}{2}\beta - 1} d\sigma_* \;\rightarrow\; P(x)\, dx$$

where

$$P(x) = \frac{1}{\sqrt{\pi}} x^{-\frac{1}{2}} e^{-x} \,, \qquad x = | \sigma_* |^\beta \tag{40}$$

$$\beta = \beta(n,h) = \frac{2(1+h)}{\delta(n) + 1 + h} \,, \qquad \delta(n) = n - h \,, \tag{41}$$

(x takes only positive values, while g_* takes either of positive or negative values). Setting $g_*(n) = k_n \sigma_*(n)$, we obtain

$$< g_*(n)^2 >= k_n^2 \int_0^\infty x^{\frac{2}{\beta}} P(x)\, dx = \frac{k_n^2}{\sqrt{\pi}} \Gamma(\frac{3}{2} + \frac{\delta(n)}{1+h}) ,$$

$$< g_*(n)^4 >= k_n^4 \int_0^\infty x^{\frac{4}{\beta}} P(x)\, dx = \frac{k_n^4}{\sqrt{\pi}} \Gamma(\frac{5}{2} + \frac{2\delta(n)}{1+h}) ,$$

where (40) and (41) are used, and $\Gamma(x)$ is the *gamma* function. Thus the flatness factor is given by

$$F_n(h) = \frac{< g_n^4 >}{< g_n^2 >^2} = \sqrt{\pi}\, \frac{\Gamma(\frac{5}{2} + 2D(n,h))}{[\Gamma(\frac{3}{2} + D(n,h))]^2} , \qquad D(n,h) = \frac{\delta(n)}{1+h} . \tag{42}$$

It is readily shown that, if $D(n,h) = 0$, F_n takes the gaussian value 3. Hence the function $D(n,h) = \delta(n)/(1+h)$ characterizes the degree of non-gaussian statistics. In the above case, we have

$$\delta(n) = n - h , \qquad D(n,h) = \frac{n-h}{1+h} . \tag{43}$$

For $h = 1$, we obtain $F_1(1) = 3$. The value $h = 1/3$ of the Kolmogorov scaling leads to $D(n, \frac{1}{3}) = (3n-1)/4$. Therefore

$$F_1(\frac{1}{3}) = \sqrt{\pi} \frac{\Gamma(\frac{7}{2})}{[\Gamma(2)]^2} \approx 5.9 , \quad F_2(\frac{1}{3}) = \sqrt{\pi} \frac{\Gamma(5)}{[\Gamma(\frac{11}{4})]^2} \approx 16.4 , \quad F_3(\frac{1}{3}) = \sqrt{\pi} \frac{\Gamma(\frac{13}{2})}{[\Gamma(\frac{7}{2})]^2} \approx 46.2 .$$

These values appear to be high, compared with the measurements quoted in (a) of the introduction. However a numerical simulation yielded $F_1 = 7.55, F_2 = 14.4$ and $F_3 = 16.1$ (Yamamoto & Kambe 1991), the first two being not far from the present results.

5. Summary and discussion

A dynamical model of intermittency in turbulence is presented in which the velocity field is represented locally in physical space and consists of three components: a regular field, a singular field with a fractional scaling exponent h and a test field. The first two are supposed to represent large-scale inertial range velocity field and the last one an intermittency which is a cascading component. Nonlinear ordinary differential equations are derived to describe growth of the inverse scale κ of the test field. Final equilibrium value of κ depends on the initial value s_0 of a velocity derivative, the fractional exponent h and the viscosity. In a particular case, the equation reduces to that of She (1991). The present analysis is based on an analytical representation of the velocity field. This suggests that the lateral derivatives of the velocity of the intermittency component scales like κ^{2-h}, while the scaling (dynamical) exponent of She is $1 - h$ (according to his notation, scaling like $J^{1-\alpha}$).

In the present formulation the longitudinal derivative σ of the cascading test field is characterized by a scaling exponent smaller by one than that of the lateral derivatives. Therefore the variable σ scales like κ^{1-h} and the corresponding pdf is given by $\exp(-\mid \sigma \mid^{\beta})$ with $\beta = 1 + h$. In the Kolmogorov scaling we have $\beta = 4/3$ which corresponds to a steeper-than exponential pdf. The present model can predict the asymmetry of the distribution function of the longitudinal derivative, which is consistent with the observed statistics.

From the analytical expression of the intermittency component, we have an estimate of the n-th order derivative. The flatness factor obtained from the pdf for the n-th order derivative increases with the order n. The estimated values (for $n = 1$ and 2) coincide with the corresponding ones of numerical simulation, but the observed values are smaller than the present one.

One of the points of the present study is that the velocity field includes a component which has a fractional scaling exponent, giving a singular behavior. This is based on the views that the fractional scaling behavior is predicted by the argument given at the end of §1, further that straining of the test field only by the regular component (Kambe 1983, 1984) is insufficient to describe the cascade in turbulence. The singular component $\mathbf{v}^{(s)}$ gives rise to a locally intensified rate of vortex stretching that could result in the intermittency. This formulation may be considered to be a realization in an analytical form of the idea of She (1991 a).

At first sight, it may appear that the present representation of the velocity is given a particular form. However the following arguments indicate that the velocity has a fairly general character more than its appearance. The regular field $\mathbf{v}^{(r)}$ has a general local expression. The direction of stretching of the field $\mathbf{v}^{(s)}$ is taken in the z direction that does not hurt its generality, but it is assumed to have a two-dimensional form that is not always the case. Another restriction is that the vorticity of the test field coincides with the z axis. In turbulent field there is of course substantial probability of such an arrangement. Once the assumed field is realized, then the cascade mechanism becomes active and a large amplitude fluctuation appears with a fair probability. In unfavorable arrangements the fluctuations will stay at small amplitudes with near-gaussian probability.

References

Bacry, E., Arneodo A., Frisch U., Gagne, Y. & Hopfinger, E. (1990) Wavelet analysis of fully turbulent data and measurement of scaling exponents, in *Turbulence and coherent structures* (eds. Métais & Lesieur, Kluwer acad. publ.).

Batchelor, G.K. (1953) *The Theory of Homogeneous Turbulence*, chap. viii, (Cambridge University Press, London).

Batchelor, G. K. & Townsend, A. A. (1947) *Proc. R. Soc.* A **190**, 534-550.

Frisch, U. & Parisi, G. (1985) in *Turbulence and Predictability in Geophysical Fluid Dynamics and Climate Dynamics* (eds. Ghil, Benzi & Parisi, North-Holland) 84-88.

Gagne, Y. (1990) Properties of fine scales in high Reynolds number turbulence, in *Advances in Turbulence* **3** (European Turbulence Symposium at Stockholm).

Hosokawa, I. & Yamamoto, K. (1989) *J. Phys. Soc. Jpn.* **58**, 20-23.

Kambe, T. (1983) *J. Phys. Soc. Jpn.* **52**, 834-841.

Kambe, T. (1984) *J. Phys. Soc. Jpn.* **53**, 13-15.

Kambe, T. (1986) *Fluid Dyn. Res.* **1**, 21-31.

Kida, S. & Murakami, Y. (1989) *Fluid Dyn. Res.* **4**, 347-370.

Kraichnan, R. H. (1990) *Phys. Rev. Lett.* **65**, 575-578.

Makita, H. (1991) *Fluid Dyn. Res.* **8** .

Monin, A.S. & Yaglom, A.M. (1975) *Statistical Fluid Mechanics* vol. 2 (MIT Press).

Proudman, I. & Reid, W.H. (1954) *Phil. Trans. Roy. Soc.* A **247**, 163-189.

Rose, H.A. & Sulem, P.L. (1978) *Journal de Physique* **39**, 441-484.

She, Z.-S. (1991 a) *Phys. Rev. Lett.* **66**, 600-603.

She, Z.-S. (1991 b) *Fluid Dyn. Res.* **8** .

She, Z.-S., Jackson, E. & Orszag, S.A. (1988) *J. Sci. Comput.* **3**, 407-434.

She, Z.-S. & Orszag, S.A. (1991) *Phys. Rev. Lett.* **66**, 1701-1704.

Sheih, C.M., Tennekes, H. & Lumley, J.L. (1971) *Phys. Fluids* **14**, 201-2215.

Van Atta, C.W. & Chen, W.Y. (1970) *J. Fluid Mech.* **44**, 145-159.

Vincent, A. & Meneguzzi, M. (1991) *J. Fluid Mech.* **225**, 1-20.

Yamamoto, K. & Hosokawa, I. (1988) *J. Phys. Soc. Jpn.* **57**, 1532-1535.

Yamamoto, K. & Kambe, T. (1991) *Fluid Dyn. Res.* **8** .

Small Scale Vortices in Turbulent Flows

Javier Jiménez

Abstract. The evidence on small compact vortex structures in turbulent flows is summarised for various experimental and numerical flow fields. It is consistent with a model of strained almost two dimensional vortices with radii of the order of the Kolmogorov scale, and circulation Reynolds numbers of a few hundred. The known alignment properties of the strain tensor are also consistent with the kinematics of this model. A possible scenario for the generation of these structures within the turbulent cascade is offered. The compact vortices are postulated to be essentially passive from the point of view of energy transfer, connected to the coherent structures observed in two dimensional turbulence

1. Introduction

It has been known for a long time that the small scales of turbulent flows are not completely random. It was first shown by Batchelor & Townsend (1949) that the statistics of the velocity derivatives are incompatible with an uncorrelated random behaviour of the velocity field at scales comparable to the Kolmogorov dissipation limit. A modern summary of the available data can be found in (Van Atta & Antonia, 1980). It was soon realised that these results implied the existence of intermittent organised structures at the short wavelength end of the spectrum, and experiments were attempted to clarify their geometry . It was Kuo & Corrsin (1972) who first presented suggestive evidence that the high vorticity loci were vortex tubes or, at most, ribbons, but it was necessary to wait for the advent of direct visualisations of numerically simulated flows before these more or less flattened vortex tubes were shown to be the dominant structures of isotropic turbulent flows at high vorticity amplitudes (Siggia, 1981, Kerr, 1985, Hosokawa & Yamamoto, 1990, She *et al.*, 1990, Ruesch & Maxey, 1991, Vincent & Meneguzzi, 1991). A re-examination of older data fields in numerically simulated turbulent shear layers, homogenous shear flows and channels, shows the presence of compact vortices of roughly similar characteristics. These will be discussed below. Finally (Douady *et al.*, 1991) produced direct experimental visualisations of strong concentrated persistent vortices in homogeneous turbulence, and (Schwarz, 1990) published pictures of organised strain structures which show up as alternating bands of consistent orientation of flakes suspended in a grid-stirred flow. It is not clear how, or whether, these structures are connected to the vortex tubes, but they are persistent and seem to have comparable dimensions.

We summarise here the experimental evidence available for these compact, small scale vorticity structures, using both published results and other available numerical data. That evidence will be discussed in the next section and shows that there is a certain homogeneity of the small structures across flows whose large scale character is very different. We will then use those general characteristics to define a simple model for the small scale vortices, and in particular to explain some of the experimental observed relations between the vorticity and the rate of strain tensor. One of our main results is that many of those properties turn out to be essentially kinematic. The evidence on isolated compact structures in two dimensional turbulence is then reviewed, and the possible relation between the two phenomena is discussed.

2. Phenomenology

The definition of a vortex tube is a subjective matter that probably varies among different authors, but it seems to refer in most cases to the structures with the highest values of vorticity in the flow (although some adjustment has to be done in wall bounded flows for the presence of very strong vortex layers at the wall). It is possible, however, to get some idea of their characteristic sizes and strengths by comparing data from different sources. This is done in Table 1, and the processing of the different cases is explained in the appendix.

There seems to be reasonable agreement that the diameter of the tubes is intermediate between the Taylor microscale, λ, and the Kolmogorov scale, η, in the range of $4-10\eta$, but, since most estimates come from numerical simulations for which $Re_\lambda \simeq 100$, where $\lambda/\eta \simeq 15$, it is difficult to distinguish between low multiples of η and high fractions of λ. The vortex length is quoted as being of the order of either a small multiple of λ or of the integral scale of the flow, L. Again, since $L/\lambda \simeq Re_\lambda^{1/2}/5 \simeq 2$, for $Re_\lambda \approx 100$, it is difficult to distinguish both possibilities.

Still harder to estimate is the intensity of the vortices, which can usually be derived only indirectly from other quantities given by the different authors. It is expressed best as a Reynolds number, $Re_\gamma = \gamma/\nu$, based on the total circulation, γ, in the tube. It is interesting that, whenever this quantity can be estimated, it seems to be in the range, $Re_\gamma \sim 150-400 \pm 100$. This is true even for flows, like the wall region in a channel, or like the turbulent mixing layer, in which the detailed turbulence dynamics

Flow type	γ/ν	ω_{max}/ω'	ρ/η	$\Delta u/u'$	
Homog. isotropic turb.	260	8.9	3.1	3.5	$Re_\lambda = 60$
Wall region of channel	100-300	4.6	4.3	3.9	$Re_\tau = 100 - 200$
Homogeneous shear	400-500	5.6	5.3	3.0	$Re_\lambda = 95$
Plane mixing layer	340	20.	2.5	6.7	$Re_\lambda = 55$

Table 1. Characteristic circulation associated to small scale vortices in various types of flows. See Appendix I for detailed references and assumptions.

are presumably quite different from those of homogeneous turbulence. Once more, since all the observations come from numerical simulations at low Reynolds numbers, and given the uncertainties in the calculation of the circulation, it is difficult to know whether the information in Table 1 should be interpreted as a range for Re_γ, or as evidence for a dependence on Re_λ. From the analysis of the only two cases in which data are available for several Reynolds numbers - the channel and the homogeneous shear flow (see appendix) - there is some reason to believe that the circulation of the cores is an increasing function of the bulk Reynolds number, at least in the range for which experimental data are available.

On the other hand, $Re_\gamma \simeq 150$ is at least a plausible value for the intensity of the smallest observable vortices. Since they are defined as loci of very high vorticity, they have to be generated by stretching of previous structures, which is the only mechanism able to amplify vorticity away from walls, but which does not modify the total circulation. As long as a vortex is being stretched, there is no limit as to how thin its core can become, and there is a well known equilibrium solution (Burgers' vortex, see Batchelor, 1967), for which the rate of strain, S, compensates viscous dissipation at a radius of order $(S/\nu)^{1/2}$. The structures in which we are interested, though, are supposed to survive even after the strain that originates them ceases to act. Otherwise, their dynamical significance would be small. In fact, all the available observations indicate that their lifetime is long with respect to their inertial time scales.

It can easily be shown that the peak vorticity of a two dimensional axisymmetric, unstrained, self similar vortex, diffusing under the effect of viscosity, decays by a factor of two in a time

$$\tau = \frac{Re_\gamma T_E}{16\pi^2}, \tag{1}$$

where $T_E = 4\pi^2\rho^2/\gamma$ is the turnover time at the $1/e$ radius, ρ, or the core . Because of the large denominator in (1), vortices with Re_γ much smaller than 150 decay too fast, and would not be identified as coherent. In fact they would probably not even have time to form as identifiable structures.

It is interesting that, in all the cases that we have studied, the radius of the cores, normalised with $\eta = (\omega'/\nu)^{1/2}$, is of the order of 3-4. This is roughly the radius of an equilibrium Burgers' vortex under a rate of strain $O(\omega')$, which can be expected to be the average fluctuating rate of strain within the flow. While the estimation of individual radii is subject to large uncertainties, the relative consistency of the values in Table 1 makes it likely that the thickness of the cores is really the Kolmogorov scale, and that they are Burgers' vortices. Note that, as explained in the appendix, the large radii for the homogeneous shear flow is probably related to numerical resolution effects.

In all the cases that we have studied, the peak vorticity at the axis of the cores is several times higher than the r.m.s. vorticity for the flow field (Table 1). The same is true for the characteristic velocity, $\Delta u = \omega_{max}\rho$, compared to the r.m.s. velocity fluctuation for the flow considered as a whole. This means that the flow in the immediate neighbourhood of the vortices is dominated by them and is relatively independent of the influence of other structures. We will show now that this can be used to understand some of the alignment properties that have been reported in recent years between the different velocity derivatives.

3. Alignment

It was first suggested by Kerr (1985), and shown later by Ashurt *et al.* (1987) that the vorticity in homogeneous turbulent flows is preferentially aligned with the eigenvector corresponding to the intermediate eigenvalue of the rate of strain tensor, $S_{ij} = (\partial_j u_i + \partial_i u_j)/2$, especially at high values of the enstrophy. This was confirmed later in other flows, both in numerical simulations (Vincent & Meneguzzi, 1990) and in laboratory experiments (Dracos *et al.*, 1989). That observation was considered surprising because the lagrangian vorticity equation can be written as $d\omega_i/dt = S_{ij}\omega_j + \nu\nabla^2\omega_i$, and it had always been assumed that the vorticity vector would be stretched along the direction of any eigenvector of the rate of strain tensor with a positive eigenvalue, and that it would eventually be aligned to the eigenvector of the most positive one. In fact, the existence of tubes was also considered initially controversial because it can be shown that the most probable state for the rate of strain tensor is to have two extensional eigenvalues (Betchov, 1956), and it was felt that this should give rise preferentially to vortex sheets.

That this is not necessarily so can be seen by considering *two dimensional* vortices and vortex sheets. In those cases the vorticity is normal to the $x-y$ plane, and generates a rate of strain tensor in which the only two non-zero eigenvalues are equal in magnitude and opposite in sign, with eigenvectors normal to the vorticity. Thus the vorticity is aligned with the eigenvector of the intermediate (zero) eigenvalue, and is not stretched by any of the other two. Since the picture that emerges from the previous survey of experimental results, at least at high enstrophy values, is one of elongated, essentially two dimensional, compact vortices, it is not surprising that the rate of strain produced by them is normal to their axes, and that it dominates the rate of strain tensor, so that any residual eigenvector is, by the orthogonality property, aligned to the vorticity. It is only the residual axial eigenvalue that does the stretching or compression of the vortex tube, while the two equatorial ones are local effects of the vorticity and do not participate in its dynamics. Note that this arrangement automatically satisfies the requirement that two eigenvalues of the rate of strain tensor be positive, while being consistent with the observation of a prevalence of tubes. This model can be checked directly by inspection of cross sections of cores in numerical turbulent fields (Jiménez, 1991). They show that the orientations of the most compressive and the most extensive eigenvectors are irregular outside the cores, but behave like those of two a dimensional flow inside them.

The argument does not depend of the detailed shape of the vorticity distribution, and can be extended to show that, in the neighbourhood of any elongated concentrated vortex that is stronger than its surroundings, the intermediate strain is aligned to the vorticity (Jiménez, 1991). In essence, the reasoning is that the strong vorticity dominates the local flow field, which becomes approximately two dimensional. It is important to realise that this property is kinematic, and that the orientation of the strains with respect to the vorticity is independent of the particular dynamical mechanism involved in the generation of the vorticity concentration.

In contrast, it was suggested in (Kerr 1985, Ashurt *et al.* 1987), that the alignment of the vorticity to the intermediate eigenvector is due to the tendency of the vorticity vector to rotate in that direction, as a consequence of the asymptotic behaviour of the solutions of a truncated local approximation to the Navier Stokes equations (Vieillefosse, 1982). While those equations might still be useful in explaining the formation of the vorticity concentrations, the present discussion suggests that the explanation of the alignment is simpler, and that it is the rate of strain tensor the one that rotates towards the vorticity, once the latter becomes strong enough.

The same model can be used to explain some of the quantitative information available on the magnitude of the strain eigenvalues at the points of maximum dissipation, which has been measured from numerical simulations to be approximately in the ratio (1:3:-4) (Ashurt *et al.*, 1987). From the model that we have developed here, we may visualise those structures as stretched vortices in which the straining and the viscous diffusion are roughly in equilibrium. This would at least be true during the formation stage, at which dissipation is maximum. It can be shown that the maximum rate of strain in those Burgers' vortices occurs in a annulus surrounding the core, slightly outside the $1/e$ vorticity radius (Jiménez, 1991). This distribution has been documented graphically in (Ruetsch & Maxey, 1991) for numerical isotropic turbulence. At the point of maximum rate of strain, the ratio of the first two eigenvalues depends on Re_γ and, for the experimental range $Re_\gamma \approx 200 - 400$, it varies between (1:2:-3) and (1:4:-5). These values are in rough agreement with the numerical ones quoted above, but they increase with vortex intensity. It would be interesting to check whether flows with a larger spread in the measured values of ω_{max}/ω' also have a larger value of the ratio of the principal strains, but the Reynolds numbers of the present numerical simulations are too low for that purpose.

4. Two dimensional turbulence

In trying to understand the origin of the compact vortices, described above for three dimensional turbulent flows, it is instructive to review the behaviour of two dimensional turbulence, where a similar phenomenon has been known for some time. Compact coherent vortices were first observed in two dimensional homogeneous turbulence by Fornberg (1977), and later by McWilliams (1984). It had been predicted by Kraichnan (1967) and Batchelor (1969) that 2D turbulence would be dominated by a self similar enstrophy cascade with a k^{-3} energy spectrum. Coherent vortices are not part of this cascade, and the spectrum of real turbulent flows tends to be steeper. It was realised by Benzi, *et al.* (1986) that the coherent vortices inhibit locally the cascade and that, in a sense, they constitute drops of laminar fluid in an otherwise turbulent flow. The distinction was formalised in (Babiano *et al.* 1987, Benzi *et al.* 1987), both of which showed that two dimensional turbulence can be understood as a superposition of two components. A turbulent one in which the two vorticity signs are intimately mixed, and which is the site of an enstrophy cascade with a k^{-3} energy spectrum, and a vortical "phase" consisting of persistent patches of vorticity of a given sign, which decay much slower. This second phase behaves almost like a Hamiltonian system of point vortices interacting only sporadically by amalgamation.

The reason for the existence of the compact vortices is only imperfectly understood. It was shown in (Brachet *et al.*, 1988) that vortices are maintained in those region in which the magnitude of the vorticity is larger than that of the rate of strain, thus probably explaining why only relatively intense vortices are observed. Regions in which the opposite is true are quickly deformed into the mixed "phase". Santangelo *et al.* (1989) traced the history of many of these structures and showed that, although some of them originate directly from the initial conditions, others emerge later from the turbulent "sea" and grow by amalgamation. While these observations tell much about how vortices are formed, they say little about why they are, and give no way of predicting how many of them can be expected to form from a given initial condition, or which would be their intensity.

Many researchers have tried to approach these questions using methods of statistical mechanics. The original idea was enunciated by Onsager (1949) who noted that, since a two dimensional gas of point vortices can be described as a Hamiltonian system, it could be treated by statistical methods. He went on to predict that the temperature of that gas would become negative at some point, and argued that this would lead to the fragmentation of the gas into individual patches of pure vorticity of one sign. This last part of the argument is difficult to understand, since the instability condition for a gas of atoms without kinetic energy, like the vortex gas, is not negative temperature but negative specific heat (Callen, 1985). Nevertheless, the accuracy of the prediction has led to a long stream of papers building on the original formulation. The latest two are (Miller 1990, Robert & Sommeria 1991) which give a fairly consistent account of the statistical mechanics of the two dimensional Euler equations. Their argument is that an initial vorticity distribution would eventually deform into a mixed state in which the mean distance between opposite vorticity signs decreases exponentially with time. After a while, the details of the distribution are no longer relevant, and a fluid particle can be characterised by the probability of finding a given vorticity level within it. The flow conserves the vorticity histogram, the total circulation, and the kinetic energy. All other flow invariants are lost to the increasingly small scales of the mixing cascade. In particular, the flow velocity and the kinetic energy can be computed using only the the mean vorticity at each point, without reference to the small scale.

A consequence is that the mixed phase is not an efficient medium in which to store the invariant total kinetic energy of the flow. Vorticities of opposite sign are so close to each other that their effects cancel and they generate no velocity. The velocity is induced by those regions in which the vorticity histogram is unbalanced towards one sign. The study of equilibrium configurations does not lead to the formation of compact intense structures. The maximum entropy criterion predicts the formation of weak extended kinetic energy features in which the mean vorticity is low. This features are observed in actual flows, where they dominate the stream function distributions. They constitute the inverse “energy cascade”. In the flow, however, these large structures are not maintained by an extended steady region of weakly unbalanced vorticity histograms, but by a neutral, highly mixed, phase which contains no energy but which houses a moving cloud of compact vortices whose average position is responsible for the large scale structure. The vortices themselves are in orbits which are bound enough to maintain on the average the large scale structures required by the statistical arguments, but the reason why they form in the first place is still open.

In fact, the statistical model can be fully solved in some simple cases, which can be used to explore its character. One such case is a periodic box in which all the vorticity is in the form of either $\omega = 1$ or $\omega = -1$, with zero total circulation. The Euler equations maintain this vorticity histogram, but mix it throughout the box. The only remaining state variable is the average vorticity at each point, which is a balance of the probability of finding pure vorticity of each sign. The question is whether the statistical mechanics would favour the formation of isolated regions where $\omega \approx \pm 1$. Using the formalism of Robert and Sommeria (1991), the problem reduces to the solution of a nonlinear elliptic equation for the stream function

$$\nabla^2 \psi = -(1/T) \tanh \psi, \tag{2}$$

subject to the condition that the total energy, $E = \int (\nabla \psi)^2 \, ds$, be given. The temperature, T, appears in (2) as an eigenvalue, which depends on the energy through the amplitude of ψ. It is easy to convince oneself that, at least for small energies, the important eigenvalue is the least negative one, and that the specific heat, $c = (dT/dE)^{-1}$ is positive. At low energies, the corresponding eigenfunction is a slow sine wave with a wavelength comparable to the size of the box. The corresponding vorticity distribution is formed by two weak extended vortices in which the mean vorticity is $O(E^{1/2})$. Since the specific heat is positive, this configuration is stable and the vorticity fluctuation remain locally small. There seems to be no place in this model for the small compact vortices but, as pointed before, the large scale average features are consistent with the results of the inverse energy cascade.

One possibility to reconcile the model with the experiments would be that the sign of dT/dE changed at higher energies, leading to a thermodynamic instability and to the segregation of a new "laminar" phase that would condense in droplets corresponding to the small vortices. While this possibility is appealing, there is little evidence for it in our calculations of the model described above, although we are continuing the study at higher energies. A second possibility is that the vortices are "frozen" features from the initial conditions that cannot decay because of the detailed local dynamics. This is consistent with the argument of (Brachet *et al*, 1988) by which the vortices are places were the rotation predominates over the strain. If one assumes that the flow velocity is mainly due to vortices of maximum vorticity, ω, it is easy to show that the maximum strain is $O(\omega)$, and occurs only in the neighbourhood of the vortices themselves. As a consequence, the only possibility to disrupt an existing vortex would be during an interaction with another one, and this event may be too rare to lead to randomisation in a reasonable time or for a box of reasonable size. What this second alternative would mean is that the basic assumption of statistical equivalence of all the microstates within a given macrostate does not hold.

In this view the vortices would be almost accidental, metastable, consequences of the initial conditions that survive because they are unable to decay. This would mean that their properties would depend on the initial condition and that they would not be easy to describe by statistical methods. Santangelo *et al.* (1989) have reported a strong dependence of the properties of two dimensional decaying turbulence on the slope of the initial energy spectrum. For a model of three dimensional turbulence, this would also suggest that the origin of the strong vortices should not be sought in an equilibrium property of the flow but in some transient effect due to local straining. Since there are more instability modes in three dimensions than in two, the three dimensional compact cores would decay faster than their two dimensional counterpart and would play a lesser part in the organisation of the flow.

It is important to realise that, when we say that the compact vortices are responsible for most of the kinetic energy, we do not mean that u^2 is large only inside the cores. A point vortex induces large velocities over a broad area which does not depend on the dimension of its core, but on its circulation. On the other hand, the total kinetic energy, $\int u^2\, ds$, can be written as $\int \psi\omega\, ds$ (Batchelor, 1967), where ψ is the stream function. That the energy resides in the vortex cores should be understood as meaning that the second integral can be extended only over them.

The kinetic energy argument can be used to predict some of the properties of the compact vortices, and Carnevale *et al.* (1991) have used it to relate successfully the evolution of several vortex properties in decaying 2D turbulence. They also need the assumptions that the interaction energy among vortices is negligible with respect to the energy due to the vortices themselves, and that the maximum vorticity in the cores is constant and roughly equal to the maximum vorticity in the initial flow. Although we have seen that the first assumption cannot be strictly true, since some binding energy has to be used to maintain the large scale structure, it seems to be a reasonable approximation. It implies that the structure of each vortex is local, independent of all the others.

5. Strained turbulence.

It is obvious that straining of vortex lines is an important process in three dimensional turbulence, and it can be argued that highly strained turbulence behaves as essentially two dimensional. If we strain a vorticity field axially by an amount $S = \Delta z(t)/\Delta z(0)$, the axial vorticity component is amplified by S while the two normal components are dampened by $S^{-1/2}$. Consequently, the ratio between axial and normal components increases by $S^{3/2}$ and, eventually, it is only the axial vorticity that is important. That is the same as saying that the turbulence becomes approximately two dimensional from the point of view of vorticity dynamics.

The effect of an axial strain on an otherwise two dimensional flow field was studied by Lundgren (1982). He showed that the strained flow is related to a fictitious unstrained two dimensional flow by a scaling transformation. If the rate of strain is defined by $dz/dt = \alpha z$, so that the total strain after time t is $S = \exp(\alpha t)$, the strained flow, u, v, ω_z (x, y, t) corresponds to a two dimensional flow, $S^{-1/2}(u, v)$, $S^{-1}\omega_z$ $(S^{1/2}x, S^{1/2}y, St/\alpha)$. In essence, the two flows behave similarly but the velocities and the vorticities of the strained flow are amplified, the distances contract, and the time, which is proportional to the eddy turnover period, runs faster. The result is that, not only a strained three dimensional turbulent flow will turn quickly into locally two dimensional, but any strained cascade will soon reach an asymptotic state corresponding to long (unstrained) times.

From the discussion in the previous section we can expect that strained turbulence will differentiate into elongated "laminar" vortices, moving in a mixed vorticity "sea". It is tempting to identify the former with the compact vortices described at the beginning of this paper, and the latter with the disorganised turbulent cascade.

It is possible to use Lundgren's transformation, plus the known results of two dimensional turbulence, to estimate the magnitude of the characteristic vorticities for the two flow components. The compact vortices are amplified by a factor S, although their circulation remains independent of time. It can be shown that assuming that the core sizes in two dimensional turbulence are of the order of the viscous cut-off, $(\nu/\omega)^{1/2}$, is equivalent to assuming that the strained laminar vortices are Burger's cores, of size $(\nu/\alpha)^{1/2}$.

The characteristic vorticity, $\Omega = <\omega^2>^{1/2}$, in the two dimensional cascade decreases inversely with time and can be described approximately by a differential equation, $d\Omega/dt \approx -5/\Omega^2$ (see Lesieur, 1990, for a summary of relevant results). Straining adds an extra term to the right hand side, which becomes,

$$d\Omega/dt \approx -5/\Omega^2 + \alpha\Omega.$$

The end result is that, at large times, the straining compensates the decay, and the characteristic vorticity tends to a constant value $\Omega \approx 5\alpha$, which is independent of the initial conditions. At the same time, the inner scale of the strained flow tends to the Burgers' scale, $\lambda \approx (\nu/\alpha)^{1/2}$. This may explain how three dimensional turbulence can undergo a large amount of internal strain without the maximum vorticity blowing up. Note that, in this process, the strain pumps energy into the vortex field, that is dissipated as the enstrophy decays.

6. Conclusions

We have seen that the small strong vortex cores observed in three dimensional turbulence are consistent with a model of almost two dimensional vortices, generated by local straining of the turbulence vorticity. We have also shown that their properties are relatively homogeneous across different types of flows.

A similar phenomenon is observed in two dimensional turbulence, in which the cores appear as coherent vortices whose vorticity is close to the most intense one observed anywhere in the flow. An attempt to derive them from the statistical mechanics of the two dimensional Euler equations failed, although further attempts are still in progress. At present the most probable explanation for their origin is that they constitute "frozen" residues of the initial vorticity distribution, whose own intensity prevents them from decaying to statistical equilibrium.

We have also seen that local straining of three dimensional turbulence generates locally two dimensional turbulence, and that the scale of the viscous vortices generated in this way are compatible with the Burgers' limit in three dimensions. We suggest that the generation of the compact cores in three dimensional turbulence might be related to this process. A similar mechanism for the collapse of strong isolated vortices in three dimensional shear layers was, of course, proposed by (Lin & Corcos, 1984, Neu, 1984).

Acknowledgements

Part of this work was accomplished at, and supported by, the Centre for Turbulence Research, at NASA Ames and Stanford University. It was also supported by the Hermes program of the European Space Agency, under contract AMD-RDANE 3/88 to the Universidad Politécnica of Madrid. Some of the data used for the phenomenological estimates were not contained in the original papers, and have been made kindly available by Drs. Moser and Rogers for the purpose of this investigation. They also provided much of the data processing software, and invaluable suggestions. Their help is warmly appreciated. I have benefited from discussions with E. Castaign, U. Frisch, E. Hopfinger, P. Moin, E. Novikov and, very especially, with P.G. Saffman on the dynamics of turbulence and with F. Higuera on statistical mechanics.

References

Ashurst, W.T., Kerstein, A.R., Kerr, R.M. & Gibson, C.H. (1987a) Alignment of vorticity and scalar gradient with strain in simulated Navier Stokes turbulence. *Phys. Fluids.* **30**, 3243-3253.

Babiano, A., Basdevant, C., Legras, B. & Sadourny, R. (1987) Vorticity and passive scalar dynamics in two dimensional turbulence. *J. Fluid Mech.* **183**, 379-397.

Batchelor, G.K. (1967) *An introduction to fluid mechanics*, Cambridge Univ. Press, pp. 271-273.

Batchelor, G.K. (1969) Computation of the energy spectrum in two dimensional turbulence. *Phys. Fluids Suppl., II* **12**, 233-239.

Batchelor, G.K. & Townsend, A.A. (1949) The nature of turbulent motion at large wave numbers. *Proc. Roy. Soc. London.* **A 199**, 238-255.

Benzi, R., Paladin, G., Patarnello, S., Santangelo, P. & Vulpiani, A. (1986) Intermittency and coherent structures in two dimensional turbulence. *J. Phys. A: Math. Gen.* **19**, 3771-3784.

Benzi, R., Patarnello, S. & Santangelo, P. (1987) On the statistical properties of two dimensional decaying turbulence. *Europhys. Lett.* **3**, 811-818.

Betchov, R. (1956) An inequality concerning the production of vorticity in isotropic turbulence. *J. Fluid Mech.* **1**, 497-504.

Brachet, M.E., Meneguzzi, M., Politano, H. & Sulem, P.L. (1988) The dynamics of freely decaying two dimensional turbulence. *J. Fluid Mech.* **194**, 333-349.

Callen, H.B. (1985) *Thermodynamics and an introduction to thermostatics*, 2nd. Ed., Wiley, pp. 203-210.

Carnevale, G.F., McWilliams, J.C., Pomeau, Y., Weiss, J.B. & Young, W.R. (1991) Evolution of vortex statistics in two dimensional turbulence. *Phys. Rev. Lett.* **66**, 2735-2737.

Dracos, T., Kholmyansky, M., Kit, E. & Tsinober, A. (1989) Some experimental results on velocity-velocity gradients measurements in turbulent grid flows. *Proc. IUTAM Symp. Topological Fluid Mech., Cambridge, August 13-18, 1989* (H.K. Moffat and A. Tsinober, eds.), Cambridge U. Press, pp. 564-584.

Douady, S., Couder, Y. & Brachet, M.E. (1991) Direct observation of the intermittency of intense vorticity filaments in turbulence. *Phys. Rev. Lett.* **67**, 983-986.

Fornberg, B. (1977) A numerical study of 2D turbulence. *J. Comp. Phys.* **25**, 1-31.

Hosokawa, I. & Yamamoto, K. (1990) Intermittency of dissipation in directly simulated fully developed turbulence. *J. Phys. Soc. Japan.* **59**, 401-404.

Jiménez, J. (1991) Kinematic alignment effects in turbulent flows, in press *Phys. Fluids A.*

Jiménez, J. & Moin. P. (1991) The minimal flow unit in near wall turbulence, *J. Fluid Mech.* **225**, 213-240.

Kerr, R.M. (1985) Higher order derivative correlation and the alignment of small-scale structures in isotropic numerical turbulence. *J. Fluid Mech.* **153**, 31-58.

Kim, J., Moin, P. & Moser, R. (1987) Turbulence statistics in fully developed channel flow at low Reynolds number. *J. Fluid Mech.* **177**, 133-166.

Kuo, A.Y. & Corrsin, S. (1972) Experiments on the geometry of the fine structure regions in fully turbulent fluid. *J. Fluid Mech.* **56**, 447-479.

Lesieur, M. (1990) *Turbulence in fluids*, (2nd ed.), Kluwer, pp. 226-283.

Lundgren, T.S. (1982) Strained spiral vortex model for turbulent fine structure. *Phys. Fluids.* **25**, 2193-2203.

McWilliams, J.C. (1984) The emergence of isolated coherent vortices in turbulent flow. *J. Fluid Mech.* **146**, 21-43.

Miller, J. (1990) Statistical mechanics of Euler equation in two dimensions. *Phys. Rev. Lett.* **65**, 2137-2140.

Moser, R.D. & Rogers,M.M. (1991) Mixing transition and the cascade to small scales in a plane mixing layer. *Phys. Fluids.* **A 3**, 1128-1134.

Onsager, L. (1949) Statistical hydrodynamics. *Nuovo Cimento Suppl.* **6**, 279-286.

Robert, R. & Sommeria, J. (1991) Statistical equilibrium states for two dimensional flows. *J. Fluid Mech.* **229**, 291-310.

Robinson, S.K. (1989) A review of vortex structures and associated coherent motions in turbulent boundary layers. *2nd IUTAM Symp. Structure of Turbulence and Drag Reduction, Zurich, July 25-28, 1989* (A. Gyr, ed.), pp. 22-50.

Rogers, M.M. & Moin, P. (1987) The structure of the vorticity field in homogeneous turbulent flows. *J. Fluid Mech.* **176**, 33-66.

Ruetsch, G.R. & Maxey, M.R. (1991) Small scales features of vorticity and passive scalar fields in homogeneous isotropic turbulence. *Phys. Fluids.* **A 3**, 1587-1597.

Santangelo, P., Benzi, R. & Legras,B. (1989) The generation of vortices in high resolution, two dimensional decaying turbulence and the influence of initial conditions on the breaking of self similarity. *Phys. Fluids.* **A 1**, 1027-1034.

Schwarz, K.W. (1990) Evidence for organised small scale structure in fully developed turbulence. *Phys. Rev. Lett.* **64**, 415-418.

She, Z-S., Jackson, E. & Orszag, S.A. (1990) Intermittent vortex structures in homogeneous isotropic turbulence. *Nature* **344**, 226-228.

Siggia, E.D. (1981) Numerical study of small scale intermittency in three dimensional turbulence. *J. Fluid Mech.* **107**, 375-406.

Van Atta, C.W. & Antonia, R.A. (1980) Reynolds number dependence of skewness and flatness factors of turbulent velocity derivatives, *Phys. Fluids.* **23**, 252-257.

Vieillefosse, P. (1982) Local interaction between vorticity and shear in a perfect incompressible fluid. *J. de Physique* **43**, 837-842

Vincent, A. & Meneguzzi, M. (1991) The spatial structure and statistical properties of homogeneous turbulence. *J. Fluid Mech.* **225**, 1-25.

Wei, T. & Willmarth, W.W. (1989) Reynolds-number effects on the structure of a turbulent channel flow. *J. Fluid Mech.* **204**, 57-95.

Appendix: Data processing for Table 1.

To compute circulations we have assumed that the vorticity distribution inside the cores corresponds to that of a viscous axisymmetric vortex $\omega(r) = \omega_{max}\exp[-(r/r_0)^2]$. It is easy to see that the total circulation of such distribution is

$$\gamma = \pi\omega_{max} r_0^2. \tag{A.1}$$

The circulation is estimated using this formula, while ω_{max} is measured directly, and r_0 is defined by $\omega(r_0)/\omega_{max} = e^{-1}$.

A.1. Homogeneous isotropic turbulence. We use data from (Ruetsch & Maxey, 1991). Their cores are described as having a mean diameter of 6η at an enstrophy level $9.5 < \omega_i^2 >$, while the maximum recorded enstrophy is given as $80 < \omega_i^2 >$. Their mean diameter can then be used directly to estimate r_0, resulting in an approximate circulation of $\gamma/\nu \simeq 250$. The aspect ratio of their vortices is given as $L/D \simeq 6$.

A.2. Wall region of turbulent channels. Kim, Moin & Moser (1987), in their analysis of their numerical simulations of channel flow at $Re_\tau = 180$, introduce a "typical longitudinal eddy" whose vorticity and diameter satisfy

$$\omega_x\nu/u_\tau^2 \simeq 0.13, \qquad u_\tau D/\nu \simeq 30. \tag{A.2}$$

They derive these parameters from the behaviour of the ω'_x near the wall. The resulting circulation is, $\gamma/\nu = (\omega_x/\nu)(\pi D^2/4) \simeq 90$. Robinson (1989), using the same data set, publishes histograms of vortex diameters and circulations. The mode and average value for the former are $u_\tau D/\nu \simeq 25$ and 35, while for the latter they are $\gamma/\nu \simeq 60$ and 160. Both analyses are in rough agreement.

Since low order near wall statistics are known to scale approximately with wall variables, we can hope that the scalings in (A.2) hold independently of Reynolds number, but a careful analysis of data both from minimal channels at different Re_τ (Jiménez & Moin, 1991), and from the full size channel in (Kim, Moin & Moser, 1987) show that, while the diameter of the strongest streamwise vortices scales reasonably well in wall units as $u_\tau D/\nu \simeq 20$, their peak vorticity and total circulation increase almost linearly with Reynolds number, from $\gamma/\nu \approx 130$ at $Re_\tau = 100$, to $\gamma/\nu \approx 270$ at $Re_\tau = 200$ (Fig. 1). Note that the large error bars in the figure are due as much to real statistical scatter in the data as to arbitrariness in the definition of what really constitutes a vortex. This arbitrariness also explains in part the difference between our values for the full channel data and those of Robinson (1989). The criteria used here among different

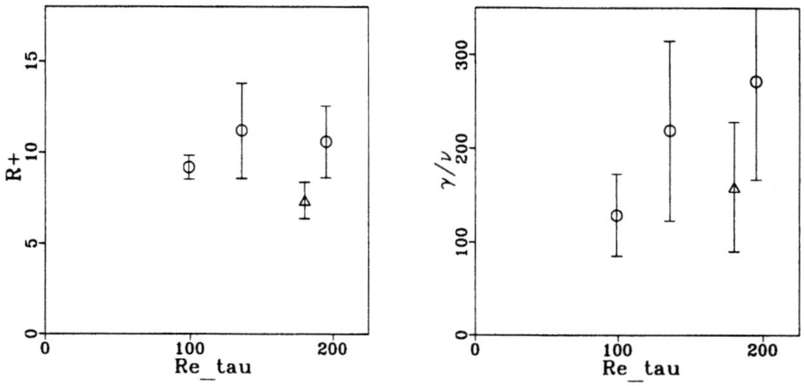

Figure 1: Radius and circulation of strong streamwise of vortices in the near wall region. Expressed in wall units. ○: Minimal channel; △: Full channel

Reynolds numbers were, however, held consistent and the trend is reliable. A similar dependence of the near wall variables on the bulk Reynolds number was documented in (Wei & Willmarth, 1989), where it was attributed to differences in the stretching of the structures of the wall region by the outer flow.

The statistics for r.m.s. vorticity near the wall are at present only available for the (Kim, Moin & Moser, 1987) channel, where $\omega_{max}/\omega' \approx 4.6$.

A.3. Homogeneous shear flow. We use data from numerical simulations of a turbulent flow in a homogeneous shear, S, developing in time from $Re_\lambda = 81$, at $t = 12$, to $Re_\lambda = 97$ at $t = 18$ (Rogers & Moin, 1987). Strong vortices appear as hairpins, whose legs are initially aligned approximately with the direction of maximum average strain. Later they rotate to become more nearly streamwise and seem to reach some sort of equilibrium. An estimation of the vortex diameter from the first zero crossing of the vorticity autocorrelation function with respect to separations normal to the plane defined by the hairpins, results in $D(S/\nu)^{1/2} \simeq 2.8$. A visual inspection of the flow fields suggest a value closer to $D(S/\nu)^{1/2} \simeq 5.8$, which does not seem to change much as time progresses. Even this larger value, however, is close to three numerical mesh spacings, and should be used with care. As maximum vorticity, we have used twice the author's estimate of three times the r.m.s. vorticity of the flow. This was checked by visual inspection of the flow field at two different times in the evolution of the flow. As the turbulence becomes stronger, the maximum vorticity increases from $\omega/S \approx 15$ to 19, but the ratio of ω_{max} to the r.m.s. vorticity stay roughly constant (~ 5.6). The resulting circulation varies from $\gamma/\nu \approx 400$ to 500.

It should be emphasised that the whole simulation corresponds to a developing stage of the flow and that the peak vorticity increases continuously, while the width of the autocorrelation function does not show signs of decreasing. It appears that the hairpins grow in intensity by accreting new vortex lines, rather than by stretching the existing ones.

A.4. Plane mixing layer. The data are from numerical simulations of a plane, three dimensional shear layer by Moser & Rogers (1991), and from some later simulations by the same authors. Two fields are analysed, one just before the second pairing ($t = 40$), and another one right after it ($t = 48$). Both fields look visually turbulent, although the outlines of the large spanwise vortex rollers are still apparent. Tubular vortices are visible and, in the edges of the two dimensional rollers, seem to be associated to the rib vortices that dominate the braid region at an earlier stage of the flow, especially in the first field. In the centre of the rollers, and in most of the second field, this association cannot be made with any certainty. A characteristic vortex diameter in both cases is $D/\delta_{\omega 0} \simeq 0.20$, where $\delta_{\omega 0}$ is the initial vorticity thickness of the layer. The peak vorticity of the cores is $\omega\delta_{\omega 0}/\Delta U \simeq 5.4$, in the second field, and slightly higher in the first one, but that difference decreases when one particularly strong rib vortex pair is separated from the sample. The total circulation per core is $\gamma/\nu \simeq 340$.

The r.m.s. vorticity at the centre of the layer is $\omega'\delta_{\omega 0}/\Delta U \simeq 0.27$, so that $\omega_{max}/\omega' \approx 20$. This value is larger than in any of the other flows, and can probably be explained by the large persistent strain induced by the two dimensional spanwise rollers, as well as by the artificially lowered value of the r.m.s. due to large scale intermittence. In the first field, the vorticity thickness of the layer is $\delta_\omega = 8.5$, while in the second it has grown to 10.0. Since the velocity difference across the layer is always $\Delta U = 2$, the peak vorticities of the cores are almost twenty five times larger than the average shear across the layer $\Delta U/\delta_\omega$, but do not seem to scale directly with it.

Address

School of Aeronautics, Universidad Politécnica
28040 Madrid, Spain
c0130001@emdupml1.bitnet
and
Centre for Turbulence Research
Stanford University

Discussion

Novikov: I am a little bit suspicious about introducing superviscosity. When you introduce superviscosity, the eigenvalues of the Stokes' operator have a property which mathematicians call Spectral Gap. This means that the distances between successive eigenvalues of the Stokes' operator are large and satisfy certain conditions. The strict theory proves that when you have a spectral gap there is an inertial manifold for two-dimensional Navier Stokes equations with superviscosity. However with normal viscosity, and in reality there is only normal viscosity, there is no gap. For a long time, well-known mathematicians, including Foias and Constantin and many others, try to prove the existence of inertial manifolds for two dimensional turbulence with normal viscosity and they did not succeed. Some kind of physics can be behind this. The introduction of superviscosity probably suppresses the part of intermittency which exists in real Navier Stokes' equations and it can lead to a formation of isolated vortices and prevents interaction and mixing of them. So what I suggest is to be careful and to compare experiments with superviscosity to experiments with real viscosity. This is of course not easy for high Reynolds numbers.

Lesieur: I just want to say that in none of the calculations that I showed hyperviscosity was used. In the cases where we used hyperviscosity, i.e. the 2-D mixing layer case, we found the same things except for one detail: The slope of the spectrum in the so-called enstrophy cascade is slightly steeper in the DNS than with hyperviscosity. We did some 3-D computations with hyperviscosity. For example starting from 2-D turbulence, perturbed by small 3-D fluctuations, you see very nice small-scale hairpin structures which develop, and eventually turbulence returns to 3-D isotropy. I don't think anybody can say that vortices are artificially created by hyperviscosity.

Zaslavsky: Why don't you consider some properties of structures, for example the distribution of sizes, alignments and so on?

Lesieur: Maybe because this is a difficult problem. No, we did not have time to look at that.

Zaslavsky: Which are the main properties for intermittency?

Lesieur: In isotropic turbulence we found low-pressure tubes whose length corresponds to the integral scale of turbulence. The average distance between these tubes is twice the integral scale. Finally, there is a third scale corresponding to the diameter of the tubes.

Moffatt: It's the temperature spectrum that you predict that will have this -1 slope? Do you know of any experiment which shows anything of this kind?

Lesieur: Well, there was a paper in Physical Review Letters where they found a k^{-1} spectrum for the longitudinal temperature fluctuation in a shear flow. Our calculation deals with isotropic decaying turbulence. Calculations were done by Chasnov, and he also found this thing. But in forced turbulence you suppress this effect. It is not surprising because you play with the phases of the vortices in the same way as in 2-D turbulence. You may suppress pairing of vortices. Actually, in grid turbulence, where Reynolds numbers are low, the temperature spectrum displays the tendency to some inertial range while the kinetic energy spectrum has no -5/3 range at all. But the slope is -1.5 and not -1. So this is what one gets out of the numerical experiment. I think it is inherently related to low pressure structures and to the exponential distribution of the low pressure.

Herring: In the example I showed earlier, at limited resolution (128^3), closure theory gave roughly k^{-1}. On the other hand, if you look at infinite Reynolds numbers, the same closure gives $k^{-5/3}$ (Herring, 1990, Phys. Fluids, A 2, 979).

Lesieur: I am quite sure that the spanwise wavelength of the perturbation is of the same order as the longitudinal wavelength. So you create longitudinal vorticity of the order of the spanwise vorticity. I am totally confident that if you evaluate spectra in the spanwise direction you will find k^{-3}. Spectra with this kind of stretching mechanism will never have slopes smaller than -3. In order to get $k^{-5/3}$ spectra you need some extra mechanism. These vortices are maybe wrapped inside the big vortices. This kind of mechanism was suggested by A. Roshko. These quasi 2-D things are wrapped into the large vorticies and then locally you have such intense velocity gradients in all directions that you get a non-linear very quick tendency to a 3-D Kolmogorov cascade. So it is during this mixing that you produce the Kolmogorov cascade, which may scatter vorticity in the flow. So I agree to most of the points of Jimenez except when he speaks of "worms". There are two types of "worms". In turbulence you have the "worms" corresponding to the vorticity in the low pressure tubes and also these -5/3 Kolmogorov "worms" which are scattered everywhere in the flow field.

Aref: I have a question of clarification to Jimenez. At the beginning of your talk it seemed to me you were saying that the rate of strain comes about due to the vorticity distribution. In the second part when you talk about the formation of the structures you were doing just the opposite. The structures come about due to the rate of strain.

Jimenez:
There is an ambient rate of strain. These "worms" are far apart. If you look near a single "worm" then there is a rate of strain which is of the order of ω, much larger than that generated by other "worms". But that is local. These other "worms" really have to come very close, if you want to approach something like the local rate. But when they are far apart, what you get is something of the order of the maximum vorticity volume ratio occupied multiplied by the "worms".

Frisch: First of all I want to say I like Javier's picture of how these things come about very much. I call them "Spaghetti". One can already put some limitations on some of the numbers by a very simple velocity argument. Working here in units where the integral scale and the RMS-velocity are taken to be one, so the viscosity is one over the Reynolds number. Let me denote by L the typical length of these things and by N the typical number of such animals per unit volume. Then you find soon in the naive way that there is a typical peak vorticity. Now I can evaluate the peripheral velocity which is of course $r \cdot \omega$. Omega is a typical peak vorticity value inside. Now calculate the dissipation coming from such tubes. It is very simple, there is the square vorticity, ω^2, and then the volume which is $r^2 \cdot L$, and then the number of such animals per unit volume and then the viscosity which is 1/Re. So you get an expression which involves the combination $r^2 \cdot \omega^2$ which is just the square of the peripheral velocity. Now, if I assume that this finite fraction of the total dissipation is contained in those tubes, which some people think is true on the basis of their numerical simulation, then I can solve this for the peripheral velocity. I mean of course the highest velocities are obtained just outside of these tubes, and what you find it goes like this: ε is the dissipation per unit volume typically people assume, that it is Reynolds number independent, or weakly dependent on Reynolds number. Now, if you take the simplest types of models, you can assume that these have a length that is comparable to the integral scales order, so that L is of the order 1 in our units and if you also assume that the number of these is such that the typical spacing between them is of order L, then you get something that possibly is an absurdity, namely that the peripheral velocity goes to infinity with Reynolds number like $\sqrt{Re}$. You cannot completely rule that out becauce these are very tiny regions and you may easily miss them, but still it seems unlikely. So here you have some constrains if you want to make dynamical models of interacting strings. Statistical mechanics of these strings is still quite a mystery even if one has some idea how they are formed. If one wants to make some kind of assessment of that, one has to keep in mind that this may have consequences for velocities which may become outrageously high in certain kinds of models. So, I consider this not at all as a criticism. It puts some limitations on the kind of further games one can play.

Novikov: You assume that essential dissipation is in these "worms"?

Lesieur: In the calculations I showed you we had this too. I think the same is true for calculations by Siggia. These are not dissipative structures. The tubes we see which have an intence vorticity and low pressure are the big vortices of the turbulence. They are stretched by mechanisms which I tried to explain. This stretching produces secondary and even tertiary vorticity, which is everywhere in the field and which may correspond to the dissipative structures.

Frisch: I recently asked Maurice Meneguzzi, who Alain Vincent had done very large scale simulations showing conspicuous vortex tubes with if he thinks these tubes contain a very sizeable fraction of the total dissipation. He said that the total disssipation had indeed a very large contribution coming from those tubes or their immediate neighbourhood. I was a bit surprised, but I am quoting him.

Lesieur: I don't think so!

Jimenez: Basically, this model was made by Tennekes in '68 and the problem is that when you try to really put all the numbers, the flatness factor is a way too high, the flatness factors go like Re or something like that. Something has to be wrong there. Not all of the dissipation is happening in this field. Actually it has to be the fraction of the dissipation which goes to zero as the Reynolds number increases, which is, I think, happening in the intermediate range when you are straining these things.

Frisch: What would you propose for a typical length? What is their radius of curvature? How does it scale with the Reynolds number? Like the integral or like the Taylor scale?

Jimenez: I think it's like the Taylor scale.

Frisch: Because one could also argue, that "worms" or "Spaghetti" form a fractal tangle, just like some people's hair.

Moffatt: The current physical picture is terribly important. If you're right, if N is of order one, then tubes are really far apart and then their motion is entirely self-induced.

Frisch: No matter, you can modify all these schemes, you can say that there is a fraction that goes like the Reynolds number to some negative power.

Moffatt: I am not sure that I am satisfied with that.

Zaslavsky: We don't know what stops the stretching of some "sausages". There are at least two reasons for this. One is dissipation and the second is instability. Which kind of estimations do you use just now, dissipation or instability?

Frisch: You can already put some constrains here. Because you see if these are being stretched by large-scale gradients, so by gradients which in those units are of order one, just by equating the stretching against the viscous diffusion you would predict a rate which scales like the square root of viscosity. But if you want these tubes to be representative of the smallest dynamically active scale, present in the flow, then you would like their radius to be comparable to the Kolmogorov scale. Then you have an inconsistency if you assume that the gradients that stretch them are of order one. They must be stronger.

Aref: I think that part of the problem here is that you are taking a very static picture of the vorticity- field and that actually what's happening is that the length of these things is set by dynamical events where different vortex-lines intersect and reconnect.

Frisch: Maybe there is a scaling range of lengths in which case probably the proper characteristic length has no meaning. Then you have a fractal tangle.
Some of you may have seen the video by Douady, Couder and Brachet (S. Douady, Y. Couder and M. Brachet, 1991, Phys. Rev. Lett. **67**, 983) where they actually claim to have observed such vortex "worms". They have a transparent cylindrical vessel with counterrotating disks at the top and the bottom. They first produced lots of bubbles induced by cavitation. Then they were hoping, probably rightly, that these very tiny bubbles will migrate to the low-pressure regions inside the vortex tubes. Watching this with a high-speed camera they did observe the formation of dark lines which look more or less like filaments.

Aref: I want to make a case for a theory that is somewhat similar to this kind of ideas done in superfluid helium, the Vinen theory, I think that might be something that you want to revisit. It works in terms of the amount of vortex-line per unit-volume. In that case all the lines are quantized. But then he basically writes a production and a decay term due to different processes. To some extent that is the integrated turbulence theory for superfluid helium. This was produced by a man called Vinen, many years ago.

Moffatt: Have you found any evidence for vortex reconnection in this computation, where you have got this marvellous technique for numerically visualizing vortices?

Lesieur: We had vortex reconnection in a mixing layer undergoing helical pairing. It is possible that these things occur also in isotropic turbulence, as was shown by Ménéguzzi. We also saw that intense helical pairing yields intense vortex stretching.
I think that if you want to develop a complete catalogue for all possible vortex interactions, it may become very complicated. This is why I still like statistical theories, because at the end it is the only way of putting all these things together.

Herring: It can not be that all is in the small scale structures. Does it in that case give the right derivative skewness which is a three-dimensional kind of thing? If that is true, could you estimate what fraction is in the straining fields in three dimensions? It seems to me there must be a lot of straining going on.

Jimenez: There's actually some statistics which were given in a paper by Ruesch and Maxey, 1991. They look how much of the dissipation is associated to the core and how much is not. In their case it is something like 0.4%, nearly nothing. There's no information there on the Reynolds number. Those strong structures are probably very important in mixing, but from the point of view of dissipation they are not, they cannot be.

Novikov: I want to mention another thing about the orientation of the vorticity and stretching. The vorticity has a tendency to aligne with the middle-eigenvector of strain. There's a very simple model which can show that such a tendency really exists. If you differenciate the Navier-Stoke equations and constitute the equation for the gradient of the velocity than you have a second derivative of pressure. If you assume that a strong vorticity is formed and that the second derivative of pressure is determined by the local gradient of velocity, you get this kind of orientation until the viscosity becomes active. This is an exact solution. (E. A. Novikov, 1990, The internal dynamics of flows and formation of singularities, Fluid Dynamics Research, V. 6, 79-89)

Moffatt: Dresselhaus and Tabor have done a calculation in which they claimed to explain the vorticity alignment that you are referring to. Tabor talked about this in Santa Barbara two weeks ago. It was based on precisely this sort of local analysis (M. Tabor, 1992, Stretching and alignment in general flow fields. In Topological Aspects of the dynamics of fluids and plasmas, ed. H. K. Moffatt et al., Kluwer, 83-111).

Frisch: I would like to make an observation regarding spurious intermittency effects which can be observed while processing Gaussian or almost Gaussian data. Suppose you have a Gaussian process and that it has a power-law spectrum and this could be multi-dimensional, but I'm considering this as a 1-D-cut, so you have a power law spectrum x^{-n} with i between 1 and 3. That's obviously a self-similar process, for instance if n = 2 it is the Brownian motion. Now, there is a theorem which tells you that if you take any smooth line, the intersections with the graph of the process form a fractal with a dimension D = (3-n)/2 (K. Falconer, Fractal Geometry, Wiley, 1990). For example, for n = 2 the dimension is 1/2. Brownian motion itself is clearly not intermittent, it's continuous, not differentiable. But if you look at the places where it achieves a given value, that's a fractal set. Now, if you have a Gaussian background field, or near-Gaussian with a Kolmogorov spectrum, something that scales, then the places where a given treshold will be exceeded or rather crossed, that is a fractal set which in 3-D space will have dimension 2 plus something. Now, this can easily lead to spurious fractality. For example, in the last few years there have been at least five papers in the Astrophysical literature where people claim to have seen fractal-magnetic structure in the sun. How do they do it? They take magnetographs and they look at the points where the magnetic flux exceeds a given value and they paint those pixels in black and then they do counting of area versus diameter or something similar. When they get a power law, they say the magnetic field is fractal. But as I just pointed out, you already get such fractals with Gaussian fields. Also in the wavelet analysis, that I was myself involved in with Argoul and others (F. Argoul et al., 1989, Nature, **338**, 51). We had a Kolmogorov spectrum and there was colour coding, which would put emphasis roughly on levels, and it was branching in a fractal way. As pointed out in later work with Bacry and others (E. Bacry et al., 1990, Proceedings of "Turbulence 89", O. Métais amd M. Lesieur eds., 203-215, Kluwer) this does not imply intermittency since such things can easily happen with Gaussian fields; maybe not exactly in the same way. To sum up: depending on how one processes data, one may see fractals without dynamical significance.

Zaslavsky: Why not, any mechanism can produce fractal structures.

Frisch: Oh, no. It's not so easy to get fractals. Fractals in dynamical systems are extremely easy to get. Fractals in the Navier-Stokes context, in physical space, are not at all so easy to get. You get them very easily when you have, for example, fingering. No doubt about that, but otherwise fractals have been speculated about, but hardly seen.

Session Three

Chairman H. Aref

Spiral structures in turbulent flow

By H.K. Moffatt

Department of Applied Mathematics and Theoretical Physics, Silver Street, Cambridge, CB3 9EW.

1 Introduction

Spiral structures are natural candidates for the role of the 'generic structures' of turbulent flow, because they are the eventual outcome of Kelvin-Helmholtz instability, an all-pervasive phenomenon associated with nearly all shear flows at very high Reynolds number. Such structures were proposed by Lundgren (1982) in a model of the fine structure of turbulence in which axial stretching of rolled-up spiral vortices played an essential role. This model could be viewed as a natural development of Townsend's (1951) model of the dissipative structures of turbulence in terms of a random distribution of vortex sheets and/or tubes, each such structure being subjected to the local rate of strain (assumed uniform and constant) associated with all other vortex structures (see Batchelor 1982, §7.4). It is known that compressive strain (with two positive principle rates of strain) is more likely in isotropic turbulence than extensive strain, so that sheets form with higher probability than tubes. However these are immediately subject to the Kelvin-Helmholtz instability which may be impeded, but not entirely suppressed, by the stretching process.

A random superposition of vortex sheets has the property that the resulting velocity field has a finite number of discontinuities per unit length on any straight line transversal. This leads to an energy spectrum function $E(k)$ proportional to k^{-2} (with an exponential cut-off at wave-numbers of the order of the reciprocal of the typical vortex sheet thickness). The Kolmogorov cascade theory, by contrast, gives $E(k)$ proportional to $k^{-5/3}$ (with an exponential cut-off at the inner Kolmogorov scale). It is natural then to enquire as to what structures in $\mathbf{x}$-space can give rise to such a power-law. This question was raised by Moffatt (1984) where it was argued that the accumulation points of discontinuities associated with spiral structures could give rise to fractional power laws $k^{-\lambda}$ with $1 < \lambda < 2$. We shall pursue this argument a little further in the present note.

The influence of spiral structures in two-dimensional turbulence has been considered by Gilbert (1988) who considered the wind-up of a weak vortex patch by a strong concentrated vortex in its vicinity, and showed that fractional power laws $E(k) \sim k^{-\lambda}$ with $3 < \lambda < 4$ could be obtained through this mechanism. This provides a convenient bridge (and reconciliation) between the vortex patch theory of Saffman (1971) and the enstrophy cascade theory of Kraichnan (1967) and Batchelor (1969) which give k^{-4} and k^{-3} respectively. The methods that we adopt in the present paper are similar to those employed by Gilbert.

2 The Kelvin-Helmholtz instability

A vortex sheet, i.e. a tangential discontinuity (of magnitude U) in tangential velocity, is well known to be unstable, on linear analysis, to perturbations of wavenumber k, the growth rate being of order kU. The fact that the shortest wavelengths grow most rapidly means that the linear problem is ill-posed, and, as shown by Moore (1979), the nonlinear development of an initially sinusoidal perturbation develops a mild singularity at a finite time t_c of order $(kU)^{-1}$, largely as a result of the singular character of the dispersion relation. The mechanism of the instability, as described by Batchelor (1967), is that the strength of the vortex sheet increases near the inflexion points where the surface is tilted in the sense favoured by the velocity difference; this tendency persists in the nonlinear regime, and at time $t = t_c$, the strength of the vortex sheet is cusp-shaped at the inflexion point (see figure 1).

Numerical procedures based on the Euler equations cannot be continued beyond this finite time singularity. If however the equations governing the evolution of the sheet are modified slightly, in a manner that mimics the fact that any vortex sheet in a real fluid would not be a perfect discontinuity, but would have finite thickness, then the integration can be pursued beyond $t = t_c$. Such a 'desingularisation' procedure has been adopted by Krasny (1986) who integrated the equations for vortex sheet evolution for various small values of a desingularisation parameter δ. When $\delta = 0$, the evolution based on the Euler equations is recovered. Figure 2 shows a reproduction of Krasny's solution for the vortex sheet, for $\delta = 0.05$, and for $t = 8t_c/3$. At this stage, five complete turns have appeared in the rolled-up vortex sheet. As δ increases, the number of turns (at fixed t) decreases; conversely, it may be supposed that, as δ decreases towards 0, the number of turns at fixed t ($> t_c$) increases without limit.

A straight line transversal passing through the centre of this spiral now samples an accumulation point of discontinuities of the velocity field. It is this property that can lead to a fractional power law in the energy spectrum. A random distribution of spiral vortices of this kind can therefore, in principle, yield a Kolmogorov spectrum (or any other power law spectrum) if the structure is right. We investigate this by an even more simplified

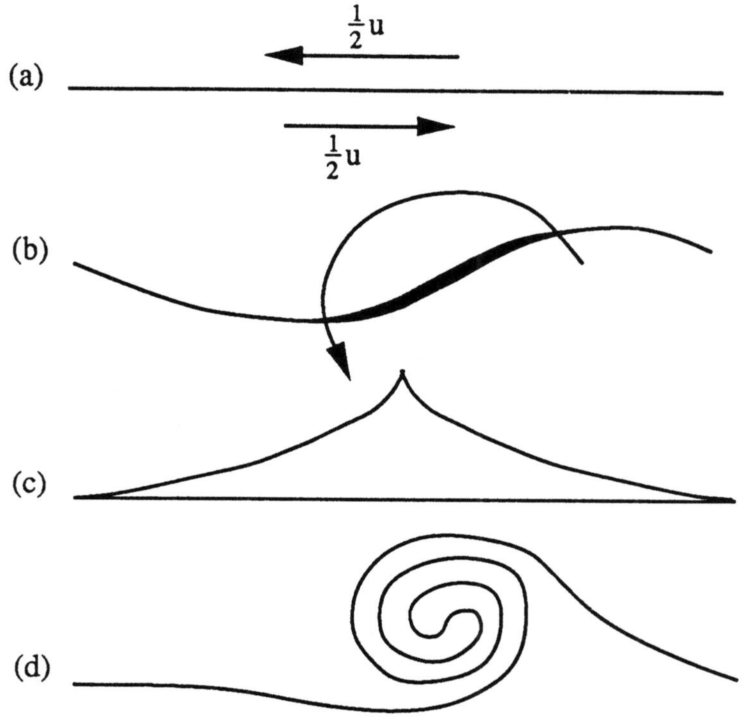

Figure 1. The mechanism of Kelvin-Helmholtz instability. (a) The unperturbed state; (b) the instability mechanism associated with accumulation of vorticity at the inflexion point; (c) vortex sheet strength, showing the cusp at the inflexion point at the instant $t = t_c$ when the curvature becomes discontinuous (Moore 1979); (d) development of the spiral singularity for $t > t_c$.

model in the following section.

3 A simple model

Consider a set of concentric circular vortex sheets of radius $x_m = \pi m^{-\alpha}$ ($m = 1, 2, 3$ etc.). Let $\delta x_m = x_m - x_{m+1}$($\sim m^{-\alpha+1}$ for large m), and suppose that the velocity in the annulus δx_m is $u_m \sim m^{\beta} \sim x_m^{-\frac{\beta}{\alpha}}$. The energy in the annulus δx_m is then $E_m \sim \int_{x_{m+1}}^{x_m} u_m^2 x_m dx_m \sim m^{2\beta-2\alpha-1}$, and the condition that the total energy be finite then implies that $\beta < \alpha$. The situation is shown schematically in figure 3.

The spectrum of the velocity field may be calculated by the method

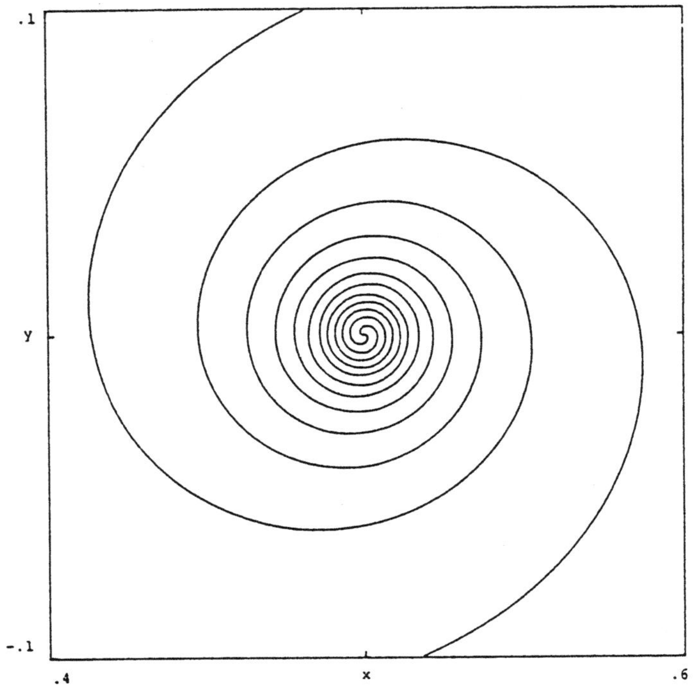

Figure 2. (From Krasny 1986) Structure of the spiral vortex computed by Krasny according to a desingularised prescription.

developed by Gilbert (1988). The Fourier coefficients of $u(x)$ are given by

$$a_n = \frac{2}{\pi}\int_0^\pi u(x)\sin nx\,dx \tag{3.1}$$

$$= \frac{4}{n\pi}\sum_{m=1}^{\infty} u_m \sin\tfrac{1}{2}n(x_{m+1}+x_m)\sin\tfrac{1}{2}n(x_{m+1}-x_m). \tag{3.2}$$

The factor $\sin(\frac{1}{2}n\delta x_{m+1})$ is effectively zero if $m^{(\alpha+1)} \gg n$; hence, provided u_m does not increase too rapidly with m,[1] the number of terms making significant contributions to this sum is $M = O(n^{\frac{1}{\alpha+1}})$. The term $R_m = \sin\frac{1}{2}n(x_{m+1}+x_m)$ is effectively a random number in the interval $[-1,1]$, and the sum can be thought of as the result of a random walk in which the step-length increases like m^β (where m labels the step). It is not difficult

[1] J.C. Vassilicos (private communication) has calculated that the condition for validity of this argument is $\beta \leq \frac{1}{2}\alpha$.

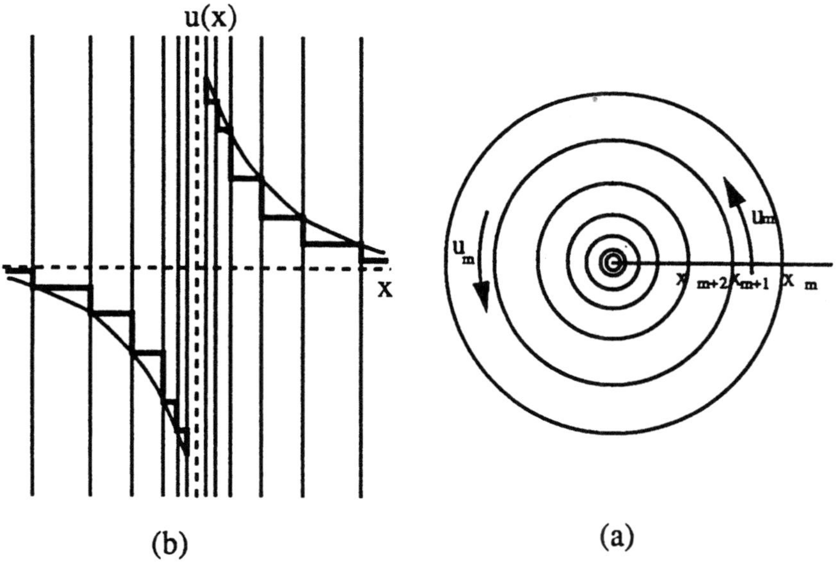

Figure 3. (a) Simplified model involving concentric vortex sheets; (b) the corresponding velocity profile on a transversal through the centre.

to deduce that

$$|a_n^2| \sim n^{-\lambda}, \text{ where } \lambda = 2 - \frac{2\beta+1}{\alpha+1}.$$

If there is a random superposition of such structures, then a straight line transversal will not pass through the centre of all of them, but rather will pass at distance y from the centre where all values of y between -1 and +1 are sampled with equal probability. Clearly, when y is small but non-zero, the infinite sum in the above equation is cut-off at a large but finite value of m. The resulting value of a_n^2 can then be averaged over all values of y, giving the modified result

$$\langle a_n^2(y)\rangle \sim n^{-\mu} \qquad \text{where} \quad \mu = \lambda + \frac{\alpha}{1+\alpha}.$$

We may note here in parenthesis the special values $\beta = 0, \alpha = \frac{1}{2}$, for which $\lambda = \frac{4}{3}, \mu = \frac{5}{3}$. These values were relevant in the work of Gilbert (1988), who considered however not the wind-up of a vortex sheet, but the wind-up of a discontinuity of *vorticity* (rather than of velocity) by a concentrated point-vortex. Consequently, in the simplest configuration considered, Gilbert found an *enstrophy* spectrum proportional to $k^{-\frac{5}{3}}$, and a corresponding energy spectrum proportional to $k^{-\frac{11}{3}}$.

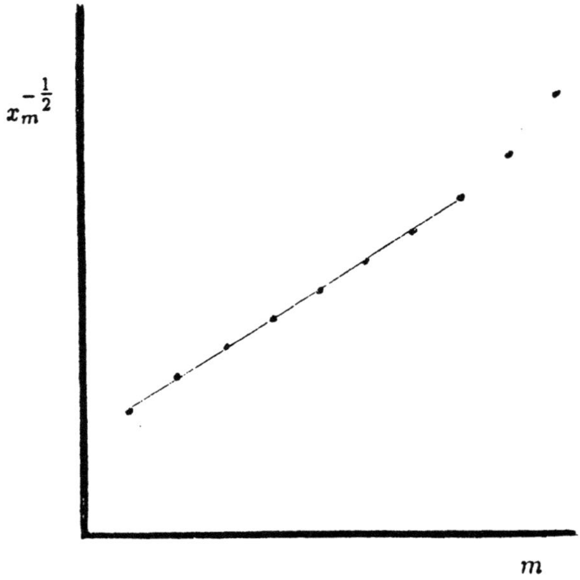

Figure 4. Plot of $x_m^{-\frac{1}{2}}$ against m for the Krasny vortex showing a linear range for $1 \leqslant m \leqslant 8$. This implies that the spiral can be represented reasonably well by the curve $r(\theta - \theta_0)^2 =$ cst., and that the circulating velocity behaves like $r^{-\frac{1}{2}}$, except in the very central region.

It is interesting to re-examine the structure of the Krasny vortex in relation to this analysis. Figure 4 shows a plot of $x_m^{-\frac{1}{2}}$, against m, where m labels the intersections of the spiral with the x-axis, as before. There is a good straight-line relationship for $1 \leqslant m \leqslant 8$, consistent with fitting to a spiral of the form $r(\theta - \theta_0)^2 =$ cst.. [2] The points for $m = 9, 10, 11$, depart from this straight-line behaviour, and this is presumably due to the fact that the vortex does not achieve an asymptotic structure in the central region (or perhaps the desingularisation procedure is at its worst in this region).

With $\alpha = 2$, as suggested by this result, a Kolmogorov exponent $\mu = \frac{5}{3}$ is obtained for the choice $\beta = 1$. For this choice, $u_m \sim x_m^{-\frac{1}{2}}$, and correspondingly the circulating velocity associated with the spiral is $u(r) \sim r^{-\frac{1}{2}}$. This slower fall-off than the velocity ($\sim r^{-1}$) associated with a point-

[2]R. Krasny (private communication) confirms this inference in further computations with desingularisation parameter $\delta = 0.03$ and a spiral with 21 turns.

vortex is of course related to the more diffuse distribution of vorticity in the spiral region. The value $\beta = 1$ means that the jump in velocity is the same across each discontinuity, i.e. that the strength of the vortex sheet is effectively constant except perhaps in the very central region. The evidence available from Krasny's computations is not conclusive on this point, but is certainly not inconsistent with this conclusion.

4 Discussion

Visual observations of turbulent flow, and equally, direct numerical simulations (see for example Kerr 1985), provide ample evidence for the existence of vortex tube structures, frequently exhibiting a spiral cross-sectional structure. The suggestion of this paper, which is a development of the argument advanced in Moffatt (1984, 1990), is that a random distribution of vortices of this kind is all that is needed to provide an inertial range spectrum proportional to $k^{-\frac{5}{3}}$. There is certainly no need to consider a hierarchy of structures on different length-scales, with a cascade of energy from one scale to the next. A single generic structure, which evolves in time due to its internal dynamics, and possibly (as in Lundgren's 1982 model) through additional straining associated with other structures, is sufficient to provide an inertial range spectrum of Kolmogorov type, although for very different reasons from those first advanced by Kolmogorov. This does not of course mean that other physical processes (e.g. vortex break-down) may not be important also in turbulent flow. But it does mean that the search for characteristic vortical structures is well worthwhile, not only to provide a better understanding of the dynamics of the energy-containing eddies, but also in understanding the process of transfer of energy to the small scales where viscous dissipation is operative.

5 Acknowledgments

I am grateful to Christos Vassilicos and Andrew Gilbert, who have made significant input in recent discussions of the problem of analysing spiral structure.

This paper was originally presented at the Conference "Wavelets, Fractals and Fourier Transforms" of the Institute of Mathematics and its Applications in December 1990, and is published with the agreement of the IMA.

6 Appendix

Everson and Sreenivasan (1992) have recently investigated spiral structures that develop in the field of a passive scalar contaminant, and have found that these spirals are better modelled by a logarithmic spiral rather than by an algebraic spiral of the type discussed above. The passive scalar problem is however rather different from the spiral wind-up associated with Kelvin-Helmholtz instability, for which the nonlinear dynamics associated with the Euler equation plays an essential role.

An important dynamical mechanism for the formation of vortex sheets has recently been identified by Pumir and Siggia (1992a,b). With any swirling flow, there is an associated unbalanced centrifugal force, which accelerates the flow outwards, and which intensifies the vorticity associated with the swirl in the accelerating region of the flow. Pumir and Siggia have suggested that shear instabilities of the developing "front" may lead, within the framework of the Euler equations, to a finite time singularity. Whether this is true or not, the mechanism is undoubtedly a potent one for the generation of vortex sheets and spiral structures, and it would seem to be a natural process in any region of the flow where the streamlines are curved (and that of course means virtually everywhere).

7 References

1. Batchelor, G.K. (1967). *An Introduction to Fluid Dynamics.* Cambridge University Press.

2. Batchelor, G.K. (1969). Computation of the energy spectrum in homogeneous two-dimensional turbulence. *Phys. Fluids Suppl.* II, 233.

3. Batchelor, G.K. (1982). *The Theory of Homogeneous Turbulence* (Cambridge Science Classics, CUP).

4. Everson, R.M. & Sreenivasan, K.R. (1992). Accumulation rates of spiral-like structures in fluid flows. *Proc. Roy. Soc. Lond.* A **437**, 391–401.

5. Gilbert, A.D. (1988). Spiral structures and spectra in two-dimensional turbulence, *J. Fluid Mech.* **193**, 475–497.

6. Kerr, R.M. (1985). Higher-order derivative correlations and the alignment of small-scale structures in isotropic numerical turbulence. *J. Fluid Mech.* **153**, 31–58.

7. Kraichnan, R.H. (1967). Inertial ranges in two-dimensional turbulence. *Phys. Fluids* **10**, 1417–1423.

8. Krasny, R. (1986). Desingularisation of periodic vortex sheet roll-up, *J. Computational Physics* **65**, 292–313.

9. Lundgren, T.S. (1982). Strained spiral vortex model for turbulent fine structure. *Phys. Fluids* **25**, 2193.

10. Moffatt, H.K. (1984). Simple topological aspects of turbulent vorticity dynamics. In *Turbulence and Chaotic Phenomena in Fluids*, ed. T. Tatsumi (Elsevier) 223–230.

11. Moffatt, H.K. (1990). Fixed points of turbulent dynamical systems and suppression of nonlinearity. In *"Whither Turbulence? – Turbulence at the Crossroads"*, ed. J.M. Lumley. *Lecture Notes in Physics* **357** (Springer-Verlag) 250–257.

12. Moore, D.W. (1979). The spontaneous appearance of a singularity in the shape of an evolving vortex sheet. *Proc. Roy. Soc.* **A365**, 105.

13. Pumir, A. & Siggia, E. (1992a). Blow up in axisymmetric Euler flows. In *Topological Aspects of the Dynamics of Fluids and Plasmas*, ed. H.K. Moffatt et al. Kluwer Acad. Publ. *Series E: Applied Sciences* **218** 293–302.

14. Pumir, A. & Siggia, E. (1992b). Development of singular solutions to the axisymmetric Euler equations. *Phys. of Fluids* A **4**, 1472–1491.

15. Saffman, P.G. (1971). On the spectrum and decay of random two-dimensional vorticity distributions at large Reynolds number. *Stud. Appl. Maths* **50**, 377.

16. Townsend, A.A. (1951). On the fine-scale structure of turbulence. *Proc. Roy. Soc.* **A209**, 418.

Solutions of exact kinetic equations for intermittent turbulence

Dr. E. A. Novikov

ABSTRACT. A new approach to the problem of turbulence is described. This approach is based on conditional averaging of equation for a local characteristic of motion, which has a mechanism of self-amplification. An exact closed equation for conditionally averaged 3-D vorticity field (with fixed vorticity at a certain point) is obtained from the Navier-Stokes equation. Corresponding closed equation is derived for conditionally averaged vorticity gradient (vg) in 2-D turbulence. Solutions of these equations are presented. The conclusion is made that the local structure of turbulent flows is not unique. The nonuniqueness is due to the phenomenon of intermittency, which depends on the large-scale structure of turbulent flows. The high order two-point moments of vorticity and vg are presented. The developed method is quite general and can be applied to a variety of physical systems with strong interaction, including magnetized plasma.

1. Introduction and small parameter

In statistical description of nonlinear systems we are faced with the problem of closure. For a system with weak interaction we can use a corresponding small parameter in order to find an appropriate closure. The first physically well motivated closure was invented in the last century by Boltzmann for the case of rare gas. For a system with strong interaction (like turbulent flow) it is very difficult to guess what kind of closure is consistent with the original (Navier-Stokes) equation. Various closures for turbulence, including quasi-normal hypotheses (Monin and Yaglom 1975), have been rejected one by one by physical analysis, comparison with measurements and, more recently, by direct numerical simulations of turbulence.

One of the difficulties in choosing an appropriate closure for turbulence is the lack of a general theory of non-Gaussian intermittent random fields. An example of such a field was constructed many years ago (Novikov 1966) and was recently republished (Novikov 1990) in more general context and with additional details. This model describes some major features of intermittent turbulence. The open problem is how to marry this kind of random field with the Navier-Stokes equation.

Another fundamental difficulty in the theory of turbulence until recently (see below) was the lack of a well recognized small parameter. We have a large parameter-Reynolds number $R = VL/\nu$ (V is a characteristic velocity, L is an external scale and ν is the kinematic viscosity). Can we somehow invert this parameter and consider an appropriate negative power of R as a small parameter? Yes, in a certain sense, we can.

In order to do this, we shall consider the internal dynamics of turbulent flow in terms of local characteristics of motion (Novikov 1990a). A local characteristic of motion must have some mechanism of self-amplification and shall not depend directly on the sources of energy. For three-dimensional turbulent flow with high R, the appropriate local characteristic of motion is the vorticity field (Taylor 1938, Novikov 1967, 1969). The mechanism of self-amplification in this case is the well known effect of stretching of vortex filaments. Statistical regime for a flow region sufficiently far from a boundary is determined by the equilibrium between vortex stretching and viscous smoothing.

Following the analysis in (Novikov 1967, 1969), let us consider the balance of vorticity in nonhomogeneous nonstationary turbulent flow:

$$\frac{\partial \omega_i}{\partial t} + v_k \frac{\partial \omega_i}{\partial x_k} = \frac{\partial v_i}{\partial x_k} \omega_k + \nu \Delta \omega_i, \ \frac{\partial v_i}{\partial x_i} = 0 \,, \tag{1}$$

$$\frac{1}{2}\frac{\partial}{\partial t}\left\langle \omega_i^2 \right\rangle + \frac{1}{2}\frac{\partial}{\partial x_k}\left\langle v_k \omega_i^2 \right\rangle - \frac{\nu}{2}\Delta\left\langle \omega_i^2 \right\rangle = \left\langle \frac{\partial v_i}{\partial x_k}\omega_i\omega_k \right\rangle - \nu\left\langle \left(\frac{\partial \omega_i}{\partial x_k}\right)^2 \right\rangle . \tag{2}$$

Here v_i is the velocity, ω_i is the vorticity and $< >$ means statistical averaging. The balance equation (2) is obtained from (1) by multiplying on ω_i, averaging and using incompressibility. For turbulent flow with high R we will use standard estimations:

$$\frac{\partial v}{\partial x} \sim \omega \sim \epsilon^{1/2}\nu^{-1/2}, \ \epsilon \sim V^3 L^{-1}, \ T \sim LV^{-1} \sim V^2 \epsilon^{-1} \,, \tag{3}$$

where ϵ is the mean rate of energy dissipation, T is a characteristic time of decay of turbulence. By using (3), it is easy to see that first two terms in the lhs of equation (2) relative to the first term in the rhs of (2) are of order of $R^{-1/2}$. This is our small parameter. The third term in the lhs of (2) is even smaller ($\sim R^{-3/2}$). Thus, self-amplification is balanced by viscous smoothing [the terms in the rhs of (2)] and the effects of nonhomogeneity and decay of turbulence are small ($\sim R^{-1/2}$). This also means that we do not have to introduce fictitious external random forces. Instead, we require natural conditions for maintaining vorticity fields at a given "temperature":

$$\nu < \omega_i^2 >= \epsilon \,, \tag{4}$$

which is, of course, consistent with (3).

The hierarchy of kinetic equations for three-dimensional turbulent vorticity fields have been obtained in (Novikov 1967, 1969) from the Navier-Stokes equation. These are the equations for n-particle (Lagrangian) and n-point (Eulerian) probability distributions of vorticity field. The equation for the characteristic functional of vorticity field has also been derived (Novikov 1969). It was indicated in (Novikov 1967, 1969) by using the same small parameter $R^{-1/2}$, that for locally-homogeneous flows with high R the kinetic equations become closed. Surprisingly, it took many years to realize that the first of these equations can be written in the simple form (7) and then can be solved. The important step is to use the procedure of conditional averaging of equation for the local characteristic with self-amplification. In Sec. 2 this method is applied to three-dimensional turbulence. The same method is applied in Sec. 3 to two-dimensional turbulence, which has a different local characteristic. Conclusions and perspective are outlined in Sec. 4.

2. Conditionally averaged vorticity field

Let us average equation (1) under the condition that ω is fixed at some point $\mathbf{x}$:

$$\overline{\frac{\partial \omega_i}{\partial t}} + \overline{v_k \frac{\partial \omega_i}{\partial x_k}} = \overline{\frac{\partial v_i}{\partial x_k}\,\omega_k} + \nu\,\overline{\Delta \omega_i} \tag{5}$$

(bar means conditional averaging). For locally homogeneous and isotropic turbulence

$$\overline{v_k \frac{\partial \omega_i}{\partial x_k}} = A(\omega)\sigma_i, \; \sigma_i = \omega_i \omega^{-1} \,, \tag{6}$$

where A is a function of absolute value of vorticity. Indeed, conditionally averaged vector in the lhs of (6) depends only on the vector ω_i and should be proportional to this vector, because there is no other distinguished direction. By multiplying (6) on $A\sigma_i$, averaging and using incompressibility, we get

$$\frac{\partial}{\partial x_k} \langle v_k B(\omega)\rangle = 0 = \left\langle A^2(\omega)\right\rangle, \; B(\omega) \equiv \int^{\omega} A(\omega) d\omega$$

Thus, the second term in the lhs of (5) is zero (for "almost all" ω_i in the statistical sense). The first term in the lhs of (5) is also zero for statistically stationary turbulence. To put it differently, the conditionally averaged angular acceleration of fluid particle, with fixed angular velocity, is zero (Novikov 1969, 1971).

After simple manipulations, equation (5) now can be written in the form:

$$\omega_\ell \, \epsilon_{ijk} \int n_k n_\ell \tilde{\Omega}_j(\mathbf{k},\omega) d\mathbf{k} = \nu \int k^2 \tilde{\Omega}_i(\mathbf{k},\omega) d\mathbf{k} \tag{7}$$

$$\tilde{\Omega}_j(\mathbf{k},\omega) = (2\pi)^{-3} \int exp\{-i\mathbf{kr}\}\, \bar{\Omega}_i(\mathbf{r},\omega) dr, \; \bar{\Omega}_i(o,\omega) = \omega_i \tag{8}$$

Here $\bar{\Omega}_i$ is the conditionally averaged vorticity field on the distance $\mathbf{r}$ from the point with fixed vorticity ω_i, $\tilde{\Omega}_i$ we will call the conditional spectrum of vorticity, $k_i = n_i k$ is the wave number vector with absolute value k and ϵ_{ijk} is the unit axial tensor. Equation (7) represents the statistical balance between self-amplification and smoothing of vorticity field. By multiplying (7) on ω_i and averaging over the ω_i, we will get (2) with zero in the lhs, as it should be. Equation (7) is a closed linear equation for $\tilde{\Omega}_i$ or $\bar{\Omega}_i$.

Let us note, for homogeneous, isotropic and statistically stationary turbulence, we can prove similarly that the conditionally averaged acceleration of a fluid particle (with fixed velocity) is zero. However, we will not get a closed equation for the conditionally averaged velocity field, because the pressure is a quadratic integral of velocity. In contrast, the first term in the rhs of equation for the vorticity (1) is a linear integral of vorticity field multiplied by vorticity. After conditional averaging, this term transforms into lhs of (7).

Equation (7) with all conditions, which we must impose on two-point statistical characteristics, formally has many solutions (Novikov 1991), because it contains closed integration over the variable $\mathbf{k}$. On the other hand, high order kinetic equations (Novikov 1967), integrated over the additional variables, reduce to equation (7). This equation is probably the only information about two-point vortex correlations, which we can get directly from the Navier-Stokes equation without resorting to "ad hoc" closure hypotheses. This leads us to the conclusion that the local structure of turbulent flow with high R is not unique. We connect this nonuniqueness with the phenomenon of intermittency, which depends on large-scale structure of turbulent flows (see the remarks of Landau and Kolmogorov about the intermittency (Monin and Yaglom 1975), recent papers (Novikov 1989, 1990) and references herein).

Having a solution of equation (7), we can calculate spectral moments of various order by the formula:

$$\Omega^{(n+1)}_{ij_1 \ldots j_n}(\mathbf{k}) = \left\langle \tilde{\Omega}_i\, \omega_{j_i} \ldots \omega_{j_n} \right\rangle = \\ (2\pi)^{-3} \int exp\{-i\mathbf{kr}\}\, \langle \omega_i(\mathbf{x}+\mathbf{r})\omega_{j_i}(\mathbf{x}) \ldots \omega_{j_n}(\mathbf{x}) \rangle\, d\mathbf{r} \tag{9}$$

Let us consider the simplest solution of equation (7), which satisfies certain necessary conditions (Novikov 1991):

$$\tilde{\Omega}_i = \frac{3\nu}{2\epsilon}\phi(k)\ \left[\omega(\sigma_i - \mu n_i) - 5\nu k^2 \mu\ \epsilon_{ijk}\ \sigma_j n_k\right], \tag{10}$$

$$\phi(k) \equiv E(k)/2\pi,\ \mu = \sigma_i n_i\ ,$$

The energy spectrum E is not universal, and therefore it is not determined uniquely by the equation (7). In the inertial range the intermittency correction to $E(k)$ is known to be small, but in the dissipation range this correction might be essential (Novikov 1963, Monin and Yaglom 1975). The indicated above small parameter $R^{-1/2}$, principally, can be used in order to extend the local description of turbulence to a global description of a particular nonhomogeneous turbulent flow. In this way, the statistical models for effects of intermittency can be connected with the Navier-Stokes equation.

From (9) and (10), after simple calculations, we get expressions:

$$\Omega^{(2)}_{ij}(\mathbf{k}) = \frac{1}{2}\phi(k)\ (\delta_{ij} - n_i n_j)\ , \tag{11}$$

$$\Omega^{(3)}_{ijk}(\mathbf{k}) = -\frac{\nu}{2}k^2\phi(k)\ (\epsilon_{ij\ell}n_\ell n_k + \epsilon_{ik\ell}n_\ell n_j)\ , \tag{12}$$

$$\Omega^{(4)}_{ijk\ell}(\mathbf{k}) = \frac{\nu < \omega^4 >}{10\epsilon}\ \phi(k)\left[\delta_{ij}\delta_{k\ell} + \delta_{ik}\delta_{j\ell} + \delta_{i\ell}\delta_{jk} - \\ -n_i(n_j\delta_{k\ell} + n_k\delta_{j\ell} + n_\ell\delta_{jk})\right] \tag{13}$$

$$\Omega^{(5)}_{ijk\ell m}(\mathbf{k}) = -\frac{\nu^2 <\omega^4> k^2}{12\epsilon}\phi(k)\Big[\epsilon_{ijp}n_p(n_k\delta_{\ell m} + n_\ell\delta_{km} + n_m\delta_{k\ell}) + \\ 3 \text{ transpositions } j \to (k,\ell,m)\Big] \tag{14}$$

These results can be checked experimentally, even if vorticity can be measured only in one point. The study of more general solutions of equation (7) is in progress, as well as the analysis of corresponding high order moments and probability distributions.

3. Statistical balance of vorticity gradient in 2-D turbulence

For 2-D incompressible flow, vorticity is not a local characteristic of motion, because there is no self-amplification:

$$\frac{\partial \omega}{\partial t} + v_k \frac{\partial \omega}{\partial x_k} = \nu \Delta \omega, \; \omega = \epsilon_{ij} \frac{\partial v_j}{\partial x_i}, \qquad \frac{\partial v_i}{\partial x_i} = 0 \tag{15}$$

Here ω is the only one component of vorticity (perpendicular to the plane of motion) and ϵ_{ij} is the 2-D unit axial tensor ($\epsilon_{12} = -\epsilon_{21} = 1, \; \epsilon_{11} = \epsilon_{22} = 0$). By multiplying (15) on ω and averaging, we get for homogeneous turbulence

$$\frac{1}{2} \frac{\partial}{\partial t} \left\langle \omega^2 \right\rangle = -\gamma \,, \tag{16}$$

$$\gamma = \nu \left\langle s_i^2 \right\rangle, \; s_i \equiv \frac{\partial \omega}{\partial x_i} \,, \tag{17}$$

where γ is the mean rate of enstrophy dissipation.

Vorticity gradient (vg) s_i is a prime candidate for a local characteristic of 2-D turbulence. Space differentiation of (15) gives:

$$\frac{\partial s_i}{\partial t} + v_k \frac{\partial s_i}{\partial x_k} = -\frac{\partial v_k}{\partial x_i} s_k + \nu \Delta s_i \tag{18}$$

Compression of fluid element in the direction of s_i produces amplification of vg. Statistical balance for homogeneous turbulence has the form

$$\frac{1}{2} \frac{\partial}{\partial t} \left\langle s_i^2 \right\rangle = -\left\langle \frac{\partial v_k}{\partial x_i} s_i s_k \right\rangle - \nu \left\langle \left(\frac{\partial s_i}{\partial x_k} \right)^2 \right\rangle \tag{19}$$

Detailed numerical experiments can check if the lhs of (19) is small in comparison with each term in the rhs of (19) for decaying turbulent flows with sufficiently high R. Such experiments can also clarify what is the corresponding small parameter for 2-D turbulence. In what follows, we assume that for statistically stationary and locally homogeneous turbulence, the "temperature" of vg is maintained (17).

Following the procedure of Sec. 2, let us average (18) under the condition that $\mathbf{s}$ is fixed at some point $\mathbf{x}$:

$$\overline{\frac{\partial s_i}{\partial t}} + \overline{v_k \frac{\partial s_i}{\partial x_k}} = -\overline{\frac{\partial v_k}{\partial x_i}} s_k + \nu \overline{\Delta s_i} \tag{20}$$

(bar again means conditional averaging). For locally homogeneous, isotropic and statistically stationary turbulence, we can prove similarly that the lhs of (20) is zero. Equation (20) can be written in the form:

$$i\, \epsilon_{jm} s_m \int k^{-1} n_i \tilde{S}_j(\mathbf{k}, \mathbf{s}) d\mathbf{k} = \nu \int k^2 \tilde{S}_i(\mathbf{k}, \mathbf{s}) d\mathbf{k} \tag{21}$$

$$\tilde{S}_i(\mathbf{k}, \mathbf{s}) = (2\pi)^{-2} \int exp\{-i\mathbf{kr}\}\, \bar{S}_i(\mathbf{r}, \mathbf{s}) d\mathbf{r},\ \bar{S}_i(0, \mathbf{s}) = s_i \tag{22}$$

Here $\bar{S}_i$ is the conditionally averaged vg at the distance $\mathbf{r}$ from the point with fixed $\mathbf{s}$, $\tilde{S}_i$ is the conditional spectrum of vg and $k_i = kn_i$ is the wave number vector with absolute value k.

Equation (22) again has formally many solutions (Novikov 1991a) and we connect this nonuniqueness with intermittency, which depends on large-scale structure of turbulent flows.

Spectral moments are defined by the formula:

$$S^{(n+1)}_{ij_1...j_n}(\mathbf{k}) = \left\langle \tilde{S}_i\, s_{j_1}...s_{j_n} \right\rangle =$$
$$(2\pi)^{-2} \int exp\{-i\mathbf{kr}\}\, \langle s_i\, (\mathbf{x}+\mathbf{r}) s_{j_1}(\mathbf{x})...s_{j_n}(\mathbf{x})\rangle\, d\mathbf{r} \tag{23}$$

Consider the simplest solution of equation (22), which satisfies necessary conditions (Novikov 1991a):

$$\tilde{S}_i = 2\nu\gamma^{-1}\, S(k)\, n_k s_k\, (1 - 4i\nu s^{-2}\, \epsilon_{\ell m} n_\ell s_m) n_i\ , \tag{24}$$
$$S(k) = \pi^{-1} k^3 E(k)$$

(E is the energy spectrum). Substitution (24) into (23) gives:

$$S^{(2)}_{ij}(\mathbf{k}) = S(k)\, n_i n_j\ , \tag{25}$$

$$S^{(3)}_{ijk}(\mathbf{k}) = i\nu k^3\, S(k) n_i \left(\epsilon_{jm} n_m n_k + \epsilon_{km} n_m n_j\right)\ , \tag{26}$$

$$S^{(4)}_{ijk\ell}(\mathbf{k}) = \frac{\nu < s^4 >}{2\gamma}\, S(k) n_i \big(n_j \delta_{k\ell} + n_k \delta_{j\ell} + n_\ell \delta_{jk}\big), \tag{27}$$

$$S^{(5)}_{ijk\ell m}(\mathbf{k}) = \frac{i\nu^2 {<}s^4{>} k^3}{6\gamma} S(k) n_i \Big[\epsilon_{jp} n_p \big(n_k \delta_{\ell m} + n_\ell \delta_{km} +$$
$$n_m \delta_{k\ell}\big) + 3 \text{ transp. } j \rightarrow (k, \ell, m)\Big] \tag{28}$$

We hope that these results will stimulate detailed numerical studies of vortex structures in 2-D turbulent flows.

4. Conclusions and perspective

A new approach to the problem of turbulence is presented in this paper. This approach is based on the conditional averaging of equation for local characteristic with self-amplification. We applied this approach to a local description of 3-D and 2-D turbulent flows. The developed method is quite general and can be applied to a variety of physical systems with strong interaction. For example, it can be shown that for magnetized strongly turbulent plasma the appropriate local characteristic with self-amplification is the gradient of electron density.

The presented results can be verified by laboratory and field measurements, as well as in numerical experiments. This new quantitative information about the local structure of turbulence, corresponding to exact Navier-Stokes equation, can be used in many different ways. We can use it in order to check any new closure hypothesis. We can use it for developing the theory of non-Gaussian intermittent random fields. We can use it for parameterization of turbulence with high R in large-eddy simulations and in stochastic models (compare with Novikov 1990). All these are separate stories. The important point is that we have a new tool, a new way of looking at problems with strong interaction.

Acknowledgement

This work is supported by the U.S. Department of Energy.

References

1. Monin, A.S. and A.M. Yaglom (1975), "Statistical Fluid Mechanics", (MIT Press, Cambridge, MA), Vol. 2.

2. Novikov, E.A. (1963), *Prikl. Mat. Mekh.* **27** (5), 944 [*Appl. Math. Mech.* **27**, 1445 (1963)].

3. Novikov, E.A. (1966), *Dokl. Akad. Nauk SSSR* **168** (6), 1279 (1966) [*Sov. Phys. Dokl.* **11**, 497].

4. Novikov, E.A. (1967), *Doklady Akad. Nauk SSSR*, 177, 299, [*Sov. Phys. Dokl.* **12**, 1006 (1968)].

5. Novikov, E.A. (1969), "Statistical Models in the Theory of Turbulence", thesis for a Doctor's degree, Institute of Oceanology Acad. Sc. USSR, Moscow.

6. Novikov, E.A. (1971), *Prikl. Mat. Mekh.* **35**, 266 [*Appl. Math. Mech.* **35**, 231 (1971)].

7. Novikov, E.A. (1989), *Phys. Fluids* **A1**, 326.

8. Novikov, E.A. (1990), *Phys. Fluids* **A2** (5), 814.

9. Novikov, E.A. (1990a), *Fluid Dyn. Res.* **6**, 79.

10. Novikov, E.A. (1991), *Phys. Rev. Let.* (submitted).

11. Novikov, E.A. (1963), *Phys. Letters A* (submitted).

12. Taylor, G.J. (1938), *Proc. Roy. Soc.*, **A164**, 15.

Address of the author

Dr. E.A. Novikov
Institute for Nonlinear Science
University of California, San Diego
La Jolla, California 92093-0402
enovikov@ucsd.edu

How important are direct interactions between large and small scales in turbulent flows?

Arkady Tsinober

"Since the energy transfer to smaller-scale motion is random, the effect of anisotropy, inhomogeneity, and nonstationarity of the mean motion on the statistical state of the turbulent fluctuations should decrease with decreasing scale." (p. 339)
"This can be translated into the language of ordinary mechanics by considering developed turbulent flow as a dynamic system with a very large number of degrees of freedom, and then isolating the degrees of freedom corresponding to small-scale (and high-frequency) components of the motion. We must then conclude that the inertial and frictional forces which correspond to the isolated degrees of freedom should be in statistical equilibrium independently of the properties of the large-scale components of the motion." (p. 345)
Monin, A. S., & Yaglom, A. M., (1975), Statistical Fluid Mechanics: Mechanics of Turbulence, Vol. 2, MIT Press.

The above question is closely related to other questions like the following ones:

- Do the small scales forget the anisotropy in the large scales, i.e. is the energy cascade an orientation loosing process?

- Are the small scales statistically independent of the large scales?

- Is local isotropy a universal property of turbulent flows at large Reynolds numbers?

It is somewhat embarrassing to ask such questions in 1991 at the 50th anniversary of Kolmogorov's theory of locally isotropic turbulence (Turbulence and stochastic processes: Kolmogorov's ideas 50 years on; a special issue of Proc. Roy. Soc. A **434**, pp. 1 - 240 (1991); Kendall (ed.) 1990, Obituary of Andrei Nikolayevich Kolmogorov. Bull. Lond. Math. Soc. **22**, 31 100). However, the existing evidence, theoretical, numerical and experimental, indicates that generally the answers to these questions are negative, though it depends on what one means by "large" and "small" scales. Usually, the large scales are associated with the maximum of turbulent energy, while the small scales are roughly of two kinds, those of order of the Taylor microscale in which most ot the dissipation takes place and scales of the order of Kolmogorov's microscale, which are most affected by viscosity.

In fact, the famous Landau remark on the influence of variation of dissipation over times and scales of large eddies should generally be valid with the term "dissipation ..." repalced by "anisotropy of large eddies".

The above questions deserve a special comprehensive review requiring considerable space. I'll limit myself to some particular comments. A partial list of relevant references on the subject is given at the end of this comment. The first publications appeared in the fifties when Townsend (1954) and Uberoi (1956, 1957) pointed out that vorticity was not (locally) isotropic in distorting flows (see also Corrsin 1958). Theoretically, most of the effort was made using RDT (Hunt & Carruthers 1990, Hunt & Vassilicos 1991, Kida & Hunt, Maxey 1982, Savill 1987, and references therein) for homogeneous shear flow. RDT is known to be limited in respect to its inability to take into account the nonlinear self interaction of turbulence, and in particular it ignores the "reacting back" (in Moffatt's terminology) of small scales on the large ones. Note that Reynolds stresses can be viewed as a result of such nonlinear self-interaction. Numerically, the main research was concentrated in homogeneous anisotropic shear-free turbulence (Brasseur & Corrsin 1987, Brasseur 1991, Lee & Reynolds 1985, and references therein) and homogeneous shear flow (Lee et al. 1990, Rogers & Moin 1987, and references therein). Recent results obtained by means of empirical eigenfunctions (Karhunen-Loeve procedure or the proper orthogonal decomposition) point to the essential non-isotropy of shear flows (Knight & Sirovich 1990, Sirovich 1991 and references therein). A related problem is the well known breakdown of computations with inadequately resolved small scales. Finally, the influence of the small scales on the large ones was demonstrated in numerical simulations with phase mixing in the small scales (Murakami, Shtilman & Levich 1992).

Both theoretical and numerical results have demonstrated that in many situations the small scales do not forget the anisotropy of the large ones. It is noteworthy that the numerical results are limited to rather modest Reynolds numbers. (Unfortunately, computers don't seem to be useful in the forseeable future for direct numerical simulations of flows at large

Reynolds numbers. Will the wind tunnel replace the computer for turbulent flows at large Reynolds numbers?)

Most of the experimental results (Antonia 1990, Antonia et al. 1986, 1987, 1988, Brown et al. 1987, George & Hussein 1990, Gibson et al. 1977, Hussein 1988) were obtained at quite moderate Reynolds numbers too. The experiments which were made at quite high Reynolds numbers deserve special mentioning (Mestayer 1982, Mestayer et al. 1976, Karyakin et al. 1991, Kuznetsov et al. 1988, 1990, Saddoughi et al. 1991). The latter have shown that small scale turbulence in shear flows is anisotropic except in the scales smaller than, say, 20 Kolmogorov scales. It is noteworthy that there is a variety of extremely serious difficulties in assessing the reliability of small scale measurements (Antonia et al. 1986, Hannart et al. 1985).

In view of the above I would like to draw your attention to the results on a "simple" turbulent flow, which provide reliable information on the existence and importance of the coupling between large and small scales. This is the developed turbulent flow in a plane channel. Results on this flow are summarized by Wei & Willmarth (1989). Rather large Reynolds numbers ($\approx 2{,}3 \cdot 10^5$) were achieved by Comte-Bellot (1965). We are interested in the distribution of the Reynolds stress over the channel cross-section, which is shown in Fig. 1. What I would like to stress is that the Reynolds stress *gradient* is essentially different from zero (also in the middle of the channel where $\langle uv\rangle = 0$). It is this quantity that enters the Reynolds equations and not the Reynolds stress. Hence, the dynamical significance of the former. On the other hand, for a developed turbulent flow in a flat channel ($\partial/\partial x \langle\ \rangle = 0$).

$$\frac{d}{dy}\langle uv\rangle = \langle(\omega \times u)_x\rangle = \langle\omega_y u_z\rangle - \langle\omega_z u_y\rangle \tag{1}$$

It is obvious from (1) that at least one correlation at the right hand side of (1) is of the order (or larger) of the Reynolds stress gradient, which together with Fig. 1 clearly indicates the importance of the interaction between large (energy containing) and small (maximum dissipation and enstrophy, but not at the same instantaneous location) scales (Tennekes & Lumley 1972, pp. 78, 79). No wonder that the small scales cannot be expected to be isotropic since the large scales are anisotropic. This seems to be true for very high Reynolds numbers too since there is every reason to believe that the behaviour of the Reynolds stress shown in Fig. 1 is independent of the Reynolds number, at least in the region where its gradient is constant. It is instructive to estimate the correlation coefficients $C_{\omega_y u_z}$ and/or $C_{\omega_z u_y}$. We will do this separately in two regions:

a) in the bulk of the flow ($0.05 \div 0.1 < y/b < 0.05$), in which

$$\frac{d}{dy}\langle uv\rangle \approx -u_\tau^2/b \tag{2}$$

b) in the region close to the wall ($0 < y/b < 0.05$), in which roughly

$$\frac{d}{dy}\langle uv\rangle \simeq 20\, u_\tau^2/b \tag{3}$$

Here u_τ^2 is the friction velocity and b the half width of the channel.

In order to estimate the correlation coefficients $C_{\omega_y u_z}$ and $C_{\omega_z u_y}$ it is necessary to know the RMS values of the corresponding quantities. These are not available for the channel flow, but attempts have been made to measure these quantities in boundary layer flows (Klewicki 1989; Balint et al. 1991, though the latter seem not to be very reliable; Vukoslavcevic et al. 1991)
Using the results by Klewicki (1989) and Balint et al. (1991) as a crude estimate one obtains

$$\langle \omega_y{}^2\rangle^{1/2}, \langle \omega_z{}^2\rangle^{1/2} \quad \begin{array}{ll} \simeq 10^{-2} u_\tau^2/\nu & \text{in the bulk of the flow} \\ \simeq 10^{-1} u_\tau{}^2/\nu & \text{close to the wall} \end{array} \tag{4}$$

It is natural also to take $\langle u_y^2\rangle^{1/2}, \langle u_z^2\rangle^{1/2} \simeq u_\tau$. Taking into account that $u_\tau D/\nu \simeq 10^4$ and using (1) - (4) one obtains $C_{\omega_y u_z}, C_{\omega_z u_y} \simeq 10^{-}$ in both cases, i.e. rather small, though the interaction between vorticity and velocity in this flow (as in any other flow with non-zero gradients of Reynolds stress) is dynamically important (if the correlation coefficient is small, it does not follow that the two quantities are not related.) The estimates (4) are rather crude, but are of the same order of magnitude as those obtained in direct numerical simulations by Kim et al. (1987) at rather low Re = 3300 (see their Fig. 14).

In view of the significance of this interaction it seems important to take it into account in attempts of low-dimensional description of turbulent flows, e.g. to perform the Karhunen - Loeve procedure (proper orthogonal decomposition) not on the basis of the Reynolds stress tensor, but using, say, the velocity-vorticity correlation tensor (it is unlikely that low dimensional description of turbulent flows can succeed withoud taking into account in some way the field of velocity derivatives, e.g. vorticity, rate of strain, etc.).This is because the dynamically significant quantities are the gradients of the Reynolds stress tensor and they are strongly related to the velocity-vorticity correlations. It seems important to study the properties of this tensor more generally since e.g. its divergence is just the vortex stretching term in the vorticity equation, its trace is the helicity-density and the off-diagonal terms form the vorticity-velocity cross-product.

Acknowledgements

The author acknowledges helpful comments by L. Sirovich.

References

Antonia, R.A., Shah, D.H. and Browne, L.W.B. 1988 Spectra of velocity derivatives in a turbulent wake. *Phys. Fluids*, **30**, 3455 - 3462.

Antonia, R.A., Anselmet, F. and Chambers, A.J. 1986 Assessment of local isotropy using measurements in a turbuelnt jet. *J. Fluid Mech.*, **163**, 365 - 391.

Antonia, R.A., Browne, L.W.B. and Shah, D.A. 1988 Characteristics of vorticity fluctuations in a turbulent wake. *Fluid Mech.*, **189**, 349 - 365

Antonia, R.A., Shah, D.A., Browne, L.W.B. 1988 Dissipation and vorticity spectra in a turbulent wake. *Phys. Fluids*, **31**, 1805 - 1808

Antonia, R.A. 1990 The role of coherent structures - Comment 3, in *Whither turbulence? Turbulence at the crossroads*, ed. Lumley, J.L., Springer, pp. 156 - 169.

Balint, J.-L., Wallace, J.M. and Vukoslavcevic, P. 1991 The velocity and vorticity vector fields of a turbulent boundary layer. Part 2. Statistical properties, *JFM*, **228**, 53 - 86.

Bradshaw, P. 1990 The role of coherent structures - Comment No. 2, in *Whither Turbulence? Turbulence at the crossroads*, ed. Lumley, J.L., Springer, 152 - 155

Brasseur, J.G. 1991 Comments on the Kolmogoroff hypothesis of isotropy in the small scales, *AIAA - 91 - 0230*, pp 9.

Brasseur, J.G. and Corrsin, S. 1987 Spectral evolution of the Navier-Stokes equation for low order couplings of fourier modes, in *Turbulence*, eds. Comte - Bellot, G. and Mathieu, J., Springer, pp. 152 - 162.

Brasseur, J.G. and Yeung, P.K. 1990 Large and small scale coupling in homogeneours turbulence:analysis of Navier-Stokes equations in the asymptotic limit, presented at the *Eighth Symposium on Turbulent Shear Flows*, Munich, Sept. 9 - 11, 1991; Preprints vol I, pp. 16-4-1 - 16-4-6.

Browne, L.W.B., Antonia, R.A. and Shah, D.H. 1987 Turbulent energy dissipation in a wake. *J. Fluid Mech.*, **179**, 307 - 326.

Champagne, F.H., Harris, V.G. and Corrsin, S. 1970 Experiments on nearly homogeneous turbulent shear flow. *J. Fluid Mech.*, **41**, 81 - 139.

Comte - Bellot, G. 1965 Ecoulement turbulent entre deux parois paralleles. *Publ. Sci. Tech. Min. l'Air*, No. 419. Ph. D. Thesis.

Corrsin, S. 1958 Local isotropy in turbulent shear flow, *Nat. Adv. Com. Aeronaut., Res. Memo.* RM 58 B 11.

Deissler, R.G. 1978 On the localness of the spectral energy transfer in turbulence. *Appl. Sci. Res.*, **34**, 379 - 392.

Domaradzki, U.A. 1988 Analysis of energy transfer in direct numerical simulations of isotropic turbulence. *Phys. Fluids*, **31**, 2747 - 2749.

Domaradzki, J.A. and Rogallo, R.S. 1990 Local energy transfer and nonlocal interactions in homogeneous isotropic turbulence. *Phys. Fluids*, **A2**, 413 - 426.

George, W.K. 1989 The self-preservation of turbulent flows and its relation to initial conditions and coherent structures, in *Advances in Turbulence*, eds. George, W.K. and Arndt, R., Springer, pp. 39 - 73.

George, W.K., and Hussein, H.J. 1990 Locally axisymmetric turbulence, in *Twelfth Turbulence Symposium*, Sept. 24 - 26, 1990, Missouri-Rolla, pp. A30 - A30-12.

Gibson, C.H. 1991 Kolmogorov similarity hypotheses for scalar fields: sampling intermittent turbulent mixing in the ocean and galaxy. *Proc. Roy. Soc. London*, **A434**, 149 - 164.

Gibson, C.H., Friehe, C.A. and McConnell, S.O. 1977 Structure of sheared turbulent fields. *Phys. Fluids*, **20**, S156 - S167.

Groth, J. Hallbäck, M. and Johansson, A.V. 1989 Measurement and modelling of anisotropic turbulent flows, in *Advances in turbulence* **2**, eds. Fernholz, H.-H. and Fiedler, H.E., Springer, pp. 84 - 89.

Hallbäk, M., Groth, J. and Johansson, A.V. 1991 Anisotropic dissipation rate - implications for Reynolds stress models, in *Advances in Turbulence* **3**, eds. Johansson, A.V. and Alfredsson, P.H., Springer, 414-421

Hannart, B., Gagne, Y. and Hopfinger, E. 1985 Domaine d'existence de l'isotropic dans une couche de milange turbulente. *C.r. Acad. Sc. Paris*, **301**, series II, 669 - 674.

Hunt, J.C.R. and Vassilicos, J.C. 1991 Kolmogorov's contributions to the physical and geometrical understanding of small scale turbulence and recent developments. *Proc. Roy. Soc., London*, **A434**, 183 - 240

Hunt, J.C.R. and Carruthers, D.J. 1990 Rapid distortion theory and the problems of turbulence. *J. Fluid Mech.* **212**, 497 - 532.

Hussein, H.J. 1988 Measuring of small scale turbulence in an axisymmetric jet using moving hot wires, *Ph. D. Thesis*, State Univ. of N. Y. of Buffalo.

Johanson, F.D. and Eckelmann, H. 1983 has a small-scale structure in turbulence experimentally verified? *Phys. Fluids*, **26**, 2408 - 2414.

Karyakin, M.Yu., Praskovsky, A.A. and Kuznetsov, V.R. 1991 Experimental verification of local isotropy assumption in high Reynolds number flows, in *Advances in Turbulence*, **3**, eds. Johansson, A.V. and Alfredsson, P.H., Springer.

Kida, S. and Hunt, J.C.R. 1989 Interaction between different scales of turbulence over short times. *Fluid. Mech.*, **201**, 411 - 415.

Kim, J., Moin, P. and Moser, R. 1987 Turbulence statistics in fully developed channel flow at low Reynolds number. *J. Fluid Mech.*, **177**, 133 - 166.

Knight, B. and Sirovich, L. 1990 Kolmogorov inertial range for inhomogeneous turbulent flows. *Phys. Rev. Lett.*, **65**, 1356 - 1359

Klewicki, J.C. 1989 Velocity - vorticity correlations related to the gradients of the Reynolds stresses in parallel turbulent wall flows. *Phys. Fluids*, **A1**, 1285 - 1288.

Kuznetsov, V.R., Praskovsky, A.A. and Sabelnikov, V.A. 1990 Turbulence fine-scale structure of intermittent shear flows, submitted to *J. Fluid Mech.*

Kuznetsov, V.R., Praskovsky, A.A. and Sabelnikov, V.N. 1988 Local structure of turbulence in free flows with strong intermittency. *Fluid Dyn.*, **23**, 843 - 850.

Lee, M.-J., Kim, J. and Moin, P. 1990 Structure of turbulence at high shear rate. *J. Fluid Mech.*, **216**, 561 - 583.

Lee, M.-J. and Reynolds, W. 1985 Numerical experiments on the structure of homogeneous turbulence, Report No. TF - 24, Thermosci. Div., Stanford Univ., pp. 253.

Maxey, M.R. 1982 Distortion of turbulence in flows with parallel streamlines. *J. Fluid. Mech.*, **124**, 261 - 282.

Mestayer, P. 1982 Local anisotropy in high-Reynolds-number turbulent boundary layer. *J. Fluid Mech.*, **125**, 475 - 503.

Mestayer, P.G., Gibson, C.H., Coantic, M.F. and Patel, A.S. 1976 Local anisotropy in heated and cooled turbulent boundary layers. *Phys. Fluids*, 19, 1279 - 1287.

Murakami, Y, Shtilman, L and Levich, E 1992 Reducing turbulence by phase juggling, Physics of Fluids A, sub judice.

Reynolds, W.C. 1990 The potential and limitations of direct and large eddy simulations, in *Whither Turbulence? Turbulence at the crossroads*, ed. Lumley, J.L., Springer, 313 - 343.

Rogers, M.M. and Moin, P. 1987 The structure of the vorticity field in homogeneous turbulent flows. *J. Fluid Mech.*, **176**, 33 - 66.

Saddoughi, S.V., Veeravalli, S.V., Praskovsky, A.A. and Bradshaw, P. 1991 An experimental investigation of the local - isotropy hypothesis in high Reynolds number turbulent flows. To be presented at the *44th Annual Meeting of the Div. Fluid Dyn., Am. Phys. Soc.*, Tempe, Arizona; also Veeravali, S.V., Saddoughi, S.G., Praskovsky, A and Bradshaw, P 1991 A note on local Isotropy in High-Reynolds number Turbulence, to be published in the Proceedings of the Monte-Verità Colloquium on Turbulence.

Safman, P.G. 1978 Problems and progress in the theory of turbulence, in *Structure and mechanisms of turbulence* **II**, ed. H. Fiedler, Springer, p. 285

Savill, A.M. 1987 Recent developments in rapid-distortion theory. *Ann. Rev. Fluid Mech.*, 19, 531 - 575.

Sirovich, L. 1991 Analysis of turbulent flow by means of the empirical eigenfunctions. *Fluid Dyn. Res.* **8**, 85 - 100

Sreenivasan, K.R. 1991 On local isotropy of passive scalars in turbulent shear flows. *Proc. Roy. Soc. London*, **A 434**, 165 - 182.

Tavoularis, S. and Corrsin, S. 1981 Experiments in nearly homogeneous turbulent shear flow with uniform shear flow with a uniform mean temperature gradient. Part 1. *J. Fluid Mech.*, **104**, 311 - 347; Part 2. The fine structure. *J. Fluid Mech.*, **104**, 349 - 367.

Tavoularis, S. and Karnik, U. 1989 Further experiments on the evolution of turbulent stresses and scales in uniformly sheared turbulence. *J. Fluid Mech.*,**204**, 457 - 478.

Tennekes, H. & Lumley, J.L. 1974 A first course in turbulence, MIT Press, pp. 78 - 79.

Teodorovich, E.V. 1990 Role of local and nonlocal interactions in the formation of the developed turbulence regime. *Fluid Dyn.*, **25**, 522 - 528.

Townsend, A.A. 1954 The uniform distortion of homogeneous turbulence. *Quart. J. Mech. Appl. Math.*, **7**, 104 - 127.

Tsinober, A. 1990 Turbulent drag reduction versus structure of turbulence, in *Structure of turbulence and drag reduction*, ed. Gyr, A., Sprigner, pp. 313 - 340.

Uberoi, M.S. 1956 Effect of wind tunnel contraction on free-stream turbulence. *J. Aero. Sci.*, **23**, 754 - 764.

Uberoi, M.S. 1957 Equipartition of energy and local isotropy in turbulent flows. *J. Appl. Phys.*, **28**, 1165 - 1170.

Van Atta, C. 1991 Local isotropy of the smaller scales of turbulent scalar and velocity field. *Proc. Roy. Soc. London*, **A 343**, 139 - 147.

Vukoslavcevic, P., Wallace, J.M. and Balint, J.-L. 1991 The velocity and vorticity vector fields of a turbulent boundary layer. Part 1. Simultaneous measurement by hot-wire anemometry. *J. Fluid Mech.*, **228**, 25 - 51.

Wei, T. and Willmarth, W.W. 1989 Reynolds-number effects on the structure of a turbulent channel flow. *J. Fluid Mech.*, **204**, 57 - 95.

Weinstock, J. & Burk, S. 1985 Theoretical pressure-strain term, experimental comparison, and resistance to large anisotropy. *J. Fluid Mech.*, **154**, 429 - 443.

Yeung, P.K. and Brasseur, J.G. 1991 The response of isotropic turbulence to isotropic and anisotropic forcing at large scales, *Phys. Fluids.*, **A3**, 884-897.

A post symposium note.

Three papers of particular interest to the subject were brought to my attention:

Bradshaw, P. 1992 Turbulence: The chief outstanding difficulty of our subject, 3rd Stewartson Memorial lecture: to be presented at the *Symposium on Numerical and Physical Aspects of Aerodynamic Flows,* Long Beach, CA.

Durbin, P.A. & Speciale, C.G. 1991 Local anisotropy in strained turbulence at high Reynolds numbers, **AMD-vol. 117**, *Recent advances in Mechanis of Structured Continua*, Eds. M. Massoudi & K.R. Rajagopal, Book No. G 00617-1991, pp 29-32

Veeravali, S.V. & Saddoughi, S.G. 1991 A preliminary experimental investigation of local isotropy in high-Reynolds-number turbulence. *Center for Turbulence Research - Stanford Annual Res. Briefs.*

Address of the author

Arkady Tsinober
Tel-Aviv University
Tel-Aviv 69978 Israel

Discussion

Aref: We now have time for some discussion and I would like to invite the speakers of this morning to come up front. And it seems to me that we have a number of things that we might discuss. One of the things is typical structures. We had the suggestion this morning that spirals are very important. Another key concept that was introduced was the kinetic equations for the vorticity field, and particularly the idea that conditional sampling might allow various kinds of progress to be made. We then had the video presentation on boundary layers and I think there was an interesting issue about - what is the word that is used by our friend in Houston - "eduction" of structures. What should one be looking at? What are some of the key aspects? Finally, a related suggestion that one of the tensors to be looked at is $U_i\omega_j$, and that's the comment in A. Tsinober's contribution. There are probably many other topics, but these were at least some of them. So I would like to invite comments from you, and maybe the authors can respond.

Zaslavsky: Can this kind of analysis be applied to another vortex model? Because such regularisation is known of course. But maybe another kind of solution can be considered.

Moffatt: I think you can do it for any vortex. The only merit of doing it for this structure is that it is a solution of the Euler equations or a near solution of the Euler equations. Now, this brings dynamics in. You can take any structure that you like that is purely kinematic and work out the associated spectrum. In a way, this is what Townsend was doing with his random vortex sheets. But random vortex sheets are not actually a solution of the Euler equation. This is a nasty problem!

Lesieur: It is not clear to me - when you spoke about the spiralling of the sheets - whether you thought you could explain the Kolmogorov spectra this way.

Moffatt: Well, yes, that was the remarkable conclusion: that the evidence of Krasny's spiral is that the best fit is given by $\alpha = 2$, $\beta = 1$, which actually corresponds to the $k^{-5/3}$ spectrum.

Now, this means that if you have a random distribution of spirals of this kind, and of course in three dimensions so they can be at any orientation you get the Kolmogorov spectrum, you don't need anything else. You don't need to think of a cascade from one scale to another. One type of structure will do. The cascade effect is produced because the spirals get tighter and tighter and that's all.

Jimenez: You seem to me to be speaking of two kinds of spirals. One was the dynamical spiral containing vorticity. And the other was the passive scalar spiral. Are you implying that they are related, that they are the same thing?

Moffatt: You're right, and I think there is a difficulty there because of course to take a picture you need some kind of scalar marker and it is true that it is not the same as the dynamic spiral that results from a vortex sheet. The Krasnay spiral is the dynamic spiral that I'm talking about. So there is some difficulty in the interpretation of experimental results in Everson and Sreenivasan (1991).

Jimenez: But the other seems to be an interface being wrapped around -

Moffatt: You're right. And particularly one of the pictures that I didn't show: the wrapping around of a rotational spiral and a smoke ring, which is so clearly a purely kinematic effect, not a dynamic effect. You may get a logarithmic spiral in situations of that kind and you may get the algebraic spiral that is controlled by the dynamics.

Malik: Vassilicos did some work on spirals (Vassilicos, 1990, Ph.D. thesis, Cambridge) and he obtained a relationship between the energy spectrum of turbulence and the associated spiral structures. One of his findings was that you need a large number of turns in a spiral, about 150, to observe this relation. But in the pictures that were shown we see only four or five at the most?

Moffatt: That's true. Of course, to make this bridge between the spiral structure and the Kolmogorov spectrum from the theoretical point of view you need an infinite number of turns; each spiral will have a singularity. Of course, in practice you observe only several turns when you have Reynolds numbers of the order of 10^3 - 10^4 typical for experiments, you see spirals with three or four turns.

Malik: You need only two turns to obtain a capacity.

Moffatt: Yes, it's absolutely miraculous, you might say, that a theory of this kind gives a spectrum of k - 5/3, or it's even more of a miracle that you get this $-^{5}/^{3}$ in experiments in which you have a finite number of turns. Now, I don't attempt to understand these things.

Wygnanski: I would like to pick up a little bit on Javier's comment regarding the spiral. First, in the visualisation of Sreenivasan there might be another aspect by taking a transsect through a spiral that might be highly three-dimensional. Now, in a forced mixing layer eddies are to a large extent two-dimensional taking the whole span of the mixing layer. We have made some phase locked measurements in a forced mixing layer where we had two components of velocity, u, v, and from that we computed streak lines and vorticity contours and got spirals which are computed and not just visual. They compare very well to what one sees as the results of streak lines produced by smoke or dye.
Now I will add another comment related to the fact that in the mixing layer you looked at the shape of a spiral that was highly distorted. In that particular case I may add that these highly distorted shapes occur when there is very little energy in those spirals. In fact, if one takes the view that these are just instability modes, then as long as the spiral is almost circular, it is amplifying up to the very limit. The roll-up is at the end of the amplification cycle. If they get distorted and tilted, they are really in a decaying stage, and the more distorted they are, the more decaying they are. There is data which shows that there is very little energy in those distorted ones, and there is lots of energy in the large almost circular ones. So that will perhaps bias the spectrum towards an α which is -1/2 and towards $k^{-5/3}$ rather than something else.

Moffatt: That's encouraging.

Lesieur: Speaking of mixing layers, Gharib in San Diego has done some 2-D mixing layer experiments in a liquid film. We had some discussions during some meetings last year because he had obtained a kinetic energy spectrum which he claimed to be $k^{-5/3}$. On the other hand, in the 2-D computations of mixing layers we did - I showed some of them yesterday - we have a spectrum which shows a slope between k^{-3} and k^{-4}, much closer to the spiral model of Gilbert. So do you think we don't have high enough resolution in the calculations?

Moffatt: Here I am really thinking of 3-D turbulence. Because if I'm thinking of an instability of a Kelvin Helmholz type, you need to have a vortex sheet in the first place. Vortex sheets don't form in 2-D turbulence unless of course you inject them as a mixing layer from the outset, that's a special situation. But in general, in 2-D turbulence you don't have vortex sheets, you have discontinuities of vorticity with weaker singularities and you get a spectral power between -3 and -4 whereas in three dimensions you get vortex sheets,

then instability of the vortex sheets, and that leads to spectral power, compatible with -5/3. So there is a big difference between 3-D and 2-D results. In a 2-D experiment with deliberate injection of vortex sheets (i.e. mixing layers), as you say was done by Gharib, I would expect a $k^{-5/3}$ spectrum to appear.
It would be fruitful to actually study the structure of these spirals by just a simple technique, by measuring the intersect, trying to see what form of spiral it is, and see if that is consistent with the spectrum you measure.

Betchov: When packing my suitcase yesterday, I instinctively took along two photographs within a box of strong turbulence in 1955 (See Fig. 1).

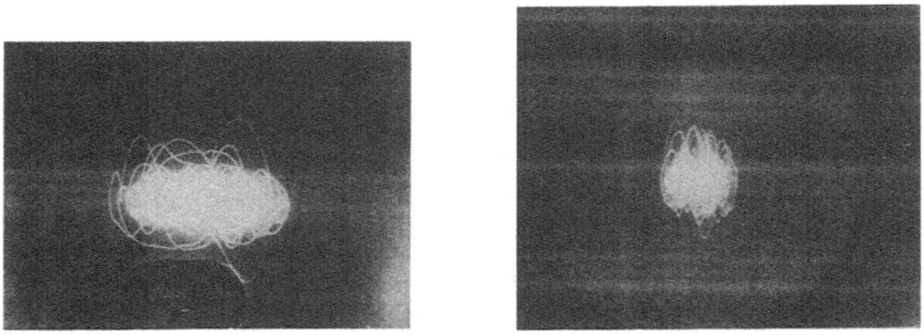

Fig. 1 Signal of u versus du/dt in "Porcupine II"

We used a cubic box of 1 meter with, on each of five sides, an array of 25 holes with cardboard tubes (See Fig. 2). The students named the box "porcupine" and accused me of

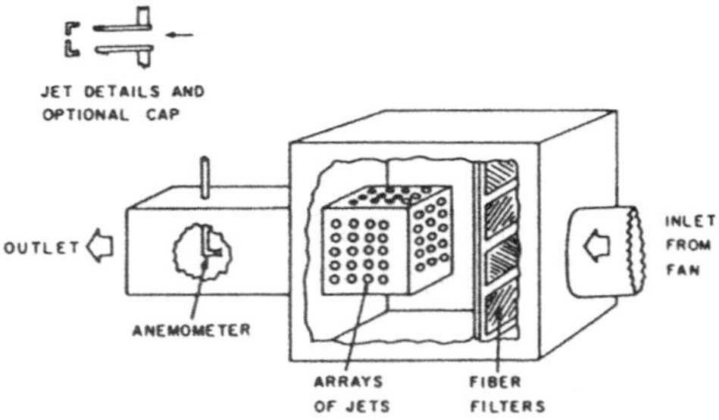

Fig. 2 Turbulence generator "Porcupine II"

picking the tubes in the university toilets. The sixth side was equipped with a large fan drawing air out of the box. This gives a mean flow with 10% isotropic turbulence.

A hot wire can determine the u velocity component along the mean flow, and its time derivative. I applied the signal u to the horizontal channel of an oscillograph, and the signal du/dt to the vertical channel. A signal free of any phase relations would give a perfectly symmetric phase space diagram. A positive mean cube of the derivative produces vertical asymmetry. You can see that the mean cube of du/dt comes mostly from isolated positive pulses. The skewness comes from few isolated strong contributions. (R. Betchov and Lorenzen, 1974, Phase Relations in Isotropic Turbulence, Physics of Fluids, Vol. 17, 1503).

I became convinced that the production of mean square vorticity is tied to an intermittent phenomenon where nature spends little time, but acts vigorously.

Several years later I made a numerical model of turbulence. I programmed the Euler equations on a computer for a box of 55*55*55 grid points with three velocities at each point. This required mobilisation of the university computer during the Christmas vacations and produced a pile of magnetic tapes. During the following months we analysed this data in depth. Initially, I had a few large random vortices in the box. When two of them collide, a region of the fluid is flattened, like making a pancake. At every point of the flow, I determined the eigen axis of the rate of deformation tensor dui/dxj + duj/dxi, and its eigenvalues α, β and γ. I also determined the vorticity components along these axes. When the flow forms a pancake, the initial vorticity has two chances out of three of falling within the pancake, where it will be stretched by the collision. I became convinced that big whirls do not generate second and third harmonics uniformly. All this cascade in Fourier space is a dream of linearized physicists. When caught in a pancake, the vorticity is stretched and contributes to the skewness of du/dx. (R. Betchov, 1975, Numerical Simulation of Isotropic Turbulence, Physics of Fluids, Vol. 18, No.10. Betchov, R., 1976, On the non-Gaussian Aspects of Turbulence, Archives of Mechanics, Warszawa, 837-845. Author's xerox on request, 53 Ave du Lignon, 1219 Geneve). In 1991 Vincent and Meneguzzi arrived to similar conclusions, using a Cray computer for a model of 240**3 grid points. (A. Vincent and M. Meneguzzi, 1991, Spatial Structure and Statistical Properties of Turbulence, J. F. M, 1-20).

Let us now look at simple isolated vortex lines. In a simple smoke ring, we have constant curvature and zero torsion. This results in simple translation of the ring. Where the curvature varies, torsion is produced. In a helicoidal vortex line, curvature and torsion stay constant. A simple equation relating curvature and torsion of such lines was obtained by Hasimoto (R. Betchov, 1965, On the Curvature and Torsion of an Isolated Vortex Filament, J. F. M., 471-479). It consists of a one-dimensional Schrödinger equation with a cubic term. This non-linear term can produce solitons, looking like a locally concentrated helicoidal perturbation on an otherwise straight tornado.

Photographs of a soliton on a full size tornado and computer experiments have been given by Aref and Flinchem. (H. Aref and Flinchem, 1984, Dynamics of a Vortex Filament in a Shear Flow, J. F. M., 477).

A thick isolated vortex line caught in a pancake becomes gradually thinner. Small perturbations can inititate the growth of helicoidal portions. These solitons will propagate along the vortex line, escaping further stretching since the vorticity bends away from the stretching direction. This process also precipitates the effect of viscosity.
If the addition of a cubic term to a Schrödinger equation produces solitons, what happens if we add a cubic term to the Dirac equations? I have now retired from the turbulent business to explore this relativistic problem. It is not a new one. Heisenberg and Durr wrote such equations around 1965. But they had no adequate computer.

Yaglom: Your picture begins with this sheet of concentrated vorticity, and then through Kelvin Helmholz instability you obtain the spiral structure. However, my feeling is that at high Reynolds numbers most of the vorticity is concentrated in some type of filaments and probably these spiral structures are something of the type which Betchov spoke about. Maybe most of the spiral structures are produced by these filaments.

Moffatt: Well, my view is that sheets form and are unstable. And the instability turns the sheets into a row of spiral filaments. But these filaments have a spiral structure, and filaments are easier to see, they are a stronger singularity than a sheet and so they are seen in numerical simulations, but that doesn't mean that there aren't sheets linking them.

Frisch: The evidence of these worms or spaghettis so far comes mostly from numerical simulations and these have relatively modest Reynolds numbers.

Moffatt: I've been teaching my students for decades that this is the Kolmogorov cascade. The very fact that we use the word "cascade" suggests many steps in the process and statistical decoupling. That comes through preoccupation with k-space. But in physical space it's not like that. If you have just one structure, one typical structure, and you distribute it at random, you can get the Kolmogorov spectrum, and you can get a transfer of energy, a systematic transfer of energy from large scales to small scales. In the case of a spiral it is through the tightening up of the spiral, that's all you need. It's a very different philosophy from the cascade.

Aref: But I would argue one has to be careful because after all the Kolmogorov theory is very much a dimensional analysis and one could argue that any structure that you propose on the basis of the dynamic equations is also going to be subject to this dimensional analysis and therefore if you do translate any structure into the language of k-space, you will inevitably get $k^{-5/3}$.

Jimenez: I want to comment on what Malik said. I tried to play along with doing spirals with vortices. You put a vortex there and you see if you get a spiral. I was trying to see not whether the spiral would entrain itself, but whether it would entrain other things across here and there. It is very difficult to get a spiral for any reasonable Reynolds number; you need vortices with Reynolds numbers of the order of many thousands; you need a big vortex. If you have things of the order of 500, you don't get a spiral, you get a growing blob of vorticity.

Moffatt: I think the separations of the spiral are typical of the inertial range when you scale. You take the global size, the radius of the whole spiral, that would be more on the scale of energy containing eddies. That's one end. You go down from that to the core, which eventully is controlled by viscosity, and that spans the inertial range.

Adrian: If you look at the low-wave number end of the inertial range, and if you apply these arguments to Reynolds numbers of infinity, what would become of a spiral vortex? It seems to me that you have instabilities inside.

Moffatt: You mean on the spiral - Well, that's not what you see. You might get a hierarchy of spirals, but you never have spirals on spirals. A spiral that is actually winding up seems to be fairly stable because of the stretching. The stretching process can stabilize it.

Betchov: If curvature and torsion are constant, the shape is constant. That's the solution of the Hasimoto equation.

Moffatt: Now that's for a filament rather than for a sheet. But a sheet that is winding up can also be stable. The actual stretching has a stabilizing effect. This is something that Saffman analysed some years ago (P. G. Saffman and G. R. Baker, 1979, Vortex Interactions, Ann. Rev. Mech., **II**, 95-122).

Aref: There is a stabilizing effect, of course, by the stretching of sheets. This is in a sense why you see the roll-up at certain places and not at others.

Zaslavsky: I have a question to Novikov. It seems to me that your arguments about the self-amplification of vortices is very reasonable, but there should be some final time of this mechanism of self- amplification. It is impossible to have amplification for infinite time. The stability works for finite time.

Novikov: Yes. The main concept is that there is a statistical balance between self-amplification and viscous smoothing.

Lesieur: In your equation you have neglected some terms which you say were of the order Reynolds to the -1/2 with respect to other terms. I agree with this assumption in the self-similar decaying turbulence, but you do not take into account the initial stage of build-up of turbulence, of stretching of vortices and possibly of singularities.

Novikov: Of course. I'm not considering the evolution in this paper. I'm considering the final statistically balanced state. Evolution is a much more difficult problem because evolution depends on the particular initial conditions, the particular boundary condition, and this is reflected in this equation because I have many solutions. In fact, my future problem is to choose which of these equations correspond to which initial and boundary conditions. And this is where I would very much like to have some people who are doing real measurements or who are doing numerical experiments, to help me to choose among all these solutions the ones which correspond to particular initial and boundary conditions, to what boundary conditions. Especially for 2-D cases - the numerical experiments available are for sufficiently high Reynolds numbers - it should be no problem.

Moffatt: I'd like a little more understanding of how a computation of this experimenter would actually determine the conditional average that you are talking about. You have a fixed point and you decide on a fixed vorticity, which is a vector, and you run the experiment and the condition is that the vector should take a given value. Now, what worries me about this is that if it was just a scalar, then we'd have no problem because a random scalar always fits a given value a given number of times, but a vector doesn't. It never repeats.

Novikov: You have to have a much bigger data-set than needed to just make correlations. But you get much more information because you have the conditionally averaged vorticity field. It's something intermediate between just a correlation which people used to measure and a two-point probability distribution, which I am now working on. In order to do this you have to choose a sub-set of data with not exactly fixed vorticity at one point, but fixed in some volume, with some degrees of freedom, and then the value of vorticity in different points averaged over this sub-set. So you get this kind of formula I presented. These formula can be compared with measurements. If it were different, I would move to a more general solution of my equation in order to be consistent with particular measurements in particular flows.

Moffatt: Well, you introduced that vital word "not exactly" the value, but in some small volume. That, I think, is important, how small - you've got to decide.

Moin: It can be done, strictly speaking, with a very small volume with the stochastic estimation techniques (R. J. Adrian and P. Moin, 1988, J. Fluid Mech., **190**). All of these are conditional averages. So in the experiment or computation you don't necessarily have to measure this conditional average by detection of the exact value and by accumulating statistics. You just measure two-point correlations of vorticity, in this case, and from that you can use the stochastic estimation technique and estimate conditional averages.

Frisch: I have a question to Evgeny Novikov. At the end of your talk you mention that you did some work in the sixties concerning the non-universal contamination of the dissipation range by the energy range. And since I'm not aware of this work, could you say a few words about it and how it relates to Kraichnan's paper (1967 Phys. Fluids **10**, 2080). I suppose yours is earlier.

Novikov: Yes, it is earlier - 1963. The story is the following. First, I published a paper in 61 on the asymptotic form of the spectrum of energy in the dissipation range based on the assumptions that in a volume of the order of Kolmogorov's scale there is some stretching and extention and the orientation of axis is not changing very quickly. If you assume this, you have some alignment of vorticity and you can then solve equations and get a spectrum of energy in the dissipation range. For example, you have an exponential decay of energy spectrum and ε, which is sitting in the exponent, is random. If you average over ε, taking into account intermittency, you can get quite a different shape of spectrum of energy, even at the beginning when it just starts to decline from the inertial range. In the inertial range you can only change the constant, and the power law a little bit. In the dissipation range - because the whole shape is moving up and down by averaging over the ε (or whatever characteristic you choose because of intermittency), you can get a completely non-universal spectrum after the averaging.

Frisch: But if you notice - the essence of Kraichnan's 67 paper was that if you have any fluctuations whatsoever in ε and therefore in $E(k)$ and you assume faster than the algebraic decrease of the spectrum, then in the far dissipation range this fluctuation will be monstrously amplified and you get arbitrary strong intermittency.

Novikov: Yes, actually it's the same thing.

Wygnanski: I have a question to Parviz. Similar data to the ones you presented were analysed by Spalart for a spot. In the boundary layer with zero pressure gradient there might be a couple of mechanisms that somehow interact in a way that make things look very random, but in the spot we have a hint that there is much more order inside, and I know that Spalart computed this using the same scheme. Do you have any information on what happens inside such a spot?

Moin: The structure of the spot was not looked at as carefully as you saw here in this picture. There is an experiment that Javier Jimenez and I did on the minimal flow unit which has just appeared in JFM. The idea was to try to isolate the minimum number of the structures that do interact with each other in a non-linear fashion as a solution of the Navier Stokes equations. We did these calculations in a duct, and we used periodic boundary conditions in the streamwise and spanwise directions. The idea was to shrink the size of the computational box to the smallest possible and still maintain a stochastic solution. We were looking for periodic solutions of the Navier Stokes equation, but with the minimum box. Presumably, having a minimum box you would have a minimum number of structures, and you could go in there and see what the essential patterns are. One streamwise vortex was always there and was responsible for the generation of patches of intense Reynolds shear stress. With just having one streak and one streamwise vortex, occasionally two, you could produce low-order statistics quite well.

Lesieur: I have another question related to the possibility of obtaining other states. If you make a very rough analogy with transition, you know, of course, that we can distinguish two secondary modes - secondary with respect to Tolmien-Schlichting waves -, which are Klebanoff's mode and Herbert's staggered mode. Well, it seems that what we have in this developed turbulent stage looks like the staggered mode. Do you think it is possible to force the boundary layer in order to obtain something closer to Klebanoff's mode?

Moin: What I just described in the terms of the minimal channel is something like forcing it. When you shrink the boundaries, you are forcing the same structure in the spanwise direction. The composite picture is a very ordered pattern of vortices and streaks etc.
Whether what we show here is a k-type instability just because they are not aligned is questionable. We show only two vortices and if you say they are not aligned, then they are staggered. I think that's stretching it too much. If you could look at the whole flow field, I don't think they are necessarily staggered. I doubt very much that they would be staggered like Klebanoff's.

Reynolds: My question is that I couldn't tell the difference between the turbulent spots' structure and the structures in the boundary layer. I think they're really very similar, and this brings me to the main point, I think that perhaps, in fact, there may be more organisation than constant contouring may show. There ought to be a method of how to do contour varying in some way to maximize the volume. Something like that would be a better way to see the structures than the constant contours show.

Moin: I feel it depends on what you call more organisation. The building block structure is organised as you will. But I also think the nature of turbulence requires this structure to be distributed in a random fashion in space and time, and when you have the whole picture, it is very complicated. So the key is not to find the right visualisation technique to be able to see everything aligned with each other because I don't think everything is aligned, everything is at a random relationship with each other. The key is to identify the minimum or the building block structure.

Reynolds: My suggestion is to do Spalart's spot simulation by the simulation you made. I bet you won't find in there the organisation that we know is there from other numerical simulations.

Wygnanski: As we know, the spot grows in x-direction and becomes huge. It always becomes commensurate with the tunnel geometry. Experiments must scale appropriately to the size of the box. However, at the boundaries of the spot, where there is fluid that just became turbulent, there is much more organisation. I think it is around those boundaries that we should really look, and then see how slowly this organisation disappears. The interior of the spot resembles the boundary layer in every respect. All the scaling rules apply there. It's only near the boundary that we can see it (the organisation), and I think those (structures) are commensurate with the streeks, etc.

Jimenez: I would like to make a comment on your question about how the ordinary turbulent spot grows. The behaviour of the spot seems fully turbulent. There is not a slow transition to turbulence.

Wygnanski: It's not slow, it's certainly not slow in a spot of zero pressure gradient, but if you go and do the same thing in a highly favourable pressure gradient, you slow the process down. In fact, I believe that in order to see these structures both in the turbulent boundary layer and in the spot we should go to pressure gradients. We would probably discover very different effects. In fact, a boundary layer in a highly adverse pressure gradient which is inviscidly unstable may produce very interesting effects, and in the case of the spot I somehow feel that a highly favourable pressure gradient will slow the process so that you can experimentally and perhaps also by regular numerical simulation see things evolving a little bit slower, so that you can sort and pick up the mechanisms one by one rather than having them all acting simultaneously.

Kambe to Moffatt: A collection of spiral vortices might give the expected energy spectrum, but I wonder how you interpret the intermittency from the model of the spiral vortices.

Moffatt: Well, I don't know and actually that's something I'd like to do or would like somebody to do. If you have a detailed solution of this kind, you can, in principle, work out all the higher order statistics, 4th- order, n-th order moments and you thus have a potential probe on the problem of intermittency (See Vassilicos, 1992).

Lesieur: I very much like the way Moffatt spoke about the formation of the spiral - the instability of these initial vortex sheets, and I think that what I called "large-scale intermittency" yesterday was exactly this kind of mechanism. So, I think what Moffatt proposed is a mechanism of formation of what I would call large eddies and a mechanism of large-scale intermittency where the initial vorticity concentrates into a smaller set, and then qualitatively you can immediately see that this produces PDF's which are much higher than Gaussian in the wing for the vorticity and for the low pressure; so I would call that large-scale intermittency. We have to be careful because when people try to measure fractal or multi-fractal structure of turbulence and see something very dramatic on their signals, it might correspond to the passage of these large vortices, and maybe not to small-scale intermittent structures.

Herring: I would simply like to say that the spacing between these elementary structures is critical and determining for the higher order statistics. The higher order statistics might be a side track or something that might be very important.

Aref: I'd like to also mention - for those interested in measurements of fractal properties of turbulence - there is a paper by Miller and Dimotakis (1991, Phys. of Fluids, V.3., pp. 168-178) that you might like to read and which gives a very different view of whether you can measure any fractality in shear layers whatsoever. In fact they find no fractality of the passive scalar.

Frisch: I want to make a comment about Marcel Lesieur's statement regarding small- and large-scale intermittency, I don't think nowadays anybody reasonable would propose having structures in a flow which are small in *all* directions, for example, the sheets, the worms or these things have one or two directions in which they are small, but they are very extended in other directions and so when one speaks about small-scale structures, one should not have in mind some kind of vaguely spherical animal; nobody takes that view seriously, so it seems that when Marcel is speaking about "large" and "small", he means that *some* dimensions are small or large. The more difficult questions are, for example, how extended are they at high Reynolds numbers. Do they have an extension or, say, radius of curvature, comparable to the integral scale or to the Taylor scale? These are really unsettled issues that present numerical simulations cannot easily answer.

Session Four

Chairman M. Lesieur

How Long is the Way from Chaos to Turbulence ?

George M. Zaslavsky

Abstract. Some problems of dynamical chaos are discussed in their relation to the problem of turbulent motion. They include the symmetry of patterns in preturbulent state, space-time chaos, Lagrangian turbulence, anomalous transport, intermittency and others. All of them are considered from a dynamical background.

1. Introduction

For a long time the onset of turbulence was a clear manifestation of how erratic behavior of a dynamical system can appear due to some instability process. From a new perspective this idea can be simply formulated that there exists a dynamical origin of turbulence. It is not necessary to explain in detail how widespread this point of view is among the community working in turbulence. Thus it seemed that a new era of turbulence would appear after the theory of dynamical chaos spread. But surprisingly there was a strong resistence of the turbulence community towards considering chaos as an explanation of the turbulence origin. In particular this community resists the idea that the theory of chaos should provide a key element in understanding the turbulence.

There is some sense in this skepticism, but it seems to me that it comes only because of an embrional form of the dynamical theory application to the theory of turbulence. The aim of the article is to give a short sketch of which part and in what way the theory of chaos could influence our understanding of turbulence. This opinion is based on several review talks given at different times (the last one was in [1]), but some part of this article concerns new considerations and new results.

2. Dynamical Origin of Turbulent Motion

This chapter will discuss two points: 1) the transition from regular (laminar) flow to irregular (turbulent) flow, and 2) how the properties of the irregular flow are consistent with the real turbulence.

2.1 Scenario of Turbulence. The appearance of irregular flow is called a scenario. The first one was proposed by L. Landau and E. Hopf. A link between chaos and irregularity of a fluid dynamic flow was demonstrated in an explicit form by E. Lorenz [2]. It was shown [3] that dimension three for the phase space of a motion is enough to produce chaos. There are several different scenarios like Pomeau-Manneville, doubling bifurcations, etc. (see review [4,5] which reflect different features of turbulence.

Although experimental data do not confirm the existence of a unique scenario of the laminar-turbulent transition, attempts to find a universal behavior are not finished. It is not clear which properties of transition should be universal and even it is not clear why such a universality should exist. The problem of the transition scenario will not be discussed below.

2.2 Spatial-Temporal Chaos. A typical "physical" way associated to the so-called "dynamical chaos" theory is to study evolution of the velocity field, $\vec{v}(\vec{r},t)$, considering its expansion

$$\vec{v}(\vec{r},t) = \sum_{\vec{k}} \vec{v}_{\vec{k}}(t) e^{i\vec{k}\vec{r}} \tag{2.2.1}$$

and to study behavior of the harmonics $\vec{v}_{\vec{k}}(t)$. The truncation operation is used to reduce the number of harmonics and just such a procedure was used by E. Lorenz. Finite-dimension system of O.D.E. for $\vec{v}_{\vec{k}}(t)$ can provide a set of bifurcations and chaos when some parameters vary.

The onset of chaos for some characteristic number R that exceeds the critical value R_c, provides a continuous spectrum of frequencies for the time expansion of $\vec{v}_{\vec{k}}(t)$. But there is no any changing in space expansion (2.2.1). Space scales (or values of $\vec{k}$) are fixed and discrete. But actually complexity of a turbulent motion should be bilateral in time and in space. Chaos should be spatial-temporal. There are several discussions of this issue [1,6,7], but there is no reliable idea to produce both space and time chaos. The most crucial property of the developed turbulence, the evolution of rather simple flows to the universal scale invariant flow, described by the Kolmogorov-Obukhov law disappears from the theory, if one uses the simple form of expansion (2.2.1). And that is maybe the most serious reason of the skepticism towards explaining the turbulence by the dynamical chaos theory.

Below a new concept will be introduced to explain the spatial chaos origin.

2.3 Patterns in Pre-Turbulence. Transition to turbulence can pass through different intermediate states which have space-organized structure. This means freezing some set of wave numbers $\vec{k}_j$ in the sum (2.2.1). There are many physical examples which provide ordered patterns before a turbulence set in (square and hexagonal cells of Rayleigh-Benard convection, zonal flows in planetary atmosphere, etc.). In the simplest cases, patterns of extended matter possess a symmetry. General considerations lead to symmetry of crystal or quasicrystal type for stationary flows [8].

Existence of a pattern with symmetry provides some general properties of Lagrangian turbulence and passive particle transport. Characteristic properties of intermittency can be established if the kind of pattern is taken into account. For large Reynolds number, properties of large scale structure should affect the small scale swirls advection. Some of these properties can be investigated with good dynamical background and they will be considered below.

3. Beltrami Flows and Symmetry

3.1 Depletion of Nonlinearity. The Euler equation

$$\frac{\partial \vec{v}}{\partial t} + (\vec{v}\,\vec{\nabla})\vec{v} = -\vec{\nabla} P \tag{3.1.1}$$

can be reduced for incompressible fluid to the form

$$\frac{\partial \vec{v}}{\partial t} + [\text{rot}\ \vec{v}, \vec{v}] = -\vec{\nabla}\left(P + \frac{1}{2}\, v^2\right) \tag{3.1.2}$$

For steady flow there are two possibilities [9]: (i) $\vec{v}$ is nonparallel to rot $\vec{v}$ and the topology of the vector field $\vec{v}$ is trivial (equations for streamlines are integrable); (ii) the condition

$$\vec{v} \parallel \text{rot}\ \vec{v} \tag{3.1.3}$$

is fulfilled and therefore on the Bernoulli surfaces ($P + \frac{1}{2}v^2 = \text{const}$) there is nontrivial topology of the field $\vec{v}$. V. Arnold suggested a model of such a field [10]

$$\begin{aligned} v_x &= A\sin z + C\cos y \\ v_y &= B\sin x + A\cos z \\ v_z &= C\sin y + B\cos x \end{aligned} \tag{3.1.4}$$

for which div $\vec{v} = 0$ and (3.1.3) is valid. Numerical simulations [11,12] showed indeed the existence of chaotic behavior of streamlines.

The flows with condition (3.1.3) are known as Beltrami flows, and (3.1.4) is called the ABC-flow (Arnold-Beltrami-Childress). Special interest to Beltrami flows has appeared during the last decade, because of two reasons. The first one is linked to the suggestion that flows with chaotic streamlines could generate magnetic field (so-called fast dynamo problem). The second one is relevant to turbulence. It is clear that condition (3.1.3) means elimination of the nonlinear term in (3.1.2). Such a reduction was discussed in [13,14] and observed in numerical simulations [15-17] (see also discussions in [18-19]). Depletion of nonlinearity or Beltramization of developed turbulence means a tendency to the alignment of $\vec{v}$ and rot $\vec{v}$. There is no convinced theory of how such a phenomena could appear. Nevertheless it seems to be reasonable to study properties of Beltrami flows as much as possible in hope of recognizing them in developed turbulence. Some other motivations, in addition to this one, will be mentioned after the concept of Lagrangian turbulence is introduced below.

3.2 Q-Flows. Let us write the condition (3.1.3) in the form

$$\text{rot } \vec{v} = \kappa \vec{v} \tag{3.2.1}$$

where κ can be arbitrary function of coordinates $\vec{r}$. The solution of (3.2.1) for constant κ is

$$\vec{v} = \kappa \text{ rot } \vec{\ell}\Psi + \text{rot rot } \vec{\ell}\Psi \tag{3.2.2}$$

where $\vec{\ell}$ is arbitrary unit vector and Ψ is an arbitrary solution of the Helmholtz equation

$$\Delta\Psi + \kappa^2\Psi = 0 \tag{3.2.3}$$

This form of solution was known very long ago as Trkalian field [20,21].

The case $\kappa = \text{const}$ (or $\kappa = \pm 1$, if it is omitted) will be considered below.

Several simple corollaries can be deduced from (3.2.1)–(3.2.3):

(i) For any Ψ of the form

$$\Psi = \psi(x,y)\cos z \tag{3.2.4}$$

we have $v_z = 0$, if $\vec{\ell} = \vec{z}^{\,0}$.

(ii) For any Ψ of the form

$$\Psi = \psi(x,y) + \phi(x,y)\cos z \tag{3.2.5}$$

we have $v_z = 0$, if $\vec{\ell} = \vec{z}^{\,0}$. The cases describe 2-D flow, depend on all three coordinates.

(iii) For $\ell = \vec{z}^{\,0}$ and

$$\Psi = \psi(x,y)$$

one finds

$$\vec{v} = (\partial\psi/\partial y, -\partial\psi/\partial x, \psi) \tag{3.2.6}$$

That is, 3-D flow, depending on two coordinates. Let $\vec{\ell} = \vec{y}^{\,0}$ and

$$\Psi = \cos z \tag{3.2.7}$$

Then

$$\vec{v} = (\sin z, \cos z, 0) \tag{3.2.8}$$

(iv) Linear combination of solutions (3.2.2) with different $\vec{\ell}$ is also a solution. So the sum of (3.2.6) and (3.2.8) gives the field

$$\vec{v} = \left(\frac{\partial \psi}{\partial y} + \epsilon \sin z, \ \frac{\partial \psi}{\partial x} + \epsilon \cos z, \psi \right) \tag{3.2.9}$$

where ϵ is a parameter and $\psi = \psi(x, y)$ is a 2-D function.

It was in [22] where so-called Q-flow was introduced in the form (3.2.9) with a special function:

$$\psi_q = \sum_{j=1}^{q} A_j \cos(\vec{r} \cdot \vec{e}_j) \tag{3.2.10}$$

where A_j are constants and $\{\vec{e}_j\}$ is a set of unit vectors, forming a regular star:

$$\vec{e}_j = (\cos(2\pi j/q), \ \sin(2\pi j/q)) \tag{3.2.11}$$

A particular case is when all $A_j = 1$:

$$\psi_q = \sum_{j=1}^{q} \cos(\vec{r} \cdot \vec{e}_j) \tag{3.2.12}$$

and a more general case is when in place of cosine in (3.2.10) there is an arbitrary function $\Psi(\vec{r} \cdot \vec{e}_j)$ where Ψ is a solution of the Helmholtz equation (3.2.3).

We will deal with the Beltrami flow (3.2.9) and $\psi = \psi_q$ defined in (3.2.12). The special form (3.2.12) possesses a unique symmetry property. For $q = 4$

$$\psi_4 = 2(\cos x + \cos y) \tag{3.2.13}$$

and (3.2.9) gives ABC-flow (3.1.4) (to within constant and phase shifts). For $q = 3$

$$\psi_3 = \cos x + \cos \frac{1}{2}(x + \sqrt{3}\, y) + \cos \frac{1}{2}(x - \sqrt{3}\, y) \tag{3.2.14}$$

The function (3.2.14) coincides with the stream function for hexagonal flow in Rayleigh-Bénard convection. The solution is the same for $q = 6$.

If trivial cases $q = 1, 2$ are excluded, then $q = 3, 4, 6$ are the only possible periodic solutions. For any others $q \neq \{q_c\} \equiv \{1, 2, 3, 4, 6\}$, ψ_q will be almost periodic function of quasicrystal-type symmetry. That is why such flows are called Q-flows.

Different properties of Q-flows were studied in [8,22,23].

The main property of Q-flows is that they form regular patterns with symmetry of different levels of complexity. Quasicrystal symmetry provides an intermediate case between periodic structures and disordered ones. They are aperiodic, but not disordered and their spectrum is pointlike. Nevertheless some properties of disordering like in a liquid state can be concluded from the form (3.2.12) [24].

This comment can explain one of the reasons why Q-flows are interesting for investigation not only large-scale behavior, but for some small-scale properties of the developed turbulence too. The Q-flow is controlled by a new parameter q as an order of symmetry, but at the same time there is no loosing of structure (even for large q) and no loosing of anisotropy of cells of the structure. And this seems to help us to understand the phenomenon of intermittency from the dynamic point of view as we will see it later.

In 2-D flows the function ψ_q in the form (3.2.10) or (3.2.12) can be considered as a stream function. Such flows with quasicrystal symmetry of pattern can survive in different dissipative media and they possess a stability in a finite range of parameters [23].

4. Origin of Spatial Chaos

4.1 Boundary Problem. There is no evidence, either from experimental data nor from computer simulation, how structure properties and Beltramization should be organized in the developed turbulence. It seems that a sensible conjecture could be considered as a starting point to understand the level of difficulties. For instance, blobs with Beltrami flow inside [14] produces a reasonable model to study some features of turbulence in a rough approximation. One should add that the level of nonstationarity of the blobs is low and blobs are embedded into the turbulent media with much smaller scales of space and time dynamics. Hence, the Helmholtz equation (3.2.3) for Ψ should be considered inside a blob with appropriate quasistationary boundary condition, on the surface G of the blob.

The absence of flow through the surface of a blob

$$\vec{v}_n(G) = 0 \tag{4.1.1}$$

is an example of such conditions. It can be rewritten as a condition for vorticity $\vec{A}$

$$\text{rot } \vec{\ell}\psi(G) = \vec{A}_{\vec{\ell}} \tag{4.1.2}$$

in the next form

$$\kappa(\vec{A}_{\vec{\ell}})_n + \text{rot}_n(\vec{A}_{\vec{\ell}}) = 0 \tag{4.1.3}$$

A more realistic way is to choose a nonlinear and maybe nonstationary condition which takes into account field exchange between a blob and outside area and moving of the boundary.

The statement that will be mentioned in the next chapter does not depend in general on the kind of boundary condition. And that is why the simplest form (4.1.1) can be used.

4.2 Connection to Quantum Chaos Problem. After the comments made above consider first the eigenvalue problem

$$\Delta\Psi_n + \kappa_n^2\Psi_n = 0\ , \quad \psi_n(G) = 0\ , \tag{4.2.1}$$

where n is a number of eigenvalue κ_n^2 and eigenfunction Ψ_n. Let the form of the boundary be simply-connected (actually this is not the case). The first eigenfunctions will have character scales $1/|\kappa_n|$ which are of order of the size of G for small numbers n. The larger n, the larger $|\kappa_n|$ and the smaller characteristic scales of solutions of (4.2.1). What can be said about the asymptotics for large n?

There is a particularly good understanding of this question [26-28]. For large n the shortwave approximation can be used. The asymptotic form of eigenfunctions can be written as a functional of classical orbits $\vec{r}(t)$

$$\Psi_n \sim \Psi_n[\kappa_n;\ \vec{r}(t)]\ . \tag{4.2.2}$$

The classical orbits for the boundary value problem (4.2.1) are the same as orbits in the free motion of a test particle inside the volume with surface G, subject to elastic collisions with boundary G (billiard problem). In the generic case all billiards (with zero measure excepted) possess ergodic properties and the motion with the mixing properties (i.e. existence of a positive Lyapunov exponent) for some part of the phase volume of a classical billiard of the form G. The example of chaos in 3-D billiard of the form of 3-D ellipsoid with two caps, sliced off by parallel planes (barrel-like form), was given in [29]. In general 3-D ellipsoid billiard is an integrable dynamic system [30]. But even a small perturbation of the form of the billiard surface generates regions of chaotic motion in the phase space.

Concerning the set of solutions (4.2.2) and $\{\kappa_n\} \equiv \{\kappa_1, \ldots, \kappa_n \ldots\}$, it is known that there exist a subset (or subsets) with quasirandom distribution of eigenvalues κ_j and quasirandom form of eigenfunctions $\Psi_n(\kappa_n, \vec{r})$, where $\vec{r}$ is coordinate vector. That is just the problem of the so-called "quantum chaos". Classical orbits of the adjoined classical dynamic system play the role of "bones" for the wave field "meat" which is the solution of (4.2.1). It is just this way that space disorder can occur in the regular boundary value problem like (4.2.1).

A more delicate problem is the space-chaos organizing and its relationship to the given boundary problem. We shall only mention this very briefly. Properties of the eigenvalues κ_n^2 can be described by their neighbor spacing distribution function $P(s)$, where

$$s_n = (\kappa_{n+1}^2 - \kappa_n^2)/\langle \kappa_{n+1}^2 - \kappa_n^2 \rangle$$

and $P(s)ds$ is density number of adjacent pairs of eigenvalues κ_n^2 for which s_n belongs to the interval $(s, s+ds)$. There is some evidence [31] for simple enough models that for $s \to 0$ the asymptotics

$$P(s) \sim s^\nu \tag{4.2.3}$$

takes place where $\nu > 0$ is an appropriate value, and for $s \to \infty$

$$P(s) \sim \exp(-\text{const.}\ s^2) \tag{4.2.4}$$

with a const ~ 1. It is unclear now how to find a connection of the result to the spectrum of turbulence.

Description of chaotic properties of eigenfunctions is more sophisticated. Different possibilities are introduced like the amplitude distribution, space-correlation functions and others [32-34,27]. No discussions will be here on the details of this kind of chaos, because the main aim was a manifestation of the reducing of the space disorder solution of (4.2.1) to the problem similar to quantum chaos.

Three small, but important comments can be made in conclusion to this section.

Comment 1. Some simple physical reasons can be introduced to clarify that initial condition (4.1.1) will generate a chaotic billiard problem as well as in the case (4.2.1).

Comment 2. Consider a simple solution of (4.2.1) in the form of ABC-flow (3.1.4). For A = B = C the right cubic symmetry specifies the flow. But if there is some difference in coefficients like A$-$C = B$-$C $\neq 0$, then a distortion of cubic symmetry will occur which leads to the equivalent classic billiard problem with chaos. Some critical distortion must appear to make the transition of a particle-in-billard motion to be chaotic in almost full phase space.

Comment 3. Even if the classical motion in the equivalent billiard is strongly chaotic, there are periodical unstable orbits, which produce strongly regular parts of some eigenfunctions. These parts are known as "scars" [27, 35-37]. They should have influence on the intermittency properties of the velocity field.

In [38] it was found a transition from regular pattern in 2-D Kolmogorov flow to a space disordered stationary pattern at the route to turbulence. Such an example may be a good model for searching the analogy to quantum chaos problem (see also forced Navier-Stokes equation model in [23]).

5. Langrangian Turbulence

There are several reviews of different levels of the topics that will be discussed here [1,8,39]. That is why they will be described in short. It was in [40], where the notion of "chaotic advection" was introduced for a scalar (passive) particle motion in a given vector field $\vec{v}(\vec{r},t)$. Considered as a dynamic system it leads to a chaotic motion or to a "Lagrangian turbulence" [12]. Different examples from theory and experiment [40–46] have led the problem to be one of the popular areas of investigation.

There exists different applications of the Lagrangian turbulence. But only one of them will be discussed here. The Lagrangian turbulence is a well formulated dynamical problem and understanding of its relation to the regular (Eulerian) turbulence could give a light on such a phenomena as intermittency. These points will be considered in the last parts of the article.

5.1 Streamlines as a Dynamic System. Passive particle equation in a given vector field $\vec{v}(\vec{r},t)$ has the form

$$\frac{d\vec{r}}{dt} = \vec{v}(\vec{r},t) \tag{5.1.1}$$

It defines the 3-D flow with unknown properties in general. Some of them were studied in a discrete version of ABC-flow [47,48], which turns out to be equivalent to a forced ABC-flow. The 2-D case for $\vec{r}$ and $\vec{v}$ can be formulated very simply in Hamiltonian form for an incompressible fluid, because of existence of a stream function. In all of these 3-D cases chaos is possible in the general case. The incompressibility condition

$$\text{div } \vec{v} = 0 \tag{5.1.2}$$

will be imposed. If the flow is stationary ($\vec{v} = \vec{v}(\vec{r})$), then a passive particle orbit coincides with the correspondent streamline. The equations for streamlines are

$$\frac{dx}{v_x} = \frac{dy}{v_y} = \frac{dz}{v_z} \tag{5.1.3}$$

which yields a dynamical system

$$\frac{dx}{dz} = \frac{v_x}{v_z}, \quad \frac{dy}{dz} = \frac{v_y}{v_z}, \quad (v_z \neq 0) \tag{5.1.4}$$

The 2-D case is integrable and is not interesting. In the 3-D case equations (5.1.4) describe a dynamical system with 3/2 degrees of freedom, if z will be considered as a time variable (the choice of z is arbitrary and x or y can replace z). So chaotic orbits are possible in (5.1.4) and this will be refered to as chaos of streamlines. Such a phenomena of the chaos of force lines of a given vector field was introduced in [49].

Even for a regular field $\vec{v}$ its streamlines can be turbulent, and this is a new topological feature of the field.

5.2 Hamiltonian Form for Streamlines. General statements for (5.1.4) will be more evident after reducing the system to a Hamiltonain form. That is available due to condition (5.1.2).

Consider, for example, Q-flow (3.2.9) and introduce a "time", τ, by the definition

$$\frac{dz}{d\tau} = \psi(x,y) \tag{5.2.1}$$

where all variables x, y, z implied to be function of τ and of the initial conditions x_0, y_0, z_0 at $\tau = 0$. Then from (5.1.4), (3.2.9) and (5.2.1), it follows

$$\begin{aligned} \frac{dx}{d\tau} &= \frac{\partial\psi}{\partial y} + \epsilon \sin z = \frac{\partial H}{\partial y} \\ \frac{dy}{d\tau} &= -\frac{\partial\psi}{\partial x} + \epsilon \cos z = -\frac{\partial H}{\partial x} \end{aligned} \tag{5.2.2}$$

where

$$H = \psi + \epsilon(y \sin z - x \cos z) \ . \tag{5.2.3}$$

So H is Hamiltonian, (5.2.2) are Hamiltonian equations, and (5.2.1) is the transformation to a new time metric [22].

A more general form of Hamiltonian description for (5.1.4) was mentioned in [50]. Conditions (5.1.2) yields

$$\vec{v} = \text{rot}\ \vec{\xi} \tag{5.2.4}$$

with some new field $\vec{\xi}$. Without loss of generality put $\xi_y = 0$. Then

$$v_x = \frac{\partial H}{\partial y}\ , \quad v_y = -\frac{\partial H}{\partial x} + \frac{\partial \xi_x}{\partial z}\ , \quad v_z = -\frac{\partial \xi_x}{\partial y} \tag{5.2.5}$$

where we set

$$\xi_z \equiv H \tag{5.2.6}$$

From (5.1.4) and (5.2.5)

$$\frac{dx}{dz} = \frac{1}{v_z}\frac{\partial H}{\partial y} \equiv \frac{\partial H}{\partial p} \tag{5.2.7}$$

with

$$p = \int^y v_z(x, y', z) dy' \tag{5.2.8}$$

Let at first $H = H(x,y,z)$. Then from (5.2.5) one has

$$H = -\int^x v_y(x', y, z) dx' + \int^x \frac{\partial \xi_x(x', y, z)}{\partial t} dx' \tag{5.2.9}$$

The last term can be transformed by the help of (5.2.5) and incompressibility condition (5.1.2)

$$\frac{\partial \xi_x(x', y, z)}{\partial z} = -\int^y dy' \frac{\partial v_z(x', y', z)}{\partial z} = \int^y dy' \frac{\partial v_x(x', y', z)}{\partial x} + v_y(x', y, z) - v_y(x', 0, z) \tag{5.2.10}$$

Substitution of (5.2.10) to (5.2.9) yields

$$H = \int^{y} v_x(x, y', z)dy' - \int^{x} v_y(x', 0, z)dx' \qquad (5.2.11)$$

The final step is to express $H = H(x, p, z)$ using (5.2.8). The second equation to (5.2.7) is

$$\frac{dp}{dz} = -\frac{\partial H}{\partial x} \qquad (5.2.12)$$

which completes the Hamiltonian form for (5.1.4).

Another kind of Hamiltonian description can be found in [39].

5.3 Stochastic Webs in Fluid. Let us come back to Q-flow in the form (3.2.9) with $\psi = \psi_q$ and ψ_q is defined in (3.2.12). The next important properties of streamlines were established in [22]. There are two different cases: $q \in \{q_c\} = \{1, 2, 3, 4, 6\}$ and $q \notin \{q_c\}$. The first case (excluding trivial cases $q = 1,2$) corresponds to 2-D periodical (crystal symmetry) function $\psi_q(x, y)$ and the second one corresponds to quasicrystal symmetry function $\psi_q(x, y)$. In the case of the crystal symmetry there exists a value

$$\psi_q(x, y) = \psi_q^{(c)} \qquad (5.3.1)$$

such that all saddles lie in the same plane $\psi_q^{(c)}$. For ψ_q in the form (3.2.12) $\psi_4^{(c)} = 0$ and $\psi_3^{(c)} = -1$. All separatrices are connected and they lie on the same plane (5.3.1), creating the patterns with square ($q = 4$) or hexagonal ($q = 3, 6$) symmetry. For any $\epsilon \ll 1$ the Hamiltonian system (5.2.1)–(5.2.3) possess the following topology of its streamlines in the real space (x, y, z). The net of separatrices is replaced by a stochastic 3-D connected web of finite width of order ϵ (Fig. 1). There are invariant cylinders inside the web. Part of streamlines represents stable orbits which wind around the invariant cylinders. Another part belongs to the web and performs a random walk. Passive particle transport is just this random walk.

A method of analytically studying the stochastic webs properties is given in [22,39] and the corresponding mapping is introduced to describe an individual streamline and its chaotic dynamic.

A more serious situation is in quasicrystal symmetry case. The same estimate that the width of the webs be of order ϵ, is obtained in [22, 39]. But it is not proved that a stochastic web exists for any arbitrary small ϵ. An example of the web of 5th order symmetry is given in Fig. 2 [8, 51]. The main structure difference for Q-flows if $q \notin \{q_c\}$ comes from the singular properties of the function $\psi_q(x, y)$. It was shown that there exists some distribution of the saddle points of $\psi_q(x)$, which does not belong to the same value of $\psi_q^{(c)}$ (see eq. (5.3.1)) [52]. There is an absolute maximum of this distribution $\psi_q^{(0)}$ which is as closer to 0 as q is larger. A connected stochastic web occurs under the perturbation ($\epsilon \neq 0$) just in the vicinity of the main maxima value $\psi_q^{(0)}$, $q \notin \{q_c\}$. For $q = 5$, $\psi_5^{(0)} \simeq 1$, and for $q = 7$, $\psi_7^{(0)} = 0$ [52].

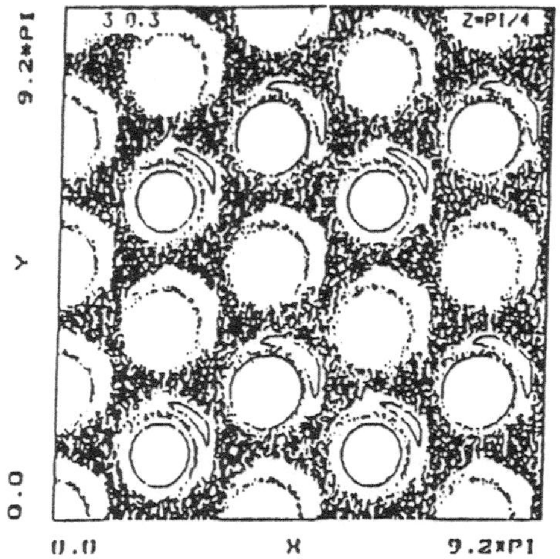

Figure 1: Poincare section of a streamline by the plane $z = \pi/4 \pmod{2\pi}$ shows a stochastic web of the hexagonal symmetry ($q = 3$; $\epsilon = 0.3$).

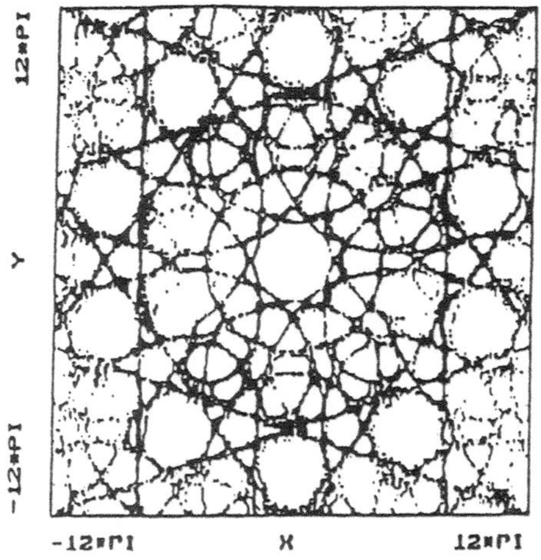

Figure 2: Example of stochastic web with 5th order symmetry ($q = 5$; $\epsilon = 0.1$).

6. Anomalous Transport and Intermittency

6.1 Flights and Jets. Transport can be investigated by consideration of the asymptotic law for displacements:

$$\langle R(\tau)\rangle \sim D_{\perp}^{1/2}\tau^{\alpha_{\perp}}$$
$$< L(\tau) >\sim D_{\parallel}^{1/2}\tau^{\alpha_{\parallel}} \tag{6.1.1}$$

where

$$R^2(\tau) = [x(\tau) - x(0)]^2 + [y(\tau) - y(0)]^2$$
$$L^2(\tau) = [z(\tau) - z(0)]^2\ , \tag{6.1.2}$$

τ is defined in (5.2.1); $x(\tau), y(\tau), z(\tau)$ are solutions of (5.2.1)–(5.2.3) and $D_{\perp}$, $D_{\parallel}$, $\alpha_{\perp}$, $\alpha_{\parallel}$ are some constants. For normal diffusion $\alpha_j = 1/2$. If $\alpha_j < 1/2$, then there are more trappings than flights and vice versa. Trapping means that a liquid particle (or streamline) spends a long time in some areas of the space. Flight means almost free propagation of a liquid particle, i.e. no random walk. In this case $1/2 < \alpha < 1$. The more flights, the closer α is to 1. Trappings and flights are responsible for anomalous transport. In a series of works this phenomena was observed for Q-flows [51, 53-55].

An example of normal and anomalous transport is given in Fig. 3 for the web with hexagonal symmetry ($q = 3$) [55]. There is normal diffusion in case "b" ($\alpha_{\perp} \simeq 1/2$) in (x, y) plane which corresponds to $\epsilon = 0.7$ and there is anomalous diffusion in the case "d" ($\alpha_{\perp} \simeq 0.75$), which corresponds to $\epsilon = 1.1$. Figures 3a,3c show Poincare sections of one streamline in one cell of the pattern, and Figs. 3b,3d are the same streamlines correspondently in unlimited space with very small size of the cells. There are flights in Fig. 3d and there is different topology of the phase space in cases 3a and 3c.

Let us describe briefly the main results concerning the anomalous transport for Q-flows.

(i) Flights and trappings are two complementary processes. A picture of transport depends on the direction of observation. For example, a flight in the z direction means trapping in the (x, y) plane. A flight appears as a winding streamline around the surface of an invariant tube. But due to instability this winding is finite and not regular. After some time the streamline leaves the vicinity of the surface and continues a random walk in the full phase volume.

(ii) Trappings appear in the vicinity of boundaries of islands in the phase space where cantori are the most effective as barriers to transport. An example of two flights is cleared in Fig. 4 [1, 53, 55].

(iii) Topology of the phase space plays an extremely important role in organizing of the transport [53, 55]. For Q-flows it depends on one parameter ϵ. Bifurcations

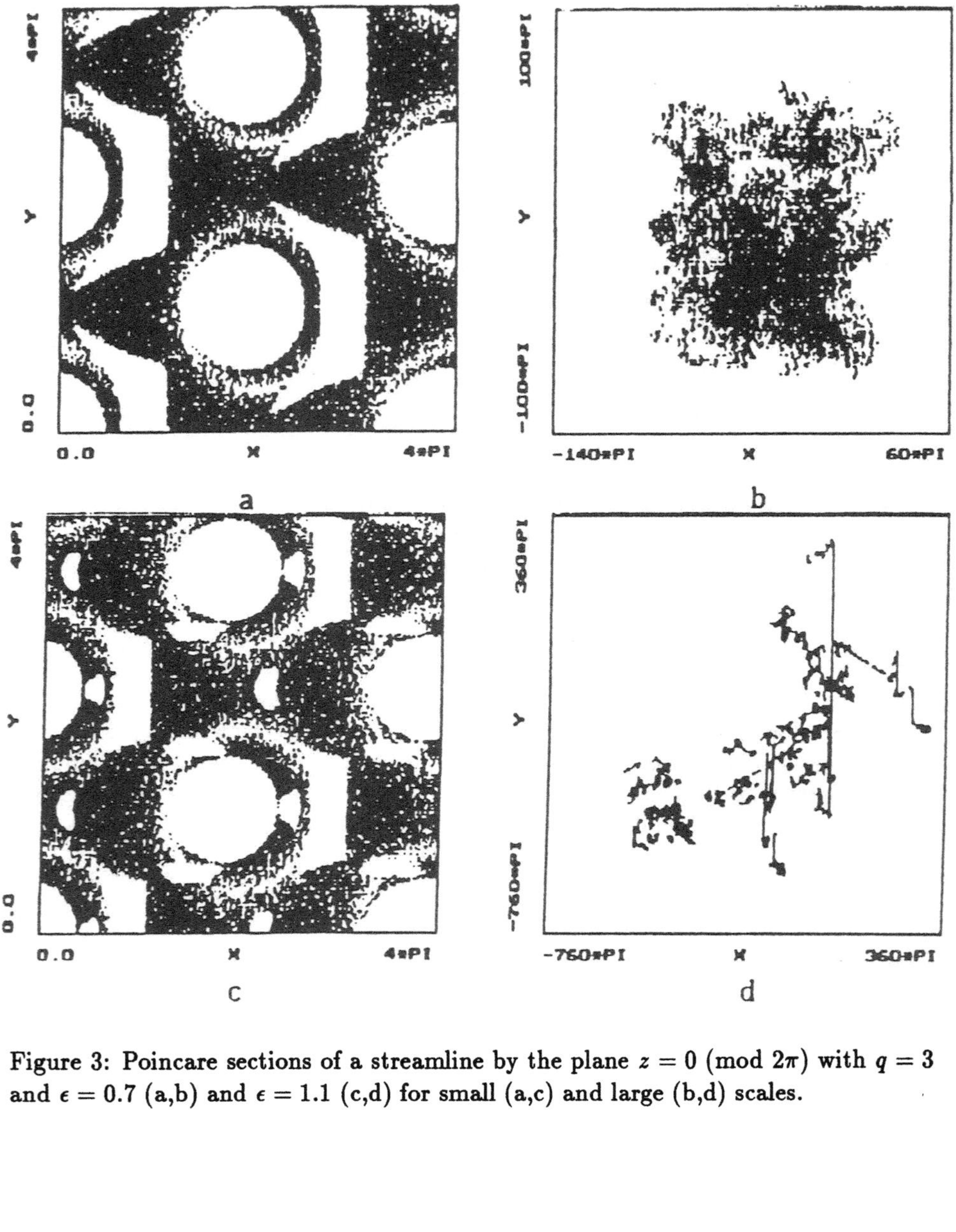

Figure 3: Poincare sections of a streamline by the plane $z = 0 \pmod{2\pi}$ with $q = 3$ and $\epsilon = 0.7$ (a,b) and $\epsilon = 1.1$ (c,d) for small (a,c) and large (b,d) scales.

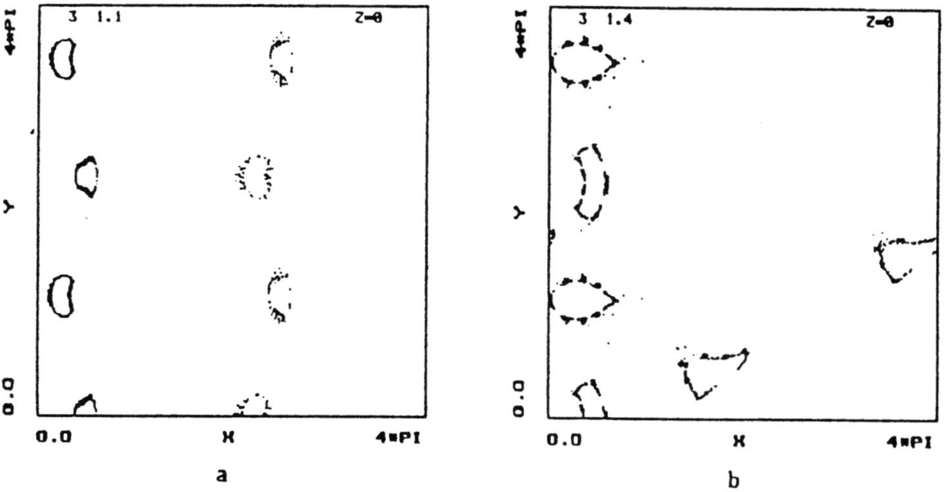

Figure 4: Flights for $q = 3$: (a) two vertical flights for $\epsilon = 1.1$; (b) one vertical and one oblique flights for $\epsilon = 1.4$.

change the phase portrait and the number and form of invariant tubes (see Fig. 3a,c). Plot of $\alpha_\perp$ and $D_\perp$ versus ϵ for $q = 3$ is shown in Fig. 5. All peaks for $\alpha_\perp$ correspond to a phase space topology change.

(iv) Extremely long flights can appear under some appropriate conditions [51]. They are called stochastic jets. Time length of jets can be comparable with the time of observation or simulation. An example of a jet is given in Fig. 6 for $q = 5$ [51]. The Poincare section of streamlines are displayed as a function of ϵ. The jet appears in Fig. 6b, when there is merger of an inner stochastic layer (see Fig. 6c) with outer stochastic web (see Fig. 2).

There are several experimental observations that indicates that the degree of anomalous transport depends on the degree of the symmetry exhibited by the pattern [56-58].

6.2 Random Walk in a Fractal Space-Time. The existence of trappings and flights shows that another kind of random walk process should be applied rather than regular Brownian motion. It seems that the Levy flight process [59] is the most appropriate to describe the observed phenomena of Lagrangian turbulence [1, 51].

The main feature of the Levy process is the absence of a characteristic space scale. That leads to divergence of moments $\langle|\vec{r}|^m\rangle$ for $m \geq 2$ for any time moment. A long

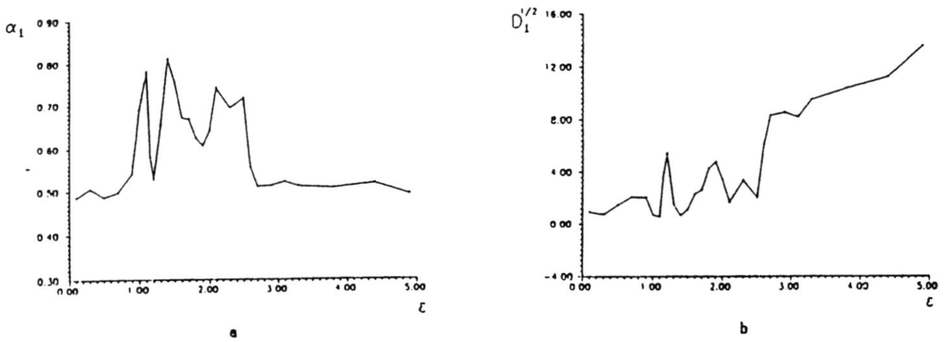

Figure 5: Plot of $\alpha_\perp$ (a) and $D_\perp^{1/2}$ (b) versus ϵ for $q = 3$.

tail density distribution, $P(\vec{r}, t)$, to find a particle in the vicinty of the point $\vec{r}$ at the moment t is in charge of the second and higher moments divergence, in contrast with Gaussian or Poissonian form of $P(\vec{r}, t)$. In [60-62] the connection of the Levy process with random walk in a fractal space and time was made and a fractal generalization for the path integral representation was applied to a typical physical process with pausing time and flights [63].

The connection of space and time fractal dimensions with an asymptotic transport law (6.1.1) for dynamic systems was introduced in [64, 51, 1] together with a new possibility in particles anomalous transport.

Let us consider two orbits with close initial conditions, separated by δr_0. Then the time of separation of the orbits δr apart is

$$t_s = \text{const } \ell n \frac{\delta r}{\delta r_0} \tag{6.2.1}$$

For big values of the const (small increments of stochastic instability or small Kolmogorov-Sinai entropy) the time t_s will be large and even may be larger than the available observation or simulation time. That means the low spreading of the initial bundle of orbits of the diameter δr_0. Just these comment leads us to introduction of the concept of stochastic jets. But not only that. Different location of the initial area δr_0 of bundles of orbits reveals different values of the exponents μ in the asymptotic law

$$\langle |r(t)| \rangle \sim t^\mu \tag{6.2.2}$$

where averaging is performed over all initial values of the bundle of orbits. Thus exponents μ should be inhomogeneous over phase space. Of course, if initial conditions cover very large area of phase space than μ will tend to a constant value and the property of inhomogeneity of the transport will be overshadowed.

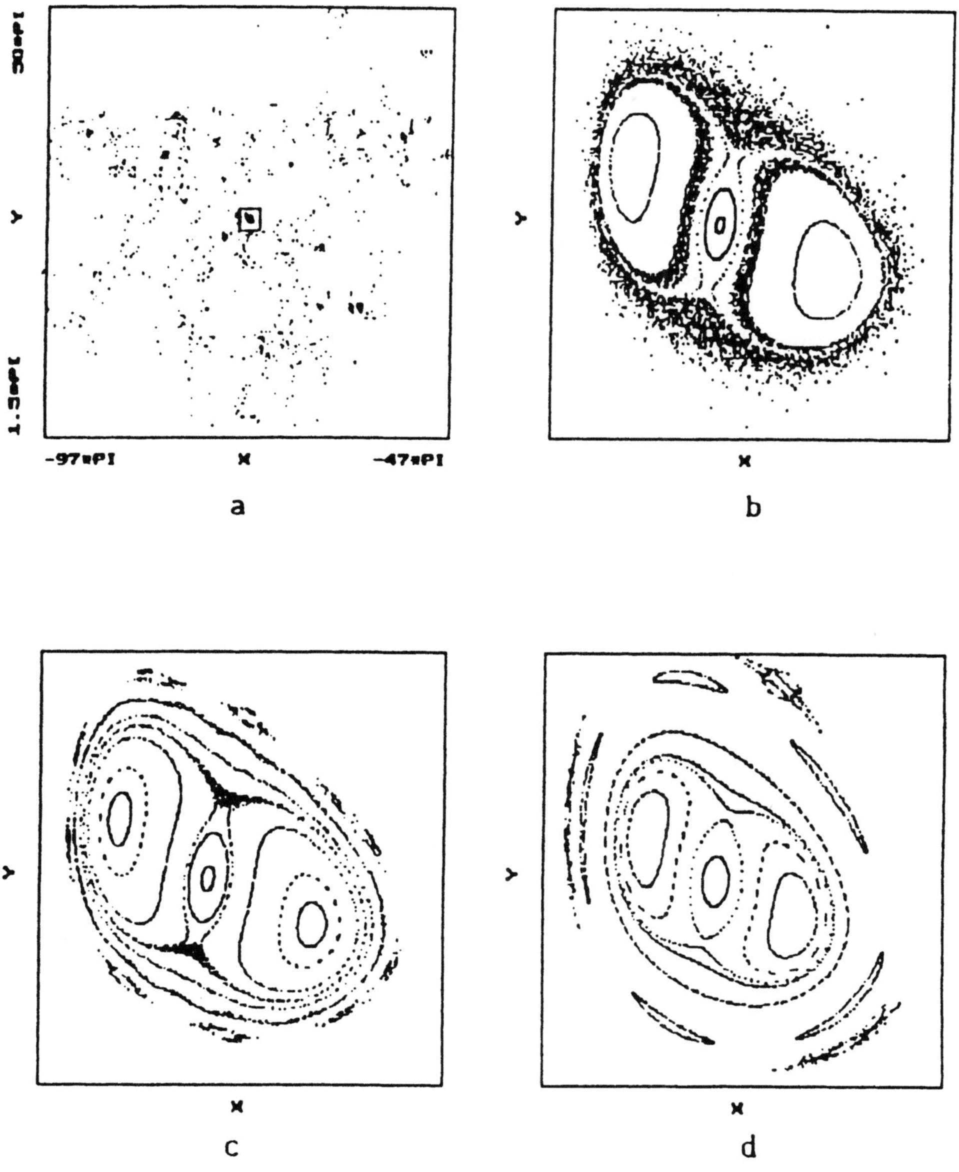

Figure 6: Poincare section in (x,y) plane of a jet for $q = 5$, $\epsilon = 0.9$: (a) – location of the jet; (b) – zoom of the square in (a); (c) – the same as in (b), but for $\epsilon = 0.8$; (d) – the same as in (b), but for $\epsilon = 0.5$. Figs. (c) and (d) explain the bifurcation of the phase portrait.

Let us consider the one-dimensional process and let $P(r,t)$ be the probability density to find particle in the position r at the moment t. Let $\Psi(r-r',t-t')$ be the transition probability density which depends only on the corresponding differences of times and positions. Then the integral equation for $P(r,t)$ can be written

$$P(r,t) = \int dr' \int_0^t dt' P(r',t')\Psi(r-r',t-t') + \Phi(t)\delta_{r,0} \tag{6.2.3}$$

where $\Phi(t)$ is the probability density not to leave position $r = 0$ during the time interval t (that is to be trapped at $r = 0$ for the time t). For $\Phi \equiv 0$ equation (6.2.3) reduces to the standard form of Smolukhovsky-Chapman equation. New term in (6.2.3) is responsible for the possibility of a time delay to make a step due to trappings.

From the definition of $\Psi(r,t)$, the probability that a transition occurs during the time t equals

$$\Psi(t) = \int dr\ \Psi(r,t) \tag{6.2.4}$$

Therefore the probability not to perform any step during the time t (i.e. to be trapped) is

$$\Phi(t) = 1 - \int_0^t dt'\ \Psi(t') \tag{6.2.5}$$

Equation (6.2.5) is the necessary important step which gives the linking between the intrinsic probabilities Φ and Ψ introduced in the main equation (6.2.3). The next steps are very formal.

Introduce the Fourier-Laplace transform of $P(r,t)$ and $\Psi(r,t)$

$$P(k,u) = \int_{-\infty}^{\infty} dr\ e^{ikr} \int_0^{\infty} dt\ e^{-ut} P(r,t)$$

$$\Psi(k,u) = \int_{-\infty}^{\infty} dr\ e^{ikr} dt\ e^{-ut}\Psi(r,t) \tag{6.2.6}$$

and Laplace transform of $\Psi(t)$ and $\Phi(t)$:

$$\Psi(u) = \int_0^{\infty} dt\ e^{-ut}\Psi(t)$$

$$\Phi(u) = \int_{-\infty}^{\infty} dt\ e^{-ut}\Phi(t) \tag{6.2.7}$$

Then (6.2.5) and (6.2.7) yield

$$\Phi(u) = [1 - \Psi(u)]/u \tag{6.2.8}$$

Equations (6.2.6) and (6.2.5) give

$$\Psi(u) = \Psi(k=0,u)\ . \tag{6.2.9}$$

Using (6.2.6)–(6.2.9) a formal solution of Eq. (3.1) can be obtained

$$P(k,u) = \frac{\Phi(u)}{1-\Psi(k,u)} = \frac{1-\Psi(0,u)}{u(1-\Psi(k,u))} \tag{6.2.10}$$

which is just the Montroll-Weiss equation [65, 66]. Only one function $\Psi(k,u)$ is in control of the final solution

$$P(r,t) = \frac{1}{4\pi^2}\int_{-\infty}^{\infty} dk \int_{c-i\infty}^{c+i\infty} du\; e^{ikr+ut}\,\frac{1-\Psi(0,u)}{u(1-\Psi(k,u))} \tag{6.2.11}$$

If the asymptotics $r \to \infty$, $t \to \infty$ are of the main interest then the integrand in (6.2.11) should be considered in the limit $k \to 0$, $u \to 0$ and the $\Psi(k,u)$ behavior is only relevant. The simplest expansion

$$\Psi(k,u) \sim 1 + t_0 u - \frac{1}{2}\,\sigma k^2 \tag{6.2.12}$$

generates the Gaussian distribution

$$P(r,t) = (4\pi\sigma t/t_0)\exp\left(-\frac{r^2 t_0}{4\sigma t}\right) \tag{6.2.13}$$

To obtain the anomalous diffusion a singular expansion of $\Psi(k,u)$ should be considered instead of (6.2.12). Let us take, for example,

$$\Psi(k,u) = 1 - A|k|^{\alpha} - Bu^{\beta} \tag{6.2.14}$$

where A, B and α, β are some constants with restrictions

$$0 < \alpha < 2, \qquad 0 < \beta < 2 \tag{6.2.15}$$

providing the positive definition of $P(r,t)$. Substitution of (6.2.14) to (6.2.11) yields

$$P(r,t) = \frac{1}{4\pi^2}\int_{c-i\infty}^{c+i\infty} du \int_{-\infty}^{\infty} dk\; \frac{Bu^{\beta-1}}{Bu^{\beta}+A|k|^{\alpha}}\; e^{ikr+ut} \tag{6.2.16}$$

or due to the definition of $P(r,t)$

$$\langle |r| \rangle \sim \frac{1}{4\pi^2}\int_{c-i\infty}^{c+i\infty} \frac{du}{u}\int_{-\infty}^{\infty} dk\; e^{-ikr+ut}\;\cdot\; |r| / \left(1+\frac{A}{B}\,|k|^{\alpha}/u^{\beta}\right) \tag{6.2.17}$$

Substitutions of

$$u' = ut\;, \qquad k' = kt^{\beta/\alpha} \tag{6.2.18}$$

into (6.2.17) gives

$$\langle |r| \rangle \;\sim\; \text{const.}\; t^{\beta/\alpha} \tag{6.2.19}$$

with $\mu = \beta/\alpha$.

The result (6.2.19) indicates the possibility to obtain different exponents to describe the anomalous transport. But there are several important consequences which will be discussed in the next section.

7. Comments "in Brief"

7.1 Fractals and Space-Time Coupling. The first comment is the direct result of the observation of the phenomena of flights and stochastic jets. There is no information about the expansion of $\Psi(k,u)$ besides the only statement that α and β reflect the fractal properties of space and time. In this case a more general expansion can be used instead of (6.2.14)

$$\Psi(k,u) = 1 - \sum_{i,j} a_{ij}|k|^{\alpha_i} u^{\beta_j} \tag{7.1.1}$$

For a given time t_μ an intermediate asymptotics has to appear which shall govern the diffusion process. The larger t_μ the smaller exponents in (6.2.20) will be dominant. It seems that a devil's staircase — like, dependence of diffusion exponents, can appear if they are considered as functions of time.

The second comment is that Eqs. (6.2.19) and (6.2.14) display the nonlinear coupling between space and time behavior of a test particle. This property can be applied to get the realistic doubt concerning the Taylor hypothesis.

The third comment is much more simpler and deeper at the same time. The result (6.2.19) is formal mapping of the microscopic properties of the space-time of flow that come from the expansion (6.2.14) or (7.1.1), to the macroscopic properties (or large scale behavior in space and time) that come from the formula (6.2.19) or its another equivalent. This mapping is inconsistent with the idea of a multifractal property of developed turbulence [67]. Actually, multifractal space for a particle wandering reflects the inhomogeneity of its local dimension. If the averaging in (6.2.2) to get an asymptotic law of transport is performed over a finite volume of initial conditions, then the macroscopic law will be a function of the location of the initial averaging volume. A more detailed description should introduce some clusters which make subspaces that are embedded into the full phase space. Every cluster has its own dimension. A random walk alternates between wandering along a particular cluster and transition from one cluster to another. Macroscopic properties of a random walk (and particularly parameter μ) will depend on the transfer probabilities.

Some ideas of the clusterization were presented in [68,69,58].

7.2 Intermittency and Drag Reduction. Two comments in this section will be very short. There is an obvious connection between anomalous transport and intermittency. The flights and trappings occur due to existence of some pattern in space or even in phase space. Particularly regular behavior of orbits due to flight or trapping gives onset of the intermittency property. The nonlinear origin of intermittency in a random field [70] is another kind of phenomena, but it is impossible to exclude some connection between these two different kinds of intermittency.

It was mentioned that intermittency and structures should imply a drag reduction [71]. New light on the drag reduction problem can arise from the dynamic theory of chaos. The possibility of the existence of lights and stochastic jets, even if the passive particles motion is chaotic, was shown in Section 6. If there is no friction or viscosity then a body could travel with such a beam of flights or jets for a long distance. The existence of viscosity and a finite mass of the body makes this distance to be shorter. But the time and length of the "free" way will be owing to the same reasons as the flights and jets origin. Trajectories (i.e. trajectories encountering little or no "drag") are "proportional" to the extent of "time" spent in flights and jets.

8. Conclusions

The study of a well formulated dynamical problem for fluid dynamics provides a possibility for deep and correct understanding of the characteristic phenomena of chaotic motion from first principles. Any of us, of course, would like to understand strong developed turbulence. The lack of possibility to approach it can be slightly compensated by the good background in dynamical methods. The only trivial comment that can be added to this is that for large Reynolds number (small diffusivity) the role of dynamical treatment should be of important value. This gives us the belief that construction of such processes as space and time disorder, intermittency and transport in the strongly turbulent case will repeat some essential properties displayed in pure dynamical models of flows with patterns.

Acknowledgements

I am grateful to Arkady Tsinober for long and stimulating discussions, to Peter Laurence who helped to improve the manuscript, and to the Courant Institute of Mathematical Sciences for its hospitality and where part of this work was done.

References

1. G.M. Zaslavsky. Dynamical Theory and Mixing. Proc. of 3d European Turb. Conf. Stockholm, 1990.

2. E. Lorenz. J. Atmos. Sci., **20**, 130 (1963).

3. D. Ruelle and F. Takens. Commun. Math. Phys., **20**, 167 (1971).

4. J.P. Eckmann. Rev. Mod. Phys., **53**, 643 (1981).

5. E. Ott. Rev. Mod. Phys., **53**, 655 (1981).

6. R.Z. Sagdeev, D.A. Usikov, G.M. Zaslavsky. Nonlinear Physics. (Warwood Acad. Publ. NY, 1989).

7. H. Aref, In: Lecture Notes in Physics No. 357, Ed. J.L.Lumley (Springer-Verlag, Berlin, 1989), comment 2.

8. V.V. Beloshapkin, A.A. Chernikov, M.Ya. Natenzon, B.A. Petrovichev, R.Z. Sagdeev, and G.M.Zaslavsky. Nature 337, **133** (1989).

9. V.I. Arnold. Mathematical Methods in Classical Mechanics. Springer, Berlin, 1978.

10. V.I. Arnold. C.R.A.C.Paris **261**, 17 (1965).

11. M. Henon. C.R.A.S. Paris **262**, 312 (1966).

12. T. Dombre, U. Frisch, J.M. Greene, M. Henon, A. Mehr, and A.M. Soward. J.Fluid Mech. 167, 353 (1986).

13. A. Tsinober, E. Levich. Phys. Lett. A, **99**, 321 (1983).

14. H.K. Moffatt. J. Fluid Mech. **159**, 359 (1985).

15. R.H. Kraichnan, R. Panda, Phys. Fluids, **31**, 2395 (1988).

16. R.B. Pelz, L. Shtilman, and A. Tsinober. Phys. Fluids **29**, 3506 (1986).

17. M.M. Rogers and P. Moin, Phys. Fluids 30, 2662 (1987).

18. U. Frisch, S.A. Orszag. Physics Today, **1**, 24 (1990).

19. L. Shtilman, W. Polifke, Phys. Fluids A, **1**, 778, (1989). A. Tsinober, Ibid, 2, 484 (1990).

20. V. Trkal. Posnamka k hydrodynamice vazkych tekutin. Casopis pro Pestovani Matematiky a Fisiky (Praha) vol. 48, p. 302 (1919).

21. O. Bjorgum. On Beltrami Vector Fields and Flows. Universitet I Bergen, Arbok (1951), Naturvitenskapelig rekke, p. 1, No. 1, p. 5. O. Bjorgum and T. Godal, part 2, ibid, No. 13, p.4.

22. G.M. Zaslavsky, R.Z. Sagdeev, and A.A. Chernikov. Sov. Phys. JETP **67**, 270 (1988).

23. V.V. Beloshapkin, A.A. Chernikov, R.Z. Sagdeev, and G.M. Zaslavsky. Phys. Lett. A, **133**, 395 (1988).

24. A.A. Chernikov, R.Z. Sagdeev, D.A. Usikov, and G.M. Zaslavsky, Phys. Lett. A, **125**, 101 (1987).

25. B.A. Malomed, A.A. Nepomnyaschii, and M.I. Tribel'skii. Zhurn. Eksp. i Teor. Fiz. **96**, 684 (1989).

26. G.M. Zaslavsky. Phys. Rep. **80**, 157 (1981).

27. M. Berry. In: "Chaotic Behavior of Deterministic System", Eds. G.Ioos, R.H.G.Helleman, and R.Stora (North Holland, 1983), p.174.

28. M. Gutzwiller. Chaos in Classical and Quantum Mechanics. Springer-Verlag, N.Y., 1990.

29. G.M. Zaslavsky, H. Strauss (to be published, 1991).

30. J. Moser. Integrable Hamiltonian Systems and Spectral Theory (Accademia Nationale dei Lincei, Pisa, 1981).

31. G.M. Zaslavsky, In: "Quantum Chaos", Eds. M.A. Cerdeira, R. Ramaswamy, M.C. Gutzwiller, and G. Casati (World Scientific, Singapore, 1991).

32. M.V. Berry, J.Phys. A 10, 2083 (1977).

33. S.W. McDonald, A.N. Kaufman, Phys. Rev. A, **37**, 3067 (1988).

34. Y.L. Bolotin, V.Y. Gonchar, V.N. Tarasov, and N.A. Chekanov. Phys. Lett. A, **144**, 459 (1990).

35. E.J. Heller, Phys. Rev. Lett. **53**, 1515 (1984).

36. E.B. Bogomolnyi, Pis'ma Zhurn. Eksp. Teor. Fiz. **44**, 436 (1986).

37. R.V. Jensen, M.M. Sanders, M. Saraceno, and B. Sundaram. Phys. Rev. Lett. **63**, 2771 (1989).

38. L. Sirovich, N. Fitzmaurice, and N. Platt. Phys. Fluids (to be published).

39. G.M. Zaslavsky, R.Z. Sagdeev, D.A. Usikov, and A.A. Chernikov. Weak Chaos and Quasiregular Patterns (Cambridge Univ. Press, Cambridge. 1991).

40. H. Aref, J. Fluid Mech. **143**, 1 (1984).

41. J. Chaiken, C.K. Chu, M. Tabor, and Q.M. Tan. Phys. Fluids **30**, 687 (1987).

42. K. Ishii, R. Iwatsu. In: Topological Fluid Mechanics. Eds. H.K. Moffatt and A. Tsinober. (Cambridge Univ. Press, 1990), p. 54.

43. E. Dresselhaus, M. Tabor. Ibid., p.75.

44. J.M. Ottino, Ibid., p.13.

45. K. Bajer, H.K. Moffatt, and F.H. Nex. Ibid., p.459.

46. K. Ishii, R. Iwatsu, T. Kambe, and N. Matsumoto. In: Nonlinear World, Eds. V.G.Baryakhtar et al. (World Scientific, Singapore, 1990) **1**, p.490.

47. M. Feingold, L.P. Kadanoff, and O.Piro. J.Stat. Phys. **50**, 529 (1988).

48. M. Feingold, L.P. Kadanoff, and O.Piro. In:"Instabilities and Nonequilibrium Structures, II,37 (1989). Eds. E. Tirapegui and D. Villaroel (Kluwer Acad. Publ.)

49. M.N. Rosenbluth, R.Z. Sagdeev, J.B. Taylor, and G.M. Zaslavsky. Nuclear Fusion, **6**, 297 (1966).

50. K.J. Whiteman. Rep. Prog. Phys. **40**, 1033 (1977).

51. B.A. Petrovichev, A.V. Rogalsky, R.Z. Sagdeev, and G.M. Zaslavsky, Phys. Lett. A, 150, 391 (1990).

52. A.A. Chernikov, R.Z. Sagdeev, D.A. Usikov, and G.M. Zaslavsky. Phys. Lett. A, 125, 101 (1987).

53. A.A. Chernikov, R.Z. Sagdeev, and G.M. Zaslavsky. In: "Topological Fluid Dynamics". Eds. H.K. Moffatt and A. Tsinober (Cambridge Univ. Press, 1990).

54. G.M. Zaslavsky, M. Tippett (1991), to be published.

55. A.A. Chernikov, B.A. Petrovichev, A.V. Rogalsky, R.Z. Sagdeev, and G.M. Zaslavsky. Phys. Lett. A, **144**,127 (1990).

56. P. Tabeling, O. Cardoso. Bul. de la societe franc. de physiq. **73**,13 (1989).

57. R. Ramshankar, D. Berlin, J.P. Gollub. Phys. Fluids A, **2**, 1955 (1990).

58. R. Ramshankar, J.P. Gollub. Phys. Fluids A (1991) to appear.

59. P. Levy. Theorie de l'Addition des variables Eleatores Gauthier-Villiers. Paris, 1937.

60. E.W. Montroll, M.F. Shlesinger. In: Studies in Statistical mechanics, v. ll, p. l. Eds. J. Leibovitz and E.W. Montroll, North-Holland, Amsterdam, 1984.

61. M.F. Shlesinger, Ann. Rev. Phys. Chem. **39**, 269 (1988).

62. H. Scher, M.F. Shlesinger, and J.T. Bendler. Phys. Today, **1**, 26 (1991).

63. J.F. Douglas. Macromolecules, **22**,1786 (1989).

64. V.V. Afanasiev, R.Z. Sagdeev, and G.M. Zaslavsky (1991) to appear.

65. E.W. Montroll, G.H. Weiss. J. Math. Phys. **6**,167 (1965).

66. J. Klafter, A. Blumen, G. Zumofen, and M.F. Shlesinger. Physica A (to appear).

67. G. Parisi, U. Frisch, In: "Turbulence and Predictability in Geophysical Fluid Dynamics and Climate Dynamics". Eds. M. Ghil, R. Benzi and C.Parisi (North-Holland, 1984), p.84.

68. G.M. Zaslavsky, R.Z. Sagdeev, D.A. Chaikovsky, and A.A. Chernikov. Sov. Phys. JETP **68**, 995 (1989).

69. L. Yu., E. Ott, Qi Chen, Phys. Rev. Lett. **65**, 2935 (1990).

70. R.H. Kraichnan. Phys. Rev. Lett. **65**, 575, (1990).

71. A. Tsinober. In: Proc. of the IUTAM Symposium on Structures and Turbulence and Drug Reduction. Zurich, 1989. Eds. A.Gyr (Springer).

Adress of the author

G.M. Zaslavsky
Courant Inst. of Math. Sciences
New York University
251 Mercer St., New York, NY 10012
and
Space Research Inst. of the USSR Academy of Sciences
Moscow 117810, USSR

Vortex Interactions as a Dynamical System

HASSAN AREF *† AND IRENEUSZ ZAWADZKI *

University of California at San Diego
La Jolla, CA 92093-0225, USA

1. INTRODUCTION

In the first announcement for this meeting the conveners, Th. Dracos and A. Tsinober, asked four questions. Their second and third question were:

> How important are kinematical properties (topology, geometry) for the dynamic behavior of turbulent flows?
>
> Are there any prospects that a dynamical systems/dynamical chaos approach will allow describing real turbulence?

The study of vortex dynamics is intimately connected with both these questions. It is connected with the first question because issues of spatial structure in turbulent flows often hinge precisely on the geometry and/or topology of the vorticity field. For example, the important concept of helicity (Moffatt, 1969) is phrased immediately in terms of the linkage of vortex lines. Furthermore, the suggestion that something as 'trivial' as kinematics would even be considered at a meeting on turbulence derives, undoubtedly, from the recent understanding that passive advection, even in laminar flows, can display the full richness and complexity of chaos in a dynamical system of low order. This phenomenon of chaotic advection was first realized in simple vortex flows, such as the three-dimensional 'ABC-flow,' which continues to haunt turbulence research by virtue of its Beltrami property, and in the advection of a scalar by three interacting point vortices in two dimensions (see Aref, 1983; 1990, for background and review).

Vortex dynamics is connected with the second question above because many simple model representations of vortex interactions can be phrased in a comprehensive and enlightening way in terms of the ideas and concepts of dynamical systems theory. In such cases we can address issues of integrable or chaotic behavior with precision, and obtain definite answers. On the other hand, vortex dynamics obviously has something to do with the evolution and

*) Affiliated with Department of Applied Mechanics and Engineering Science
†) Affiliated with Institute of Geophysics and Planetary Physics, and San Diego Supercomputer Center.

development of so-called 'coherent structures,' which have been identified in virtually every kind of turbulent flow, from the simpler shear flows, such as mixing layers and wakes, to boundary layers, to that epitome of disorder: homogeneous, isotropic turbulence. Thus, if dynamical systems methods and concepts are to be used to solve problems of turbulence, the systems to which these techniques should be applied almost certainly must derive from vortex dynamics.

In this paper we first give a brief survey of instances in which problems of vortex interactions have been cast 'cleanly' in the framework of dynamical systems theory (§2). Although much of this material is familiar, it is useful to recount it. The 'flavor' of the description that one is led to in this approach is quite different from, and, in fact, almost complementary to the Fourier-mode-amplitude description that pervades the theory of turbulence as currently presented.

With this survey behind us, we argue (in §3) that a key element in vortex models relevant to turbulence phenomena is *topology change*. This takes many forms. In two dimensions like-signed vortices can 'merge,' opposite-signed vortices can 'pair,' and, on occasion, large vortices can split or 'tear.' In three dimensions vortices 'reconnect.' In all cases these processes can be quite complicated. Sometimes considerable 'debris' is left behind by the main vortices of the process. Invariably, the degrees of freedom before and after a topology change are quite different: Initially independent vortices may have fused to single structures, or pieces of the original vortices may make up new vortices in the final state.

We give a few examples of topology changes in three-dimensional flows based on recent computer simulations (Aref & Zawadzki, 1991; Zawadzki & Aref, 1991a,b). Then we argue that this is really the main new feature that needs to be addressed in adapting dynamical systems to complex flows. The relevant dynamical systems for turbulent flow are *punctuated,* in the sense that the topology changes, which are assumed to occur on a short timescale, divide the evolution of the system into 'eras' characterized by essentially 'free' evolution of a fixed set of vortices, an evolution that is, in some sense, easy to describe. This picture is elaborated towards the end of the paper (§4).

An earlier version of the ideas and results contained herein was presented at the workshop on *Coherent Structures and Intermittencies in Turbulence,* Arizona State University, May 20-24, 1991.

2. INTEGRABILITY AND CHAOS IN VORTEX DYNAMICS

In this section we consider several cases that have arisen over the years in which a flow is represented successfully by a dynamical system, the degrees of freedom of which are the vortical structures in the flow. Such systems deal ultimately with one of the most elegant equations in all of fluid mechanics, Helmholtz' equation for the evolution of the vorticity vector field,

$$\frac{\partial \omega}{\partial t} + (\mathbf{V} \cdot \nabla)\, \omega = (\omega \cdot \nabla)\, \mathbf{V} + \nu \Delta \omega. \tag{2.1}$$

Here $\mathbf{V}$ is the fluid velocity field, $\omega = \nabla \times \mathbf{V}$ is the vorticity, and Δ denotes the Laplacian operator. The vorticity equation is one of the 'most Lagrangian' dynamical equations in all of fluid mechanics, requiring analysis in terms of individual fluid particles to bring out its full physical significance. This comes about by noting the very similar form of the *dynamical* equation (2.1) and the purely *kinematical* equation governing the stretching and reorientation of an infinitesimal line element. Expanding the advection equation,

$$\frac{d\mathbf{x}}{dt} = \mathbf{V}(\mathbf{x},t), \tag{2.2}$$

for two nearby points, at $\mathbf{x}$ and $\mathbf{x} + \delta\mathbf{s}$, respectively, gives

$$\frac{\partial(\delta s)}{\partial t} + (\mathbf{V}\cdot\nabla)\delta s = (\delta s\cdot\nabla)\mathbf{V}. \tag{2.3}$$

Comparison of (2.1) and (2.3) leads to the well-known interpretation that the material derivative of the vorticity changes due to two processes: *stretching* and *reorientation* of the vorticity vector, embodied in the term $(\omega\cdot\nabla)\mathbf{V}$, and *diffusion* of vorticity, due to the term $\nu\Delta\omega$. The profound result that the kinematic viscosity, ν, can be interpreted as the diffusion coefficient for vorticity is a corollary to this argument. In plane flow $(\omega\cdot\nabla)\mathbf{V}$ vanishes, so the single component of vorticity perpendicular to the flow plane is a constant of the motion for each particle in inviscid flow.

The Lagrangian interpretation of the vorticity equation is useful both in analysis and for the construction of numerical methods. The above well-known material, and the observation that (2.2) may have chaotic solutions even for simple flow fields, $\mathbf{V}$, contains the gist of the connection between kinematics, vorticity dynamics and chaos (a theme elaborated on by Aref *et al.*, 1989).

2.1 *Formats of vorticity*

Vortex interaction models have employed a number of analytically convenient, idealized vorticity 'formats.' These are relatives of the many other idealizations used in mechanics, such as the point mass, the unstretchable string, and the frictionless constraint. In this subsection we mention some of these formats, their governing equations of motion, and their key dynamical-systems properties.

Point vortices

A point vortex of circulation Γ may be thought of as a δ-function singularity of the single surviving component of the vorticity field in two dimensions,

$$\zeta = \Gamma\, \delta(\mathbf{x} - \mathbf{x}_0). \tag{2.4}$$

This induces a flow field given by the streamfunction $\Gamma G(\mathbf{x},\mathbf{x}_0)$, where G is Green's function for the domain in question:

$$\Delta G(\mathbf{x},\mathbf{x}_0) = \delta(\mathbf{x} - \mathbf{x}_0); \quad G(\mathbf{x},\mathbf{x}_0) = 0 \text{ for } \mathbf{x} \in \partial\mathcal{D}. \tag{2.5}$$

Here $\partial\mathcal{D}$ designates the boundary of $\mathcal{D}$. The total streamfunction from an assembly of point vortices within $\mathcal{D}$ is given by superposition, since the relation between ψ and ζ is linear (Poisson equation):

$$\psi(\mathbf{x},t) = \sum_{\alpha=1}^{N} \Gamma_\alpha \, G(\mathbf{x},\mathbf{x}_\alpha(t)), \tag{2.6}$$

where $\mathbf{x}_\alpha(t)$ gives the position, and Γ_α the circulation, of vortex $\alpha=1,\ldots,N$.

In the vicinity of any given vortex, α, Green's function takes on the asymptotic form

$$G(\mathbf{x},\mathbf{x}_\alpha) \approx \frac{1}{2\pi} \log |\mathbf{x} - \mathbf{x}_\alpha|. \tag{2.7}$$

This corresponds to the streamfunction of a point vortex of unit circulation on the unbounded plane. Thus, to compute the fluid velocity due to all vortices except α itself, but including the images of vortex α in the boundary of the domain, we should use the streamfunction

$$\psi_\alpha(\mathbf{x},t) = \sum_{\beta=1}^{N}{}' \, \Gamma_\beta \, G(\mathbf{x},\mathbf{x}_\beta) + \Gamma_\alpha \, g(\mathbf{x},\mathbf{x}_\alpha), \tag{2.8}$$

where

$$g(\mathbf{x},\mathbf{x}_\alpha) = G(\mathbf{x},\mathbf{x}_\alpha) - \frac{1}{2\pi} \log |\mathbf{x} - \mathbf{x}_\alpha|. \tag{2.9}$$

Helmholtz's theorem that the vortex moves with the local fluid velocity at its location now demands that the vortex velocity be given by

$$\frac{dx_\alpha}{dt} = -\left(\frac{\partial\psi_\alpha}{\partial y}\right)_{\mathbf{x}=\mathbf{x}_\alpha} ; \quad \frac{dy_\alpha}{dt} = \left(\frac{\partial\psi_\alpha}{\partial x}\right)_{\mathbf{x}=\mathbf{x}_\alpha}. \tag{2.10}$$

Explicitly,

$$\frac{\partial\psi_\alpha}{\partial \mathbf{x}} = \sum_{\beta=1}^{N}{}' \, \Gamma_\beta \frac{\partial}{\partial \mathbf{x}} G(\mathbf{x},\mathbf{x}_\beta) + \Gamma_\alpha \frac{\partial}{\partial \mathbf{x}} g(\mathbf{x},\mathbf{x}_\alpha). \tag{2.11}$$

The gradients on the right hand side are to be evaluated at $\mathbf{x}=\mathbf{x}_\alpha$ for use in (2.10). The gradients of Green's function are $\frac{\partial}{\partial \mathbf{x}_\alpha} G(\mathbf{x}_\alpha,\mathbf{x}_\beta)$, the gradient of g is $\frac{1}{2}\frac{\partial}{\partial \mathbf{x}_\alpha} g(\mathbf{x}_\alpha,\mathbf{x}_\alpha)$. Hence, (2.11) may be written

$$\Gamma_\alpha \frac{dx_\alpha}{dt} = -\frac{\partial \mathcal{H}}{\partial y_\alpha} \quad ; \quad \Gamma_\alpha \frac{dy_\alpha}{dt} = \frac{\partial \mathcal{H}}{\partial x_\alpha}, \tag{2.12}$$

where

$$\mathcal{H} = \sum_{1 \le \alpha < \beta \le N} \Gamma_\alpha \Gamma_\beta \, G(\mathbf{x}_\alpha, \mathbf{x}_\beta) + \frac{1}{2} \sum_{\alpha = 1}^{N} \Gamma_\alpha^2 \, g(\mathbf{x}_\alpha, \mathbf{x}_\alpha). \tag{2.13}$$

Another way of writing this arises by substituting for G in terms of g from (2.9). Then

$$\mathcal{H} = \mathcal{H}_0 + \frac{1}{2} \sum_{\alpha,\beta = 1}^{N} \Gamma_\alpha \Gamma_\beta \, g(\mathbf{x}_\alpha, \mathbf{x}_\beta), \tag{2.14}$$

where $\mathcal{H}_0$ is the Hamiltonian for point vortex motion on the unbounded plane,

$$\mathcal{H}_0 = \frac{1}{2\pi} \sum_{1 \le \alpha < \beta \le N} \Gamma_\alpha \Gamma_\beta \log |\mathbf{x}_\alpha - \mathbf{x}_\beta|. \tag{2.15}$$

The above summarizes the elegant Kirchhoff-Routh-Lin theory of point vortex motion (see Lin, 1943). The Hamiltonian may also be written in terms of the Riemann mapping function of the domain in question (Hasimoto *et al.*, 1984). It is advantageous to think of the xy-plane in which the motion is taking place as a complex z-plane, where $z = x + iy$. The equations of motion (2.12) for the unbounded plane then take the particularly elegant form

$$\frac{d\bar{z}_\alpha}{dt} = \frac{1}{2\pi i} \sum_{\beta = 1}^{N}{}' \frac{\Gamma_\beta}{z_\alpha - z_\beta}, \tag{2.16}$$

where the overbar denotes complex conjugation.

Some of the main results of the theory presented here are:

- The motion of one vortex in any domain (with slip boundary conditions) is *integrable*. (This follows from Routh's statement of (2.12) in the special case of one vortex.).

- Motion on the unbounded plane is *integrable* for N = 1, 2, 3 and any values of the vortex circulations. (That this is true in principle was known to Kirchhoff and Poincaré. Details for particular cases were worked out by Gröbli in 1877. The special case of identical vortices was rediscovered by Novikov in 1975. A classification of motions according to the values of the vortex strengths was obtained by Synge in 1949 and, independently, by Aref in 1979. The special case of three vortices with zero net circulation was treated by Rott and Aref in 1989. For this material see the review by Aref, Rott & Thomann, 1992.)

- Motion on the unbounded plane of 4 identical vortices is *chaotic*. (This was established numerically by Novikov & Sedov, and by Aref & Pomphrey; see Aref, 1983, for review.)

- Motion on the unbounded plane of 4 vortices with $\Gamma_1 = -\Gamma_3$, $\Gamma_2 = -\Gamma_4$ displays *chaotic scattering* unless Q = P = 0 in which case the system is integrable. Integrability holds more generally for $\Gamma_1 + \Gamma_2 + \Gamma_3 + \Gamma_4 = 0$ and Q = P = 0. The main reference is Eckhardt & Aref, 1988; see also Aref *et al.*, 1988; references to the integrable cases are given in Aref, Rott & Thomann, 1992.)

- Motion in a singly or doubly periodic domain of 3 vortices is *chaotic* unless $\Gamma_1 + \Gamma_2 + \Gamma_3$ vanishes.

- Motion of an advected particle in the field of 3 interacting, identical vortices on the unbounded plane ('restricted' 4-vortex problem) is *chaotic*. (This was established numerically by Aref & Pomphrey, and analytically by Ziglin in 1980; see Aref, 1983.)

We have gone into this considerable level of detail for two reasons: First, the point vortex system, and the understanding we have of it from a dynamical-systems point of view, at least in broad outline, may be indicative of what is possible for vortex interactions in general. Second, the point vortex system is related to other interesting 2D vortex dynamics problems, such as the motion of vortex sheets and vortex patches, that arise from it by taking appropriate continuum limits. These systems provide the first infinite-degree-of-freedom vortex systems to which the application of dynamical-systems concepts should be attempted. Finally, certain aspects of the phenomenology uncovered in 2D vortex interactions, of which point vortices are one extreme, have surprising analogs in the interaction of 3D localized vortices such as vortex rings and tubes.

Vortex sheets and vortex patches

The continuum limits of the point vortex theory that have been considered in the literature pertain primarily to the case of unbounded flow, although as the above developments suggest there is no particular reason that these problems could not be written out for arbitrary domains.

In the vortex sheet limit one assumes that point vortices populate a plane curve, C, ever more densely, leading in the limit to a vortex sheet strength, $\gamma(s,t)$, defined such that the amount of circulation captured along the curve between arclength s and s + ds is $d\Gamma = \gamma ds$. The resulting equation of motion is known as the *Birkhoff-Rott equation:*

$$\frac{\partial \bar{z}(s,t)}{\partial t} = \frac{1}{2\pi i} \mathcal{P}\int_C \frac{\gamma(\sigma,t)\, d\sigma}{z(s,t) - z(\sigma,t)} . \tag{2.17}$$

Clearly, this is what one gets by taking the continuum limit of (2.16). The principal value integral is consistent with the $\beta \neq \alpha$ omission in (2.16).

Similarly, if we consider a continuum limit in which the point vortices gradually populate some finite region $\mathcal{A}$ ever more densely, Eq. (2.16) becomes

$$\frac{\partial \bar{z}}{\partial t} = \frac{1}{2\pi i} \int_{\mathcal{A}} \frac{\zeta' \, dx'dy'}{z - z'} . \tag{2.18}$$

Here ζ' is the vorticity at the field point z'. If the limit is taken such that the vorticity within the region $\mathcal{A}$ becomes uniform – in which case we speak of a 'vortex patch' – then (2.18) can be transformed into

$$\frac{\partial z}{\partial t} = \frac{\zeta_0}{2\pi} \oint_{\partial \mathcal{A}} \log|z - z'| \, dz' \tag{2.19}$$

(Deem & Zabusky, 1978; Su, 1979), where ζ_0 is the uniform value of ζ', and $\partial\mathcal{A}$ is the boundary of $\mathcal{A}$.

For both (2.17) and (2.19) the issue of chaotic motion (which may manifest itself as singular structure for complex time; see Kimura, Zawadzki & Aref, 1990, for the potential relevance of such singularities to problems in vortex motion and acoustics) has been overshadowed by the issue of singularity formation after a finite (real) time. For vortex sheets, where singularity formation occurs very early in the evolution of an initially straight-line sheet, this topic has a long literature starting with the asymptotic analysis by Moore (1979), and the numerical investigations of Meiron, Baker & Orszag (1982). Desingularized versions of (2.17), e.g., where a cut-off parameter is inserted in the denominator (Krasny, 1986), have not been explored from a dynamical systems viewpoint. The solutions of such desingularized equations often simulate the physically expected roll-up of a vortex sheet in a fluid at large but finite Reynolds number.

The question of whether (2.19) is ill-posed has recently been the subject of controversy. Buttke (1989) suggested that (2.19) did, indeed, produce singularities after a finite time when started from smooth initial conditions. This claim was refuted by Dritschel & McIntyre (1990), who did not find such singularities in their computations, and argued that the singularities observed by Buttke (1989) derived from the numerical method that he used, and were not an intrinsic property of the integro-differential equation (2.19).

In this context the work of Abrashkin & Yakubovich (1985) should be mentioned (see also Zeitlin, 1991). These authors approached (2.18) by asking: Under what conditions is the motion determined by this equation in a Lagrangian representation given by at most two frequencies? They found a number of solutions corresponding to 2D, finite vortices of various shapes and various internal vorticity distributions. Among them is the rotating ellipse of uniform vorticity first found by Kirchhoff more than a century ago. This investigation is interesting in the context of dynamical systems theory because it suggests an approach to the problem of deciding integrable behavior in 2D vortex dynamics. What Abrashkin & Yakubovich (1985) have tried is, basically, an 'integrable *Ansatz*' for the motion of individual fluid particles. What their analysis appears to be suggesting is that one can indeed identify solutions in which the motion contains at most two frequencies, but beyond that complexity sets in. The pursuit of a postulated form in terms of a Fourier series in time for the spatial coordinates of fluid particles may be the best tool we have for addressing integrability or chaos in many-degree-of-freedom vortex flows such as sheets and patches. It is reminiscent of the approach adopted by Moore (1979) in his analysis of the singularity problem for the vortex sheet, although he worked with collective modes for the entire sheet rather than individual particle coordinates.

A single vortex filament

In 3D it is natural to consider the idealization of a space curve C, usually called a 'filament' in this context, on which vorticity is concentrated. If the filament is closed, or goes to infinity, the solenoidality of the vorticity field is satisfied, and the evolution of the space curve is governed by:

$$\frac{\partial \mathbf{x}}{\partial t} = -\frac{\Gamma}{4\pi}\, \mathcal{P}\oint_C \frac{(\mathbf{x}-\mathbf{x}')\times d\mathbf{s}}{|\mathbf{x}-\mathbf{x}'|^3}. \tag{2.20}$$

To derive (2.20) one first writes a general formula for the velocity field about a vortex filament of circulation Γ. This is similar to the Biot-Savart formula giving the magnetic field about a current filament in electromagnetic theory. Then, one uses Helmholtz's theorem to equate the velocity of a point on the filament to the induced velocity at its location, just as we did in (2.10) in the simpler, 2D case.

There is an interesting asymptotic analysis of (2.20), which shows that the leading order contribution to the motion of a point on the filament is a local one, proportional to Γ, and to the local curvature, κ, and directed along the filament binormal vector, $\mathbf{b}$. The coefficient of $\Gamma\kappa\mathbf{b}$ contains a factor $\log(\kappa\sigma)$, where σ is a measure of the vortex cross-section, and, thus, tends to zero in the analysis (cf. Batchelor, 1967). The so-called *localized induction approximation* (LIA) for vortex filament motion, that arises by replacing (2.20) by

$$\frac{\partial \mathbf{x}}{\partial t} = C\Gamma\kappa\mathbf{b}, \tag{2.21}$$

where C is assumed to be *a finite constant,* was considered already by Da Rios (1906; cf. Ricca, 1991). It was rediscovered and promoted by Hama (1963, 1988). Since (2.21) is phrased entirely in terms intrinsic geometric quantities, it is possible, as realized by Da Rios and rediscovered by Betchov (1965), to rewrite the evolution (2.21) in terms of the curvature and the torsion, τ, of the filament. Scaling time by $C\Gamma$, and denoting derivatives with respect to time by an overdot, and with respect to arclength by a prime, these equations are

$$\dot{\kappa} + (\kappa\tau)' = -\kappa'\tau \;\; ; \;\; \dot{\tau} - \left(\frac{\kappa''}{\kappa} - \tau^2 + \frac{1}{2}\kappa^2\right)' = 0. \tag{2.22}$$

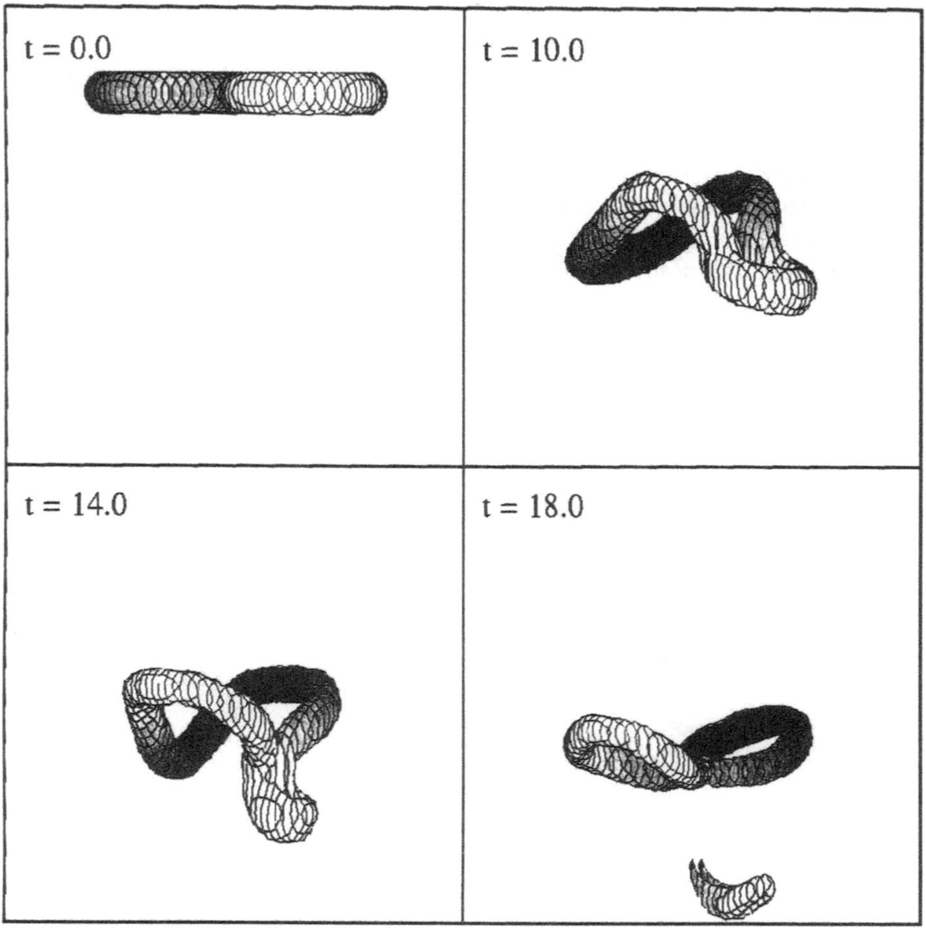

Figure 1: Interaction of two circular rings with circulation ratio 0.9:1.0, initially moving side by side. Reconnections similar to those for identical rings occur, but the 'excess' vorticity is shed in a small 'ringlet.'

Hasimoto (1972) noticed that the 'Madelung transformation,'

$$\psi(s,\theta) = \kappa(s,\theta) \exp\left\{ i \int^{s} \tau(\sigma,\theta)\, d\sigma \right\}, \tag{2.23}$$

where θ denotes the scaled time variable, reduces (2.22) to the *nonlinear Schrödinger equation* (NLSE):

$$\frac{1}{i}\frac{\partial \psi}{\partial \theta} = \frac{\partial^2 \psi}{\partial s^2} + \tfrac{1}{2}|\psi|^2 \psi. \tag{2.24}$$

It is not our intention to pursue this topic in all its ramifications, which include experimental observations of solitary waves on concentrated vortices, extensions of the model to include axial flow, and several developments related to finding the spatial structure of the filament directly from Eq. (2.24). Our main purpose is to point out that, once again, a specific vortex dynamics model has led to a clear characterization in terms of integrability or chaos, and that the methods of dynamical systems theory in this case allow considerable insight and guidance into the physics of a single, concentrated, strong vortex. (See Aref & Flinchem, 1984, for an attempt at connecting solitary waves on vortices to turbulent structure.)

We mention that steady configurations of a vortex filament evolving under LIA exist in which the filament is knotted, and, thus, has non-vanishing helicity (Kida, 1981; 1982).

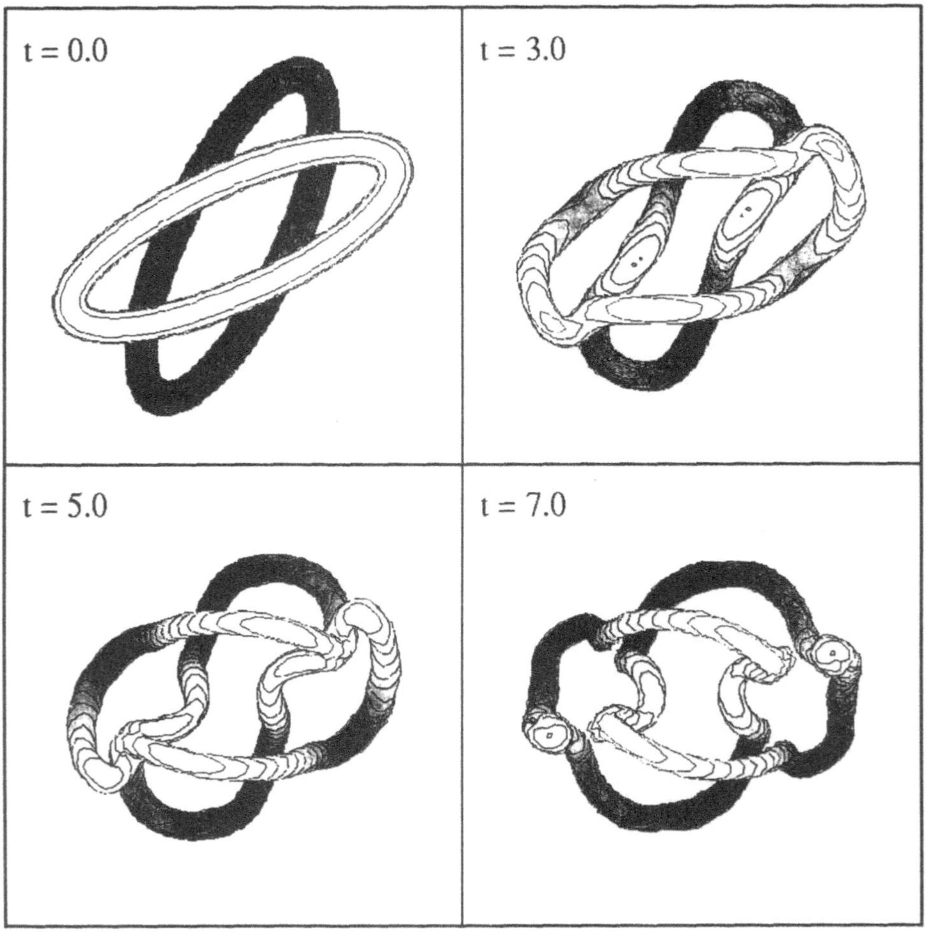

Figure 2: Interaction of two identical, elliptical vortex rings with major axes tilted by 45°. Initially one ring is slightly behind the other. They are travelling in the same direction. Two reconnections take place, and an intermediate linked state is produced.

3. TOPOLOGY CHANGE

Changes in 'topology' or connectivity are a common feature of fluid flows. Most of the 'lines' that we are used to thinking about, such as streamlines, timelines and pathlines, show a rich variation in topology with changes in flow parameters and initial conditions. Chaotic flows have intrinsically complex patterns of flow lines. In an overall sense, 'topology change' is something with which we are intimately familiar in fluid flows.

The type of topology change that we have in mind in connection with vortex flows is, however, quite special, primarily because of the correspondence in the inviscid limit of vortex lines and material lines. While many of the topology changes in the configuration of streamlines, etc., are essentially *kinematic* and purely *geometric,* topology changes in the pattern of vortex lines typically involve important *dynamical* information. For example, in 2D the merging of vortices to form ever larger structures is associated with the important phenomenon of an 'inverse' energy cascade in 2D turbulence. Similarly, the linking of vortex lines in 3D is associated with the production of helicity.

In 2D the various processes leading to topology change in the vorticity field are relatively well understood. The most important process is vortex merging, which can take a continuum of forms depending on the relative strengths of the vortices, but which in essence consists of like-signed vortices coming together in such a way that the number of distinct maxima in the vorticity distribution is reduced with time. Vortices of opposite sign pair rather than merge, forming tightly bound dipole structures that propagate through the fluid for long distances in a solitary wave type motion. The simple point vortex model discussed in §2 does not allow merging *per se,* although like-signed vortices can become trapped in bound states, orbiting endlessly, and effectively behaving as a single vortex. Point vortex dynamics does readily admit pairing, and the study of point vortex pair collision dynamics provides an interesting if somewhat formal example of the phenomenon of chaotic scattering (Eckhardt & Aref, 1988; Aref *et al.,* 1988).

In 3D the situation is much less clear cut. Certain 'canonical' cases of vortex reconnection have been studied in great detail, both experimentally and numerically, but a global view of the various possibilities of reconnection is only now being pursued. We shall make a few remarks about the reconnections that we have observed for interacting vortex rings using a newly developed, three-dimensional vortex-in-cell code (Zawadzki & Aref, 1991a).

For circular rings one of the key mechanisms is elucidated in the experiments on collisions of identical rings by Schatzle (1987), refining earlier work by Kambe & Takao (1971), and Oshima & Asaka (1975, 1977). The rings undergo an 'exchange interaction' in which half of each original ring ends up with half of the other ring, forming two new rings that propagate away from the interaction region. The entire process is symmetric, and at high Reynolds number is very close to being time-reversible, in the sense that if a state after reformation of the two rings is 'frozen,' and the vorticity reversed, it will upon 'release' retrace the forward process step by step modulo the appropriate rotation in space.

For identical rings moving towards one another along parallel directions the essential parameter to vary is the offset of the axes of propagation. For pure head-on collision indefinite expansion and stretching of the rings take place (possibly cut short by instability). When the rings are offset, however, two reconnections take place roughly at the points of intersection of the two initial circles projected onto a single plane. The resulting vortices now consist of pieces of the original vortices according to this geometrical ratio (see Zawadzki & Aref, 1991b, where the mixing of a scalar due to this process is studied).

If the scenario just described is perturbed, other interesting phenomena occur. For example, if instead of having two identical rings, we let the ratio of the circulations be 0.9:1.0, an 'exchange interaction' as for identical rings will still take place, but the 'excess' vorticity in one ring will be 'shed' in the form of smaller ring (Fig.1). If the ratio of circulations is smaller, e.g., 0.7:1.0, the stronger ring is so dominant, that the evolution of the weaker is reminiscent of the winding up of a passive scalar.

Richer behavior is seen in the cases of elliptical rings, where the internal dynamics of the rings, and its ability to 'resonate' with the reconnection process can play an important role. With elliptical rings it is possible to produce linking of initially unlinked vortex rings. The simplest initial condition that leads to this consists of two elliptical rings started quite close to one another and with major axes tilted by approximately 45°. The key feature of this initial state is to have the rings touch first at two points (rather than four), setting off reconnections there. In this way linking is achieved for a brief interval of time, and the helicity in the flow increases dramatically (Fig. 2).

We have explored some of the phase space of two identical, elliptical vortex rings computationally, and find a complex mix of reconnection with linking, reconnection without linking, and 'leapfrogging' motions (the latter have been known for circular rings for a long time, see, for example, the experiments by Yamada & Matsui, 1978).

A Hamiltonian theory of 'weak' interactions between circular vortex rings was given by Roberts (1972). With the insight that we now have into reconnections between colliding circular rings, it should be possible to construct an approximate dynamical description, analytically clumsy but fully programmable on a computer, that would follow a 'gas' of vortex rings. Small deflections of distant rings would be handled by Roberts' (1972) theory, and vortex rings that actually intersect would be reconnected 'by hand,' using one of the scenarios identified by detailed collision experiments, to form new rings. This, in spirit, would be similar to the procedure adopted by Schwarz (1977) in his simulations of superfluid turbulence, and would provide a realization of a model similar to the one considered long ago by Synge & Lin (1943), where homogeneous, isotropic turbulence was represented as a collection of interacting Hill's vortices. Further refinements would have to face the full topological complexity of the flow field, where vorticity is certainly not confined to a family of rings, in general. While such approximations in terms of specific vortical structures may seem rather drastic, it is not clear that they are any more artificial than sweeping assumptions made in statistical theories of turbulent flow.

4. 'PUNCTUATED' DYNAMICAL SYSTEMS

For a dynamical systems approach a topology change implies a reassignment of degrees of freedom to spatial structures. In some cases the number of degrees of freedom is changed as when two interacting rings of slightly different circulations shed a third ring during reconnection, or when two finite area vortices merge to one in a 2D flow. These phenomena suggest that the type of dynamical system to be explored is what Carnevale, Pomeau & Young (1990) refer to as 'punctuated' Hamiltonian systems. The basic idea is that the evolution of the system proceeds in two modes. One is a 'weak interaction' mode in which relatively simple propagation of well defined entities takes place. This mode is 'slow' and takes up most of the time of the evolutionary development of the system. The second is a 'strong interaction' mode during which realignment of degrees of freedom, topology changes, etc., take place. This mode is intermittent and fast.

'Punctuated' systems were considered in the context of colloidal aggregation by Smoluchowski (see Chandrasekhar, 1943, for a review; the original work is from 1915-16). In this case the slow mode is Brownian motion of colloidal particles. The fast mode is the actual aggregation process in which larger clumps are formed from smaller ones. An even simpler example is the gas of hard spheres. Here the spheres propagate ballistically between instantaneous collisions, which constitute the only interactions. Another example with completely different physics but conforming to a similar 'punctuated' phenomenology is the evolution of a 2D foam (Weaire & Rivier, 1984; Aref & Herdtle, 1990; Herdtle & Aref, 1991). Here the slow mode is the diffusion of gas from bubble to bubble, while the fast mode is the reconnection of soap films and subsequent elastic relaxation. Two-dimensional turbulence appears to provide a third example (Carnevale *et al.*, 1991), with the slow mode being the propagation and advection of individual vortices and vortex pairs, and the fast mode being the processes of merging, pairing and tearing.

Vortex dynamics does not, of course, conform to this picture as cleanly as, say, hard spheres. The fast mode is not instantaneous, and the fusion or reconnection of vortices can leave behind considerable 'debris' in the form of low-intensity vorticity. Much emphasis has been placed on elucidating this structure (cf. Melander & Zabusky, 1988; Kida, Takaoka & Hussain, 1989). In three dimensions this vorticity may become amplified due to vortex stretching. However, we know from general considerations on three-dimensional turbulence that the small scales arise by a cascade of energy from larger scales. This both sets bounds on their energy content, and suggests that tracking the large scales, which ultimately feed the small scales through the cascade, is a legitimate approach. Such arguments on the dynamical significance of the 'large scales' have an impassioned history in the literature on coherent structures in shear flow turbulence.

One general observation worth recording is that topology changes in 2D flows tend to reduce the number of degrees of freedom, in the sense that one has a coarsening of the structures present in the flow field. In 3D by contrast one finds a number of processes that increase the number of degrees of freedom. This is in general agreement with the overall sense of energy transport in the prevailing 'cascades' of 2D and 3D homogeneous, isotropic turbulence.

We conclude with a quote from the paper by Saffman & Meiron (1986) on the potential of capturing turbulent flows in 3D with a finite number of vortex degrees of freedom: 'But if the interest lies in the possibility of representing, with at least qualitative accuracy..., the dynamics of a general, three-dimensional, rotational, incompressible velocity field (e.g., the motion of coherent structures) by a system with a few degrees of freedom, and this is the hope for applications of recent work in dynamical systems to turbulent flow, then it appears that the vorton representation is seriously flawed.' The quote is admittedly taken somewhat out of context, since the paper deals with a specific type of singularity (3D δ-functions of vorticity, which they call 'vortons'). It is also, in our view, overly pessimistic. A stumbling block for the application of dynamical systems to turbulent flows has been the issue of identifying a small number of degrees of freedom. The notion of punctuated systems may relax this stringent requirement considerably. Now the turbulence is at each instant dominated by a certain set of degrees of freedom, but we are prepared to allow this set to change repeatedly as the flow evolves. In fact, the frequency of such changes is now promoted as an essential ingredient in the dynamics.

Acknowledgements

This work and participation in the Colloquium were supported by National Science Foundation PYI award CTS 84-51107, and by DARPA/ACMP URI grant N00014-86-K-0758. The computations were performed at the San Diego Supercomputer Center (SDSC). SDSC is supported by the National Science Foundation.

References

Abrashkin, A. A. & Yakubovich, E. I. 1985 Planar rotational flows of an ideal fluid. *Sov. Phys. Dokl.* **29**, 370-371.

Aref, H. 1983 Integrable, chaotic and turbulent vortex motion in two-dimensional flows. *Ann. Rev. Fluid Mech.* **15**, 345-389.

Aref, H. 1990 Chaotic advection of fluid particles. *Phil. Trans. Roy. Soc. (London) A* **333**, 273-289.

Aref, H. & Flinchem, E. P. 1984 Dynamics of a vortex filament in a shear flow. *J. Fluid Mech.* **148**, 477-497.

Aref, H. & Herdtle, T. 1990 Fluid Networks. In *Topological Fluid Mechanics*, H. K. Moffatt & A. Tsinober eds., Cambridge University Press, pp.745-764.

Aref, H., Jones, S. W., Mofina, S. & Zawadzki, I. 1989 Vortices, kinematics and chaos. *Physica D* **37**, 423-440.

Aref, H., Kadtke, J. B., Zawadzki, I., Campbell, L. J. & Eckhardt, B. 1988 Point vortex dynamics: Recent results and open problems. *Fluid Dyn. Res.* **3**, 63-74.

Aref, H., Rott, N. & Thomann, H. 1992 Gröbli's solution of the three-vortex problem. *Ann. Rev. Fluid Mech.* **24**, 1-20.

Aref, H. & Zawadzki, I. 1991 Linking of vortex rings. *Nature* **354**, 50-53.

Batchelor, G. K. 1967 *An Introduction to Fluid Dynamics*. Cambridge Univ. Press. Ch. 7.

Betchov, R. 1965 On the curvature and torsion of an isolated vortex filament. *J. Fluid Mech.* **22**, 471-479.

Buttke, T. F. 1989 The observation of singularities in the boundary of patches of constant vorticity. *Phys. Fluids A* **1**, 1283-1285.

Carnevale, G. F., McWilliams, J. C. & Pomeau, Y., Weiss, J. B. & Young, W. R. 1991 Evolution of vortex statistics in 2-dimensional turbulence. *Phys. Rev. Lett.* **66**, 2735-2737.

Carnevale, G. F., Pomeau, Y. & Young, W. R. 1990 Statistics of ballistic agglomeration *Phys. Rev. Lett.* **64**, 2913-2916.

Chandrasekhar, S. 1943 Stochastic problems in physics and astronomy. *Rev. Mod. Phys.* **15**, 1-89.

Da Rios, L. S. 1906 Sul moto d'un liquido indefinito con un filetto vorticoso di forma qualunque. *Rend. Circ. Mat. Palermo* **22**, 117-135.

Deem, G. S. & Zabusky, N. J. 1978 Vortex waves: Stationary 'V states,' interactions, recurrence, and breaking. *Phys. Rev. Lett.* **40**, 859-862.

Dritschel, D. G. & McIntyre, M. E. 1990 Does contour dynamics go singular? *Phys. Fluids A* **2**, 748-753.

Eckhardt, B. & Aref, H. 1988 Integrable and chaotic motions of four vortices II: Collision dynamics of vortex pairs. *Phil. Trans. Roy. Soc. (London) A* **326**, 655-696.

Hama, F. R. 1963 Progressive deformation of a perturbed line vortex filament. *Phys. Fluids* **6**, 526-534.

Hama, F. R. 1988 Genesis of the LIA *Fluid Dyn. Res.* **3**, 149-150.

Hasimoto, H. 1972 A soliton on a vortex filament. *J. Fluid Mech.* **51**, 477-485.

Hasimoto, H., Ishii, K., Kimura, Y. & Sakiyama, M. 1984 Chaotic and coherent behaviors of vortex filaments in bounded domains. In *Turbulence and Chaotic Phenomena in Fluids*, T. Tatsumi ed., IUTAM/Elsevier Science Publ., pp.231-237.

Herdtle, T. & Aref, H. 1991 Numerical experiments on two-dimensional foam. *J. Fluid Mech.* (In Press).

Kambe, T. & Takao, T. 1971 Motion of distorted vortex rings. *J. Phys. Soc. Japan* **31**, 591-599.

Kida, S. 1981 A vortex filament moving without change of form. *J. Fluid Mech.* **112**, 397-409.

Kida, S. 1982 Stability of a steady vortex filament. *J. Phys. Soc. Japan* **51**, 1655-1662.

Kida, S., Takaoka, M. & Hussain, F. 1989 Reconnection of two vortex rings. *Phys. Fluids A* **1**, 630-633; Corrigendum, *ibid.* **2** (1990) 638.

Kimura, Y., Zawadzki, I. & Aref, H. 1990 Vortex motion, sound radiation, and complex time singularities. *Phys. Fluids A* **2**, 214-219.

Krasny, R. 1986 Desingularization of periodic vortex sheet roll-up. *J. Comp. Phys.* **65**, 292-313.

Lin, C. C. 1943 *The Motion of Vortices in Two Dimensions*. Toronto Univ. Press.

Meiron, D. I., Baker, G. R. & Orszag, S. A. 1982 Analytic structure of vortex sheet dynamics. Part 1. Kelvin-Helmholtz instability. *J. Fluid Mech.* **114**, 283-298.

Melander, M. & Zabusky, N. J. 1988 Interaction and 'apparent' reconnection of 3D vortex tubes via direct numerical simulations. *Fluid Dyn. Res.* **3**, 247-250.

Moffatt, H. K. 1969 The degree of knottedness of tangled vortex lines. *J. Fluid Mech.* **35**, 117-129.

Moore, D. W. 1979 The spontaneous appearance of a singularity in the shape of an evolving vortex sheet. *Proc. Roy. Soc. (London) A* **365**, 105-119.

Oshima, Y. & Asaka, S. 1975 Interaction of two vortex rings moving side by side. *Nat. Sci. Rep. Ochanomizu Univ.* **26**, 31-37.

Oshima, Y. & Asaka, S. 1977 Interaction of two vortex rings along parallel axes in air. *J. Phys. Soc. Japan* **42**, 708-713.

Ricca, R. L. 1991 Rediscovery of Da Rios equations. *Nature* **352**, 561-562.

Roberts, P. H. 1972 A Hamiltonian theory for weakly interacting vortices. *Mathematika* **19**, 169-179.

Saffman, P. G. & Meiron, D. I. 1986 Difficulties with three-dimensional weak solutions for inviscid incompressible flow. *Phys. Fluids* **29**, 2373-2375.

Schatzle, P. R. 1987 *An Experimental Study of Fusion of Vortex Rings*. Ph. D. thesis, California Institute of Technology.

Schwarz, K. 1977 Theory of turbulence in superfluid ^{4}He. *Phys. Rev. Lett.* **38**, 551-554.

Su, C. H. 1979 Motion of fluid with constant vorticity in a singly-connected region. *Phys. Fluids* **22**, 2032-2033.

Synge, J. L. & Lin, C. C. 1943 On a statistical model of isotropic turbulence. *Trans. Roy. Soc. Canada* **37**, 45-79.

Weaire, D. & Rivier, N. 1984 Soap, cells and statistics - Random patterns in two dimensions. *Contemp. Phys.* **25**, 59-99.

Yamada, H. & Matsui, T. 1978 Preliminary study of mutual slip-through of a pair of vortices. *Phys. Fluids* **21**, 292-294.

Zawadzki, I. & Aref, H, 1991a Numerical experiments on three-dimensional hydrodynamics using the vortex-in-cell method. *J. Comp. Phys.* (submitted).

Zawadzki, I. & Aref, H. 1991b Mixing during vortex ring collision. *Phys. Fluids A* **3**, 1405-1410.

Zeitlin, V. 1991 On the backreaction of acoustic radiation for distributed two-dimensional vortex structures. *Phys. Fluids A* **3**, 1677-1680.

Finite-dimensional dynamics and chaos in fluid flows

Tom Mullin

1. Introduction

We present a discussion of the role of finite-dimensional dynamics and chaos in interpreting nonlinear fluid mechanical motion. The discussion will be restricted to two examples of fluid flows which have been studied by the author and which appear to be understandable in terms of ideas based in modern thinking in dynamical systems. The specific examples we have chosen are the flow between concentric rotating cylinders commonly called the Taylor-Couette problem and the flow through a nominally two dimensional sudden symmetric expansion. The first of these is a so called 'closed flow' problem and the second is an example of an 'open flow' where disturbances can grow as they are carried down stream. The aim of the present article is to focus attention on the practicalities of applying the abstract concepts of finite-dimensional dynamics to the experimental study of fluid flows. We will show how a careful consideration of the important symmetries of these problems can lead to the uncovering of structurally stable local organising centres for the global low-dimensional dynamical behaviour.

2. Flow Through a Sudden Expansion

An example of a pitchfork symmetry-breaking bifurcation in an fluid mechanical system is that of flow through a two dimensional sudden symmetric expansion. Details of a numerical and experimental study can be found in Fearn, Mullin and Cliffe (1990). At small Reynolds numbers ($Re \leq 35$) the flow is symmetric in that the recirculation regions after the expansion are the same size. When the Reynolds number is increased above this range then the flow is no longer symmetric and one of the recirculating regions is always much longer than the other. There must now be two stable flow states on the symmetric boundary conditions and these have arisen through a symmetry breaking bifurcation. Each stable flow is a mirror image of the other and the symmetric flow is unstable. We show in figure 2.1 a flow visualization photograph of one of the pair of flows along with one of the respective streamline plots at Re =80. The asymmetry in the first recirculations after the expansion is clear and continues to grow at higher Reynolds numbers but then saturates when other asymmetric recirculations grow downstream.

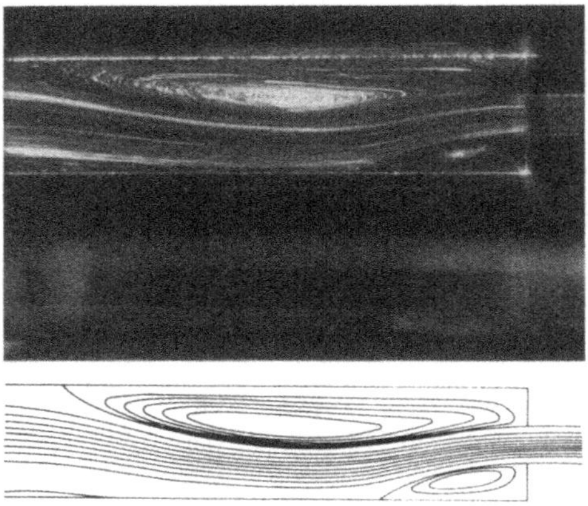

Figure 2.1 Flow visualisation photograph and streamline plot for an asymmetric state in the sudden expansion at Re = 80.

The form of the bifurcation has been calculated using numerical bifurcation techniques. We show the results of this calculation in figure 2.2 where it can be seen that the bifurcation has the form of a simple pitchfork. At low Re there is a unique trivial solution which corresponds to the symmetric flow state. There is then an exchange of stability between this and a pair of asymmetric flows through the pitchfork bifurcation at Re_{crit} =40.5. Experimental estimates of the bifurcation can be made by measuring some feature of the flow which distinguishes between the different flow states. Here we choose to measure, using laser-Doppler velocimetry, the vertical velocity on the mid-plane as the functional which discriminates between flow types. For a perfectly symmetric flow the vertical velocity component will be zero, and positive or negative for the respective asymmetric states. The results of the measurements are shown superposed on the numerical results in figure 2.2. There is clearly some agreement far away from the bifurcation point but obvious disagreement close to the numerically determined critical point. The reason for the disagreement is the effect of imperfections in the experiment which disconnects the bifurcation. In the present case the numerical calculations are carried out on perfectly symmetric boundary conditions which cannot be realised in the physical experiment.

The imperfection in the experiment can be modelled by adding a small offset to the inflow boundary in the numerical scheme. This was done for the results shown in figure 2.3 and the overall agreement between calculation and experiment is now greatly improved. The numerical offset is equivalent to ≈ $1/20\ mm$ in the experiment which is on the limit of the accuracy of the construction of the apparatus. Surprisingly, the disconnection is ≈ 13% in *Re* which is an order of magnitude larger than the corresponding effects in Taylor-Couette flow (Mullin (1991)). As a consequence of the presence of this imperfection, only one of the two flow states can be reached by continuous increase in *Re*. The other state can only be reached by a sudden start of the experiment above *Re* ≈ 50. Thus after the flow-control valve is opened suddenly above this value there is a 50/50 chance of achieving the disconnected state after an initial transient. Once the disconnected solution has been reached, the Reynolds number can then be reduced and an estimate of the limit point of the fold made. The above numerical and experimental results conclusively show that there is a pitchfork bifurcation enthuse open flow system.

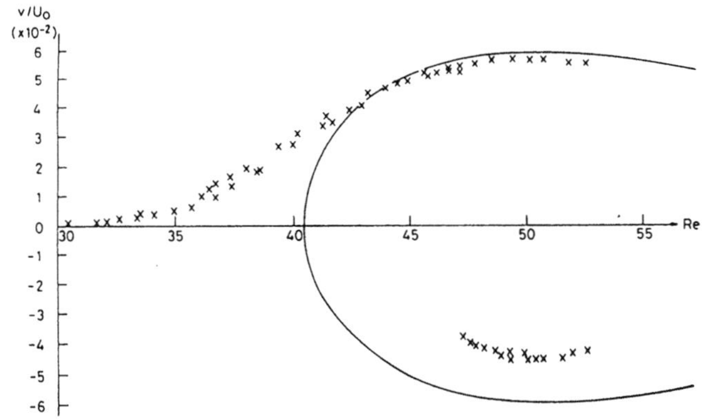

Figure 2.2 Pitchfork bifurcation in the sudden expansion flow. The crosses are experimental points and perfect symmetry has been assumed in the numerics.

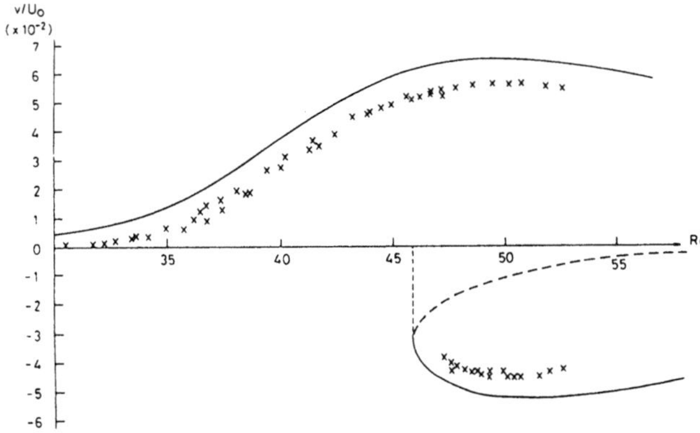

Figure 2.3 Experimental and numerical comparison for the disconnected pitchfork. The imposed numerical imperfection is equivalent to ~ $1/20\ mm$ in the experiment.

If the Reynolds numbers is now increased further then time-dependent motion sets in above $Re \approx 152$ as shown by Fearn et al. Just prior to the onset of the time-dependence the steady flow field in the experiment has some obviously three-dimensional features. In addition there is numerical evidence of Sobey and Drazin (1986) for the presence of a subcritical Hopf bifurcation in the related problem of two-dimensional smooth expansion. However, it has not yet been proven if the time-dependent flow arises through such a mechanism in the experiment. The time-dependence which is experimentally observed is not strictly periodic as can be seen from its power spectrum presented in figure 2.4. There is some width to the spectral peak indicating a variation in the frequency about a mean value. In addition, if we reconstruct the phase portrait from the experimental time series, then we obtain the result shown in figure 2.5. There is no obvious structure in the phase portrait and so we might conclude that low-dimensional dynamical behaviour is not present in this system as it undergoes 'natural' transition to turbulence. Finally, there is

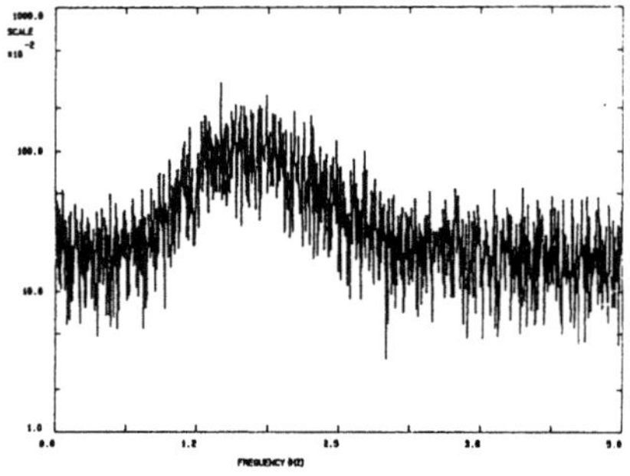

Figure 2.4 Power spectrum for the vertical flow component in the sudden expansion taken when the flow has just become time-dependent.

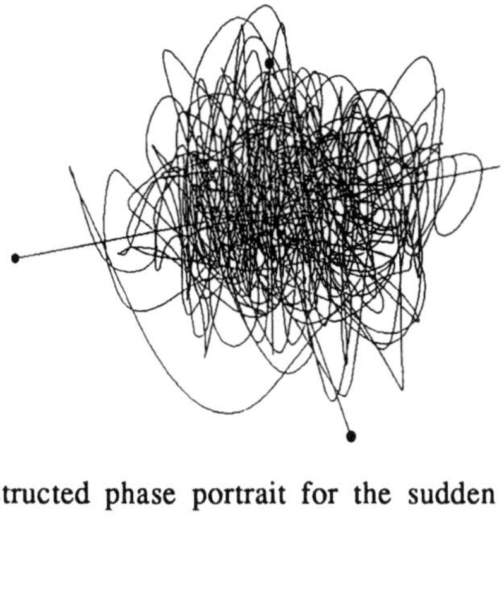

Figure 2.5 Reconstructed phase portrait for the sudden expansion just above transition.

no evidence for hysteresis in the experiment as there is a sharp transition between steady and time-dependent flows for both increase and decrease of *Re*. We will return to this experiment later and see that there is in fact evidence for low-dimensional chaos in a slightly modified version of the system.

3. Taylor-Couette flow

The Taylor-Couette flow problem is concerned with fluid motion in the gap between concentric rotating cylinders. Here we will consider the case of the inner cylinder rotating only. One feature of the Taylor-Couette system is that it exhibits a rich variety of hydrodynamic instabilities. The lowest order instability results in the formation of cellular structure in the flow and was first studied by Taylor in 1923. There have been many papers written on the topic since that time and a good review of the subject is to be found in DiPrima and Swinney (1981). In more recent years, the problem has been extensively studied in relation to the application of the ideas of finite-dimensional dynamics to fluid flows. It is this latter issue that we will address here but before proceeding with discussion of these complicated flows we will first review the fundamental problem of the steady bifurcation set associated with the cellular structure.

3.1 The Appearance of Cells

There are three control parameters which govern the Taylor-Couette system. These are the radius ratio $\eta = r_1 / r_2$, the aspect ratio $\Gamma = l / d$, where l is the length of the annular domain and d the gap width, and the Reynolds number which is defined as follows;

$$Re = \frac{\Omega r_1 d}{\nu}$$

where Ω is the angular speed of the inner cylinder, radius r_1, d is the width of the cylindrical gap and $\nu = \mu / \rho$ is the kinematic viscosity of the fluid. A schematic diagram of the experimental arrangement is presented in figure 3.1. The radius ratio is fixed in any experiment and the latter two parameters define a control plane. They are precisely maintained in the experiment by accurate construction of the apparatus, fine control of the motor speed and close regulation of the temperature. This is important because critical bifurcation sequences can be covered by very narrow spans of control space. Observations are either made using flow visualization or a point measurement of a velocity component obtained using a laser-Doppler-velocimeter.

At small *Re* the fluid is dragged around by the motion of the inner cylinder so that there is a shear across the fluid layer. If particles are added to the flow which reflect the incident light, then this flow regime will give an almost uniform reflectance except for some three-dimensional effects at the ends of the cylinders. When *Re* is increased above a critical

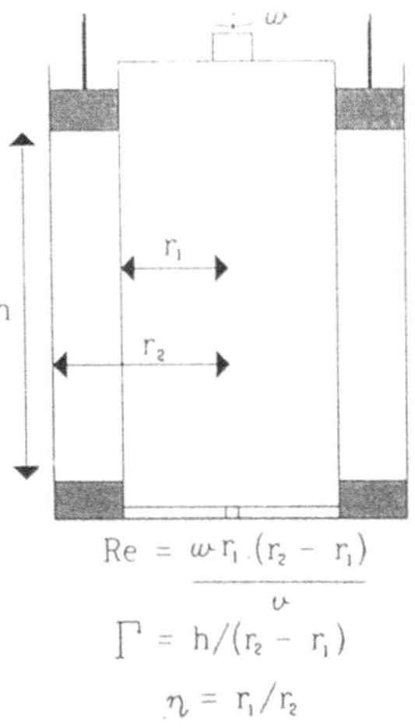

Figure 3.1 Schematic of the Taylor-Couette flow apparatus.

range a banded or cellular structure appears along the length of the cylinder as shown in figure 3.2 and the flow has the appearance of a series of doughnuts stacked upon each other.

Figure 3.2 Front view of Taylor cells.

In a brilliant combination of classical analysis on the Navier-Stokes equations and fluid experiment, G. I. Taylor (1923) gave an explanation for the first appearance of the cells as a hydrodynamic instability. In his model he considered the cylinders to be infinitely long so that periodic boundary conditions could be applied. Then the cellular instability can be represented by simple trigonometric functions which together with the assumption of an arrow gap between the cylinders makes the problem amenable to analysis.

In modern bifurcation parlance, the situation can be thought of as a pitchfork bifurcation from the trivial rotary shear state to an axisymmetric cellular flow as shown in figure 3.3.

Each branch of the pitchfork would correspond to a half wavelength shift with respect to the other along the length of the cylinder as indicated. This is of course easily achieved on the periodic boundary conditions of the model where there is translational invariance in the assumed periodicity but it is a situation which cannot be realised in the laboratory. Thus, the central portion of the flow may appear to have an approximately periodic structure in the experiment but there will always be departures from this at the ends of the physical system, no matter how large the apparatus.

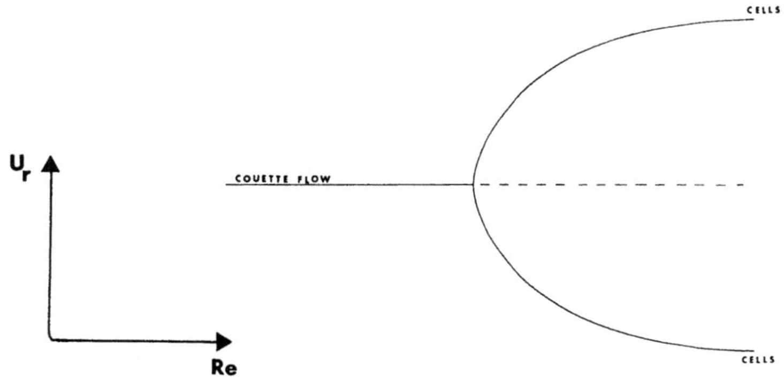

Figure 3.3 Pitchfork bifurcation for the Taylor-Couette model on periodic boundary conditions.

We have already seen how the presence of physical imperfections can be accommodated in bifurcation theory in the example of the flow through the sudden expansion. Indeed, as mentioned above, there are three dimensional flow effects at the ends of the flow domain which are precursors to the appearance of cells. Thus it might seem that we could represent the effects of the ends as a simple softening of the original pitchfork. We can measure the effects of the imperfection by obtaining estimates of the lower limits of

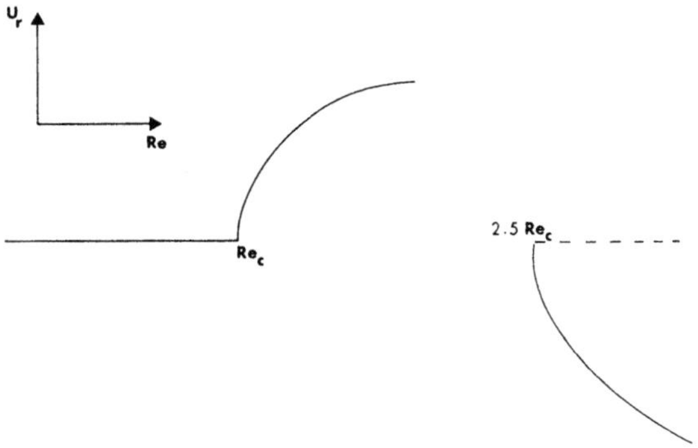

Figure 3.4 Schematic of the disconnected bifurcation for Taylor-Couette flow with realistic boundary conditions.

stability of the disconnected portion of the pitchfork. This was first carried out by Benjamin (1978) and later extended both numerically and experimentally in a number of investigations. (See Mullin (1991) for a review of this work.).

The central result of all of these studies is that the bifurcation is disconnected by a factor of ≈ 2.5 as shown schematically in figure 3.4. Thus the present case is not an example of the simple softening of a pitchfork such as we saw in the sudden expansion results. The reason is that the essential symmetry of the model which has been lost in the experiment is that of translational invariance whereas in the previous example it is the loss of geometrical mirror-plane symmetry that is important. In other words, in the Taylor-Couette problem the physical experiment is a major perturbation on the pitchfork in the model.

We stress the importance of this fundamental issue here because it has very important consequences for the dynamical behaviour. It is well known that low-dimensional dynamical behaviour is often organised by the presence of double bifurcation (codimension-2) points in a system (see Guckenheimer and Holmes (1983)). An example of this could be the interaction of a pitchfork with a Hopf bifurcation which is structurally stable in the presence of a small disconnections of the pitchfork. This particular situation has been examined in-depth by Healey, Broomhead, Cliffe, Jones, and Mullin (1991) for the case of a nonlinear electronic oscillator where they found evidence for 'Silnikov' homoclinic chaos. In the Taylor-Couette problem, if the Reynolds number is increased beyond the first appearance of cells then a travelling wave instability arises through a Hopf bifurcation as shown by Davey, DiPrima and Stuart (1968). Therefore, one might expect an interaction between the disconnected pitchfork and the Hopf as the control parameters are varied in the experiment.

However, since the pitchfork bifurcation has to all intents and purposes been destroyed then we might expect that any associated finite-

dimensional dynamics would also disappear. Nevertheless, there is in fact a simple geometrical pitchfork symmetry-breaking bifurcation as shown by Mullinand Cliffe (1986), and the flows which arise there break the up-down mirror symmetry of the physical boundary conditions. This new bifurcation helps organise the dynamics and gives rise to low-dimensional dynamical behaviour. It cannot be accommodated in the periodic model which is often used to interpret the Taylor-Couette experiment and requires a proper account of the realistic boundary conditions in any numerical investigation. Thus it is only by considering this pitchfork bifurcation that progress can be made in understanding the origins of global low-dimensional dynamical behaviour.

3.2 Low-dimensional Chaos

We now give an outline of a study whose primary aim was to obtain a better understanding of the low-dimensional dynamics arising out of steady state-Hopf interactions. The full details of the experimental and numerical study are given in Tavener, Mullin and Cliffe (1991). We will concentrate on a particular codimension-2 point where there is an interaction between Hopf bifurcations and the limit point of a fold. A sketch of the relevant sequence of bifurcations which is derived from the numerical and experimental results is shown in figure 3.5.

The folded surface is one half of a subcritical bifurcation to a pair of asymmetric four-cell flows. This is a realistic representation of the physical situation as the bifurcation will be disconnected in the experiment. Along the top surface of the fold the asymmetric flow firstly loses stability to a travelling wave through a Hopf bifurcation at H. The detection of this bifurcation is the limit of the numerical work which at present cannot follow the three-dimensional time-dependent flows which arise at such points. Experimental evidence is then found for a secondary Hopf bifurcation and the subsequent state can now be represented as motion on a torus in phase space. If the second parameter in the problem, the aspect ratio Γ is varied then the Hopf bifurcation point H moves from the stable to the unstable surface. Thus there is a codimension-2 point in (R, Γ) space where the two bifurcation points coincide.

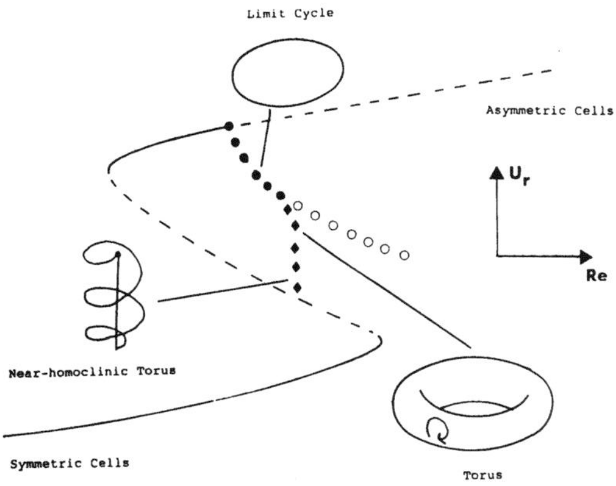

Figure 3.5 Sketch of the bifurcation sequence near a codimension-2 point in Taylor-Couette flow.

So far the sequence of events is consistent with the algebraic normal forms discussed in Chapter 7 of Guckenheimer and Holmes. In addition Cliffe (1991)has performed a centre manifold reduction of the discretised Navier-Stokes equations about the codimension-2 point and evaluated the coefficients in the normal form. He has established that the present situation corresponds to CASE III in Guckenheimer and Holmes and so we should expect to find homoclinic tori in the neighbourhood of the codimension-2 point. We show in figure 3.6 a series of experimental Poincare maps which were recorded in a parameter range very close to the codimension-2 point. It is evident that there is a period doubling of one of the frequencies on the torus before chaos appears. This cannot be accounted for in the normal form which is three-dimensional and the period-doubling on the torus must take place in a minimum of four dimensions.

If we now consider the reconstructed phase portraits shown in figure 3.7 it is clear that there is very strong evidence for homoclinic tori. We show the closest approach to homoclinicity in figure 3.7(a) when the motion is regular. This reconstructed phase portrait is qualitatively the same as the one presented in Chapter 7 of Gickenheimer and Holmes yet it is very away in parameter space from the codimension-2 point. Thus the apparent agreement maybe coincidental. Finally, we show a chaotic phase portrait in

Figure 3.6 Series of experimental Poincare maps taken near the codimension-2 point showing period-doubling on a torus.

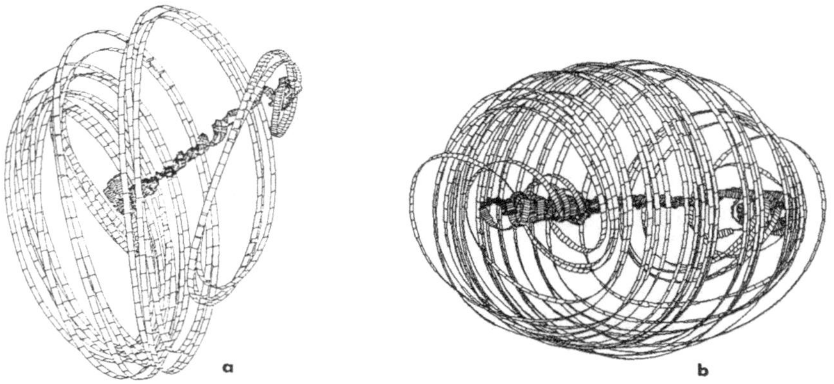

Figure 3.7 Reconstructed phase portraits at point of closest approach to homoclinicity (a) Regular behaviour, (b) Chaotic motion.

figure 3.7(b) which was measured in a neighbouring parameter range. Further discussion of these results can be found in Mullin and Price (1989).

4. Concluding Remarks

We have shown that the notions of finite-dimensional chaos are relevant to the understanding of the transition to weakly irregular states in a carefully controlled closed fluid flow system. The nontrivial dynamical motion is organised by the underlying bifurcation structure so that the global behaviour can be understood in terms of some local events in parameter space. This observation makes the possibility of establishing a connection between the governing equations of motion and the considerably simplified dynamical models tractable. Despite these obvious successes there are considerable difficulties to be overcome even for this simple system as there are other routes to chaos in neighbouring parameter ranges including intermittencies as discussed by Price and Mullin (1991).

If the above approach is applied to the more general problem of open fluid flows then high-dimensional behaviour is encountered at even modest Reynolds numbers. This is probably because only a few spatial modes are allowed to participate in the interactions which produce the chaos in the Taylor-Couette problem. However, we end on a note of optimism for future application of the ideas of low-dimensional dynamical behaviour to open fluid flows. As discussed above, the flow through a sudden expansion appeared to show no evidence for low-dimensional behaviour in the transition phase. However, recent work by Madden and Mullin (1991) has shown that finite-dimensional dynamics including period-doubling, homoclinicity and intermittency can be induced by adding a small periodic disturbance to the inflow. We show an example of this in figure 4.1 in the

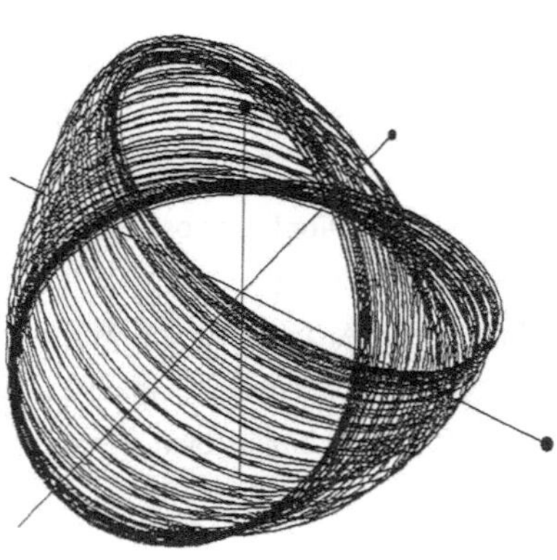

Figure 4.1 Three dimensional Poincare map obtained from the reconstructed phase portrait for modulated flow through a sudden expansion showing motion on a torus.

form of the reconstructed phase portrait of a torus. The two frequencies on the torus can be made to lock by either changing the frequency of the applied oscillation or the Reynolds number. In either case, finite dimensional chaos results which we believe is one of the first experimental observations of its kind. We note that this low dimensional behaviour is different to that observed by Olinger and Sreenivasan (1988) or Williams-Stuber and Gharib (1990) where there is a natural periodicity in the flow field. In the present case there is no obvious periodic motion in the unforced flow. Thus low-dimensional dynamical behaviour may well be relevant to the understanding of a much wider class of weakly disordered flows.

Acknowledgements

The main part of our research programme is supported through the SERC 'Nonlinear Initiative' to whom we are greatly indebted. I would also like to thank J. Kobine, F. Madden and T. Price for their help with the preparation of this manuscript.

References

Benjamin T.B. 1978. Bifurcation phenomena in steady flows of a viscous liquid. 1. Theory. 2. Experiments. Proc. Roy. Soc. Lond. A359, pp 1-26 and 27-43.

Davey A.,DiPrima R.C. and Stuart J.T. 1968. On the instability of Taylor vortices. J. Fluid Mech. 31, pp 17-52.

DiPrima R.C. and Swinney H.L. 1981. Instabilities and Transition in the flow between concentric rotating cylinders. In, Hydrodynamic Instabilities and the Transition to Turbulence (ed. H.L. Swinney and J.P. Gollub: Springer)

Fearn R.M., Mullin T. and Cliffe K.A. 1990. Nonlinear flow phenomena in a symmetric sudden expansion. J. Fluid Mech. 211, pp 595-
608.

Guckenheimer J. and Holmes P. 1983. Nonlinear oscillations, dynamical systems, and bifurcations of vector fields. (Springer-Verlag)

Healey J.J.,Broomhead D.S.,Cliffe K.A.,Jones R. and Mullin T. 1991. The origins of chaos in a modified Van der Pol oscillator. PhysicaD 48,pp 322-339.

Madden F. and Mullin, T. 1991. Finite-dimensional dynamical behaviour in an open channel flow. (In preparation)

Mullin T. and Cliffe, K.A. 1986. Symmetry breaking and the onset of time-dependence in fluid mechanical systems. In, Nonlinear Phenomena and Chaos(ed. S. Sarkar: Hilger)

Mullin T. and Price T.J. 1989. An experimental observation of chaos arising from the interaction of steady and time-dependent flows. Nature 340 pp 294-296.

Mullin T. 1991. Finite dimensional dynamics in Taylor-Couette flow. IMA J. Applied Math. 46, pp 109-119

Olinger D.J. and Sreenivasan K.R. 1988. Nonlinear dynamics in the wake of an oscillating cylinder. Phys. Rev. Lett. 60, pp 797-799.

Price T.J. and Mullin T. 1991. An experimental observation of a new type of intermittency. PhysicaD 48 pp 29-52.

Sobey I.J. and Drazin P.G. 1986. Bifurcations of two-dimensional channel flow. J. Fluid Mech. 151 pp 395-410

Tavener S.J., Mullin T. and Cliffe K.A. 1991. Bifurcation phenomena in a rotating annulus. J. Fluid Mech. 229 pp 483-498.

Taylor G.I. 1923. Stability of a viscous fluid contained between two rotating cylinders. Phil. trans. Roy. Soc. A223 pp 289-343.

Williams-Stuber K. and Gharib M. 1990. Transition form order to chaos in the wake of an airfoil. J. Fluid Mech. 213 pp 29-57.

Tom Mullin
Clarendon Laboratory
University of Oxford
U. K.

Discussion

Lesieur: I will ask the first question. Your title is: "How long is the way from chaos to turbulence?" I didn't see many perspectives in what you presented.

Zaslavsky: The Lagrangian approach gives us a possibility to get crucial information about intermittency and scaling. It can be studied in a deep and detailed way. It is necessary to consider some kind of flows, treat them as dynamical systems and try to understand what is happening. The Lagrangian approach is one of the appropriate ways to study intermittency and the interaction between large and small scales. There is also some possibility to transfer these properties from Lagrangian chaos to Eulerian turbulence. It should also be possible to observe structure decomposition from large scales to small ones in the Lagrangian frame. This statement can be made because of self-similarity or quasi self-similarity of the turbulent process.

Frisch: I've always been advocating to do somewhat less statistical studies in place of somewhat more deterministic studies. I'm still a bit skeptical about the applicability of such ideas for the questions which are at the core of this meeting.
In discussing your talk there is just one point perhaps I would like to mention. It was noticed by Arnold that if you set up 3-D flows by applying a force, you can sometimes get an exact solution. The ABC flow can be seen as an exact solution of the 3-D Navier-Stokes with a superimposed force, which itself has the same structure as the ABC flow. Then you increase the Reynolds number and study bifurcations. Arnold conjectured that flows which have Lagrangian chaos are going to become chaotic in the Eulerian sense sooner (in Reynolds numbers) than flows which do not already exhibit Lagrangian chaos. This has not yet been checked, it may also not be true. What's your feeling?

Zaslavsky: I don't think that that's the only possibility. I am fairly optimistic about studying the Lagrangian picture as far as possible.

Frisch: You say one should not just consider things like intermittency from a phenomenological view point. We've seen enough models and we want some real theory that was your message. I certainly share that view, but I don't have the feeling that we are close to such a theory.

Zaslavsky: Let us be in consistence with the multi-fractal concept of the turbulence. That means that there are many fractal clusters imbedded into real space. Each passive particle or small eddy starting its orbit in some cluster should spend with a high probability a long time at the same cluster. There are different ways to describe the particle's orbit. One of them is to say what kind of diffusion it is. To simplify the answer, that is to say what μ is. Another cluster gives another value of μ. That leads us to the inhomogeneous distribution of the diffusional exponent μ in space, which describes the large scale asymptotics. At the same time, each cluster has the fractal dimension α, and the problem is to find a connection between α and μ. I call this connection a small-scale - large-scale mapping. This mapping can be described by some functional. That's a real picture of turbulence, but there is some simplification of it because it is not Eulerian turbulence. It is a Lagrangian turbulence. My position is to study these phenomena as far as possible.

Lesieur: This is a stupid question. Can you give me the answer or something about this μ for the problem of particle pair dispersion?

Zaslavsky: A possibility to find two particles separation does exist. There are several intermediate asymptotics for this μ. One of them is $\mu = \alpha/\beta$, where α is related to the fractal dimension and to separation of orbits in space; and β is the same thing, but in time. I cannot give the values of α and β right now, but there is a way of finding them the dynamical considerations. On the pictures that I showed, there are some islands and stochastic chaotic behaviour of the orbit outside the islands. There are not only these islands, but there are sub-islands. Any of these sub-islands has its own similar structure. So if the Hamiltonian is given, that means if the velocity field is given, this structure can be desribed in principle. As a result, one can get α and β, and consequently obtain μ.

Lesieur: I repeat, suppose you are asked to solve some problem related to the dispersion of some pollution in the atmosphere or some accidental release, then you have to know what α and β are. We are not speaking the same language. I still do not know how much the polution will desperse.

Zaslavsky: I cannot answer your question in such a form, but I would like to repeat that there is a possibility of finding the answer by using dynamical methods.

Moffatt: We're not talking about dynamics, we're talking about eddy diffusion of marked particles. The answer to Marcel's question was touched upon by Novikovs t^3 law of dispersion. And this sort of ABC-type of flow doesn't contain the essential structure of turbulence, which is eddies spread on many scales, ingredients of the flow. You have one scale. That's why Q-flows are characterised by that one scale and therefore your exponent μ always works out something between 1/2 and 1.

Zaslavsky: The picture is not so simple. The ABC-flow has three coefficients. Suppose two of them are equal and the coefficient C is the parameter. So the picture strongly depends on this parameter only. By changing the parameter we will change the phase portrait. But at the same time there are invisible changes in small scale structure. Hence, one has many scales, and small- sclae properties are governed by only one parameter C. The example with the ABC-flow demonstrates how to use dynamics in more complicated cases.

Moffatt: What are the bounds on the value of μ?

Zaslavsky: From one at the boundary to zero, but it may be narrower.

Moffatt: For turbulence with particles separating in the inertial subrange it's 3/2.

Zaslavsky: Yes, that is because in the Q-flows and particularly in the ABC-flow the nonlinear terms are absent.

Herring: In 2-D, pair dispersion is supposed to proceed by an exponential law because pair dispersal is sensitive to turbulent scales near the pair separation and because the energy spectrum is $\sim k^{-3}$. So that may be an entirely different kind of diffusion.

Zaslavsky: The 2-D stationary case has some special features. At the same time it can be considered in the same way as the 3-D case from the dynamical point of view. Turbulence occurs not only due to random forces. There is definitely an influence of the mechanisms of dynamical chaos, which provides relevance of the structural properties of the flow and correlations. The exponent in diffusional law reflects these correlations. The origin of flights is the same as the origin of such correlations. There is no exponential distribution of particle displacements. On the contrary, there are power-like tails in the distribution function. Due to these tails such exponents μ appear. All this can be studied in some regular way.

Moffatt: I would like to comment only on the reconnection of vortex tubes. There is a very nice argument, given by Michael Freedman and Mitch Berger recently, explaining why there are two kinds of reconnection: one starts with two tubes like that (Figure (a)), and they reconnect in this way the figure shown therein, and Hassan showed examples of this.

(a)

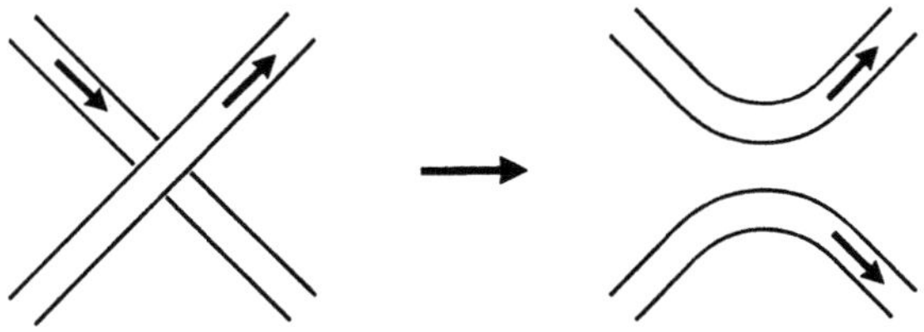

(b)

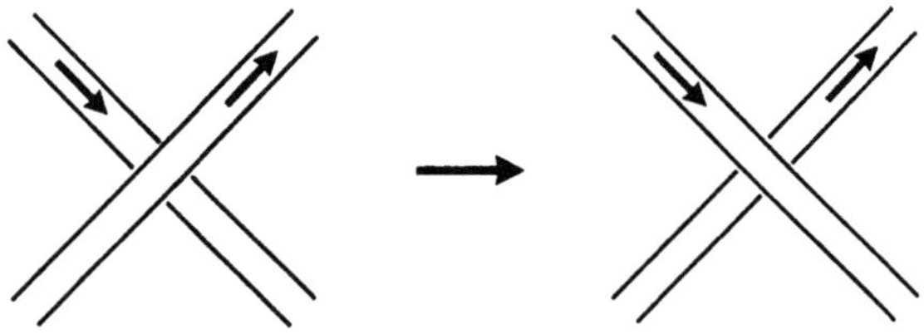

The other is more fundamental in a way: you start with one tube and another crossing it (Figure (b)), and it goes to one in which the tubes have reconnected in the way shown in this figure, and that is involved in your final example where the vortex tubes become linked. The Freedman-Berger argument is that the first type of reconnection occurs much more easily than the second type. The argument is very simple. You imagine the vortex tube has a finite cross-section. Let's split it up into a large number n of constituent tubes, n being chosen so that the scale of each constituent tube is under the scale at which viscous diffusion takes place on a time scale of order one. In order to achieve the reconnection (a), you can reconnect these constituent elemental tubes one at a time, and you'll need n reconnections in order to achieve that, perhaps 2n. Here in Figure b you've got n tubes in each one, and each one of the n has to cross each one of the other n, and so there are n^2 like n people shaking hands with everyone else. I'm not sure whether to accept this. It's so simple. But when n is large, take n to be of the order of 10^6 or so, and you can see that the first type of reconnection will occur with much greater facility than the second type of reconnection. You buy that argument? It's consistent with the fact that you find some difficulty in achieving this sort of thing (b) whereas the other type of reconnection (a) occurs without much effort.

Aref: It would seem to me that the one you have on top is dependent on essentially parallel vorticity and would be something that would happen on a diffusive time scale whereas the one on the bottom is an inertial-type thing and would happen maybe more quickly. I'm not sure if I completely understand the counting.

Herring: What is the importance of the adverse helicity. In reality if you do the numerical simulation, you will not find the adverse helicity unless you make some kind of artificial situation.

Aref: You're asking what's the importance of the contents of a fully developed turbulent flow. I don't know. Let me say one thing, something I didn't mention: You find that the helicity maximum when these things link is at best 70% of the topological linkeage. And that's interesting. It looks from the pictures, which are at 20% of the maximum vorticity, as if there's a clean linkage, but there is a lot of helicity that is associated with two other things. One is that the tubes inside the vortex are twisted and that gives you a helicity contribution, and there are still vortex lines that run around both tubes, so it's not just from the topological linking that you get some. Whether this is relevant to the real flow, I don't know. There aren't very many turbulent flows that are dominated by two vortex tubes. What happens when these things reconnect in that way is not quite so simple that they just cut in and do this. There is actually a tendency that they will become parallel right before the intersection point.

Jimenez: Could you comment a little on when two vorticies intersect at different angles? Some of them are parallel, some are perpendicular to each other, what happens in the intermediate range? What happens with two randomly oriented vorticies?

Aref: It depends on where you are in the phase space of two identical circular rings, and also two elliptical rings. Some answers to these questions we know. For example, we understand that the reconnection process for two identical rings depends on the angle of inclination. The Schatzle and Coles experiment was at about 15^0. You get very similar phenomena for 0^0,15^0, 30^0; but at 45^0 to 60^0 you don't get reconnection. There are similar things for interacting elliptical rings. I don't have it so precise that I can say if the two pieces of vorticity coming together are at such and such an angle, then you get one or the other, but I have some global results.

Jimenez: What is easier for two vortex lines? To pair or to reconnect?

Aref: I think it is easier for them to reconnect. At least it is faster.

Jimenez: That makes some sense because the pairing is adapted to larger, scales whereas the reconnection is adapted to smaller scales.

Moin: Getting back to the theme of the conference, I was wondering what are the prospects for a low dimensional description of turbulence. You mentioned about looking at the proper degrees of freedom.

Aref: I think that one could make progress with things dominated by coherent structures. I think that, for example, if I took a mixing layer and allowed 10 point vortices with periodic boundary conditions, and I did the dynamics with them; every time they merged I would reduce the number of degrees of freedom. I am allowed to rescale and replenish. So I am always allowed about 20 degrees of freedom, but not the same 20 degrees forever. I think I could do a reasonable calculation to obtain things like the Reynolds stress, etc. In a sense it is very similar to what you were trying to do in the boundary layer with your minimum box. You believed somehow that there is some minimum unit that you could take and identify, and then you could repeat that and you had a pretty good boundary layer. That's also the reason that people like Don Coles study, what they call "synthetic turbulence" by shooting vortex rings at each other.

Moin: To follow up on that: suppose you did put your 20 vortical structures in a box and let's say try to make this question of dimension concrete. Suppose you used the standard Liapunov exponent, what do we get for that? You still have to try to connect these degrees of freedom on a complete set of basic functions. You are considering these 20 vortices as 20 basic functions?

Aref: Right. The big stumbling block between chaos and turbulence, as I see it, is that the chaos problems deal with a finite number of degrees of freedom and turbulence is a multi-degree of freedom phenomenon. Now, the only hope of reaching that seems to me to say turbulence may be a multi-degree of freedom phenomenon, but not all the degrees of freedom are equally important at all times. So that there may be some intervals where I can reduce the description tremendously down to a small number of degrees of freedom, but they won't be the same degrees of freedom all the time; and that's part of the reason that I think deciding that these five modes are going to be the ones I describe, can not be true forever. Maybe the way around that is to consider sort of a changing of degrees of freedom from area to area and actually using this change in the degrees of freedom as part of the dynamical description.

Frisch: Is there hope to reduce the number of degrees of freedom? After hearing Hassan's last comment, I want also to take issue with that now. If you look at the way the chaos research is evolving now, if you look at what is happening at the leading edge of the field, you notice the following: 15 - 20 years ago people were studying very low dimensional systems, mostly zero-dimensional ones (i.e. ODE's), with very interesting results, that are now very well understood. Then the game was to increase the number of degrees of freedom. Finally, the studies shifted to extended systems. Such systems, were they not extended, would have just 1, 2 or 3 degrees of freedom. Because they are extended they have infinitely many degrees of freedom, and instead of being governed by, say, a Landau equation, they are governed by a Ginzburg-Landau equation. So there is a lot of work on the generic behaviour of the Ginzburg-Landau equation. One finds, for example, defects: spirals, sinks, sources, etc. Next, in studying the interactions of defects one recovers a problem looking a bit like the Helmholtz vortex problem. In this way, dissipative dynamics are sometimes reduced to conservative Hamiltonian dynamics, but with infinitely many degrees of freedom.

Aref: Once you start studying defects, you are really half playing the game of the finite-dimensional system.

Frisch: Yes, but you have many. And then you end up doing statistical mechanics.

Zaslavsky: There is of course the possibility of reducing the number of degrees of freedom. This is very important when deeling with extended systems. But there is a possibility of enhancing the number of degrees of freedom for chaos, and that is just the situation in so-called quantum chaos. Any kind of finite boundary problem in general produces many degrees of freedom as many as the number of eigenvalues.

Aref: Actually, one thing I omitted to say is that in looking at vortex interactions it seems that you mostly look in two dimensions. The tendency is that the number of effective degrees of freedom actually goes down because of the dominant processes of either merging or pairing. In 3-D there is actually a tendency for the effective degrees of freedom to go up because there tend to be significant pieces of vortices that get shed off. It's only in some of these very symmetrical cases, like identical rings coming together, that you have two rings in the initial state, two rings in the final. Make the rings a little bit different, and they start shedding significant pieces. Also, let me make a suggestion. People in the turbulence community would like to know what chaos really might contribute to turbulence in some sense. I think that's a very interesting thing to look at whether chaos ideas could in fact contribute to the Strouhal number/Reynolds number relations. This type of thing is in the "flavour" of chaotic dynamics. It's probably not something that anybody's turbulence theory will yield very quickly. It might be the sort of thing that chaotic dynamics could contribute to.

Mullin: I agree that one possible use for the ideas of low-dimensional dynamics when applied to fluid flows is in the understanding of the chaotic motion of well-defined coherent structures. The central aim of our work is to try and establish a connection between experimental observations of relatively simple flows and the governing equations of motion, the Navier Stokes equations. The problems we choose, such as Taylor-couette flow, are situations where one can solve the full equations on the appropriate boundary conditions using a supercomputer. At least nowadays one can solve any complicated steady flow, but finding a solution for even simply periodic flows remains a formidable task. One feature of the steady flows which plays a central role in determining the dynamical behaviour is the rich multiplicity of steady states. However, an equally important property seems to be that when the Navier Stokes equations are put on symmetric boundary conditions, there is a tendency for the flow to break the imposed geometrical symmetry at modest values of the Reynolds number. Thus, if we start with a symmetric flow in a problem which has a simple mid-plane symmetry, then this flow will exchange stability with a pair of asymmetric states above a critical Reynolds number. Any subsequent development of time-dependent behaviour on the asymmetric branches can produce an interaction between the time-dependent asymmetric flow and the unstable solution corresponding to the original symmetric solution. This in turn can give rise to homoclinic behaviour of the Silnikov type, where chaos may arise through a familiar route such as period doubling. Thus there is an organising centre present for the chaotic motion, which can be exploited using Lyapunov Schmidt reduction, for example, to establish a link between ideas from low-dimensional dynamical systems and the Navier Stokes equations.

Frisch: I have a specific question concerning the experiment where you observe shedding. You tried your best to make this thing 2-D. Do you get parallel shedding or symmetric shedding?

Mullin: The motion that's seen at the end of the final vortex is three-dimensional. Numerical investigations of the problem provide conflicting evidence for the presence of a 2-D Hopf bifurcation, and this is still under investigation. However, the experiments suggest that there may be a 3-D Hopf instability. The calculations we carried out are 2-D, and there is good agreement between them and the experiment for the first bifurcation. In addition, the power spectrum of the experimental signal for the first time-dependence suggests that there is a nearly periodic wave which looks three-dimensional when observed using flow visualisation.

Frisch: Does it become 3-D because of the boundaries which is a trivial effect or because of spontaneous symmetry breaking in the third direction?

Mullin: My speculation is that it will become 3-D through a symmetry-breaking bifurcation in the span-wise direction. After all, symmetry-breaking is rather common in fluid flows in my experience.

Reynolds: The experiment looked like you had one side wall being a free surface.

Mullin: No. The flow is totally enclosed, there is no free surface.

Reynolds: We always have a strong flow in the free circulation region. It doesn't take much to produce it.

Mullin: The Reynolds number is very small in these experiments and ranges from 10 to 200. We tried to measure a cross-stream component, but found that there was no detectable component within experimental error. We have tried to be very careful in setting up this flow in the laboratory, and details of the experimental arrangement can be found in the original article in JFM and the thesis referred to therein. The most significant observable 3-D effect is the growth of boundary layers along the side walls above Reynolds number 40.

Lesieur: Prospects not based on the model equations, which as you say are difficult to relate to Navier Stokes equations, but prospects that are based on the fluid dynamics?

Zaslavsky: Some common language is necessary to discuss all these laws. This means that chaos is a phenomenon together with tools with the help of which it is investigated. This means we have to understand this phenomenon with the help of these tools from Navier Stokes equations. This is absolutely clear. Now the next question is what kinds of tools can be applied to Navier-Stokes equations to get something useful for turbulence. Something can be applied of course, I absolutely avoided the formation of turbulence. It seems it is clear, by Mullins results and other, how turbulence appears. I mentioned that it seems that it is possible to study the appearance of space chaos, but the main problem is the spatio-temporal chaos, especially for high Reynolds numbers. It is a huge problem, and it is between the beginning of turbulence and the end of turbulence for high Reynolds numbers. Phenomena like intermittency and phenomena like basic transport may be understood with the help of the dynamic tools.

Aref: The RNG that I was alluding to was not the RNG by the Princeton people, but it was actually again an attempt to apply RNG ideas that are similar to the real-space RNG that's been quite successful in critical phenomenon studies and which is, in fact, nothing but a systematic procedure for reducing the number of spatial degrees of freedom. I think to apply that kind of ideas to, for example, flow dominated by vortex structures, would be very worthwhile and, in fact, something that E. Siggia and I worked on. The 2-D case is the easiest to think about. Assume that you have a large number of vortices running around. Every once in a while some of them end up in a bound state or in a sort of merge. Just think about point vortices, so they end up in a bound state, either plus-plus or plus-minus, and you eliminate those degrees of freedom and put in effective degrees of freedom that are, say, two plus one, which is a plus-two vortex and a plus-one-minus-one pair, and so on. And now you consider the interaction-tracking system consisting of these, and again when it coarsens, you put new degrees of freedom. And then you try to do a systematic reduction procedure and try to study that transformation rather than just doing a simulation. And the hope is that by studying that transformation certain things will start to converge to an equilibrium distribution, for example vortex strength, or the spatial distribution, and in that way you might be able to understand this turbulent state. It is very similar to what is done in critical phenomena. Basically, what you are trying to do is look at two systems that are somehow related by some kind of exchange. And just like you have a relationship, for example, in space across something like a shock, and you use those to jump across the shock in space, you need tools to jump across the change in degrees of freedom. What information do the dying degrees of freedom give to the new degrees of freedom, that you must retain in order to make the dynamics consistent? For example, if you work on vortices, you want to conserve the circulation.

Novikov: One general comment on the connection between vortex dynamics and real turbulence. I myself have in many stages tried various approaches, from pure statistical, pure dynamical, and now I'm trying to combine both of them. I think that if we are talking about the real turbulence which is observed in nature with high Reynolds numbers, we cannot avoid statistics. We cannot just consider vortex dynamics only without thinking about on which background this vortex dynamics exist. Because there is a very strong connection of a coherent structure with the background turbulence, and actually these coherent structures are born by a kind of mechanism by which turbulent flow is getting rid of some degrees of freedom and excess of energy is put into this coherent structure. This is a very delicate problem. You cannot simplify too much. So now my personal point of view: first of all, we have to understand the statistical background of turbulent flow and then how a coherent structure interacts with this statistical background. Without having in mind this picture, just considering artificial simplified dynamical systems with a small number of degrees of freedom, probably will not help much. We have somehow to include this statistical background of turbulence.

Zaslavsky: It may be that such statistical behaviour doesn't exist without coherent structures. Statistical diffusion cannot avoid the existence of coherent structures.

Novikov: These coherent structures are born inside. They appear as manifestation of excess of energy. This was pointed out by Onsager many years ago. He considered a system of point vortices and in the state where the temperature is negative, the system is getting rid of the excess of energy in such a way that it creates this coherent structure. It's exactly the same thing, but it is for very simple systems of point vortices. For 3-D real turbulence everything is much more complicated.

Moin: Marcel asked to what extent proper decomposition can help. Would it be an idea to think of these proper decomposition modes as the degrees of freedom that you were talking about.

Aref: That's what we have been trying as you know.

Moin: Now the next step is to ask to what extent that model has been successful.

Aref: There was an interesting suggestion made this morning that maybe the quantities to look at are not the eigenfunctions of $\overline{u_i u_j}$ but $\overline{u_i \omega_j}$. Maybe those are more interesting basis functions in which to expand.

Moin: The only reason for using these eigenfunctions would be that they contain modes with high energy. What it turns out actually, experience shows, is that it is only the first eigenfunction that carries significant energy and that is only if you can find a domain extremely near the wall.

Session Five

Chairman J. R. Herring

The Origins of Turbulence

Michael Gaster

Abstract. The process of transition from a laminar to a turbulent boundary layer flow is discussed. From detailed experimental measurements and dye filament observations it appears that transition often occurs in a very localised way involving the formation of distinct coherent structures. Mostly experiments as well as analysis are carried out on purely periodic disturbances and these features are not apparent. It turns out that naturally occurring waves, excited by a random turbulence field, for example, behave in a quite different manner from periodic excitation. This aspect of the transition process is explored.

1. Introduction

A great deal of current activity is focused on the role of coherent structures in turbulent flows. It seems that a proper understanding of these structures and the way that they interact with one another can be used to help explain some of the observations made in fully developed turbulence. Even linear interactions have been useful in this context.

My interests are with the initial formation of turbulent flows from laminar ones, and especially with the laminar/turbulent transition of boundary layers. From experimental measurements that have been made of the breakdown process of the boundary layer on a flat plate I have formed a picture of the sequence of events that lead to the creation of high frequency bursts of turbulence in that flow. The processes that have been found to cause this breakdown involve coherent structures and it seems very likely that there is some common ground between them and the coherent structures that have been identified in fully developed turbulence. The behaviour patterns in these two situations will, of course, not be quite the same, but maybe the observations made in the relatively simpler transitional flows offer further clues as to the formation of structures in the turbulent state. Experiments performed in a laminar flow have the advantage that events can be artificially triggered to create deterministic patterns of behaviour, whereas in the turbulent case the structures have to be inferred from statistical measurements.

2. Instability

If one watches a dye- or smoke-laden laminar flow undergoing transition to turbulence, one sees an almost explosive local breakdown of the streaky flow pattern. The marker mixes
with the surrounding fluid very rapidly indeed in certain regions of physical space/time of the flow. The mixing in these regions is far greater than that arising from viscous diffusion alone. The process appears initially to be a very localised phenomenon that must be linked to some form of instability mechanism.

Over the last hundred years or so, since Reynolds' (1883) observations of transition to turbulence in a pipe flow were reported, a very solid theory of flow stability has been developed. This theory considers the temporal evolution of disturbances imposed on a given base flow. Of particular interest has been the determination of the long term behaviour of initially very small disturbances. If the amplitudes of the flow perturbations are small enough the equations of motion can be linearised. Even with this simplification it turns out that the determination of the stability of any given flow is difficult, and only simple base flow geometries that enable the linear partial differential equations defining the perturbation to separate can be solved analytically. For example, boundary layers are modelled as infinite parallel flows with a specified streamwise velocity profile. Only then can the system of eigenmodes be readily determined. It turns out that the eigensolutions appear as travelling waves such that for a given wavenumber the complex frequency parameter arises as the eigenvalue. The flow is deemed to be stable or unstable depending on the sign of the maximum value of the real part of the exponent. If this quantity is less than zero the flow is considered to be stable because all these Fourier modes eventually decay. In general the unstable waves predicted are long and their growth rates slow. At first sight, this global type of flow instability does not seem to relate at all to flow visualizations of the transition process, that do not show any wavy structures. In fact one of the strongest critics (Taylor (1938)) of the Tollmien-Schlichting model of boundary layer transition made the point that no one had observed or detected such waves, even with sensitive hot-wire anemometers then becoming available.

Nevertheless, the basic linear theory of hydrodynamics stability, defined through solutions of the Orr-Sommerfeld equation, was formerly shown to apply to boundary layers by the experiments of Schubauer & Skramstad (1947). Their experimental work was carried out in a specially developed low turbulence wind tunnel at the National Bureau of Standards in a flow with far lower free-stream "noise" than had been used before. They created controlled periodic wavetrains in the boundary layer by means of a two-dimensional ribbon wavemaker. The introduction of regular waves in this way enabled detailed measurements to be made without undue interference from those created naturally by the free-stream turbulence or the acoustic field in the wind tunnel.

Over the last 50 years both experimental and analytical work has followed the idea that Tollmien-Schlichting (Schlichting (1933)) waves are the prime instabilities that lead to transition and hence turbulence. Great emphasis has therefore been placed on the determination of the conditions for these waves to grow, defining flow stability. Account can now be taken of the slow growth in boundary layer thickness with distance from the leading edge, of the fact that wavemakers create spatially evolving waves rather than temporally growing ones used in the original theoretical model. Also weak nonlinear effects can be taken account of in the predictions of the behaviour of imposed waves. In fact the waves that Schubauer and Skramstad created were spatially evolving waves *forced* by the wavemaker. These waves can certainly increase in amplitude as they propagate downstream, but this process does not constitute true *instability*. Transition to another flow state is not necessarily achieved locally where the appropriate instability parameters are exceeded, as is the case in flows like Taylor flow between concentric rotating cylinders that do exhibit true instability.

Stability theory still only describes the evolutions of low frequency travelling waves in the boundary layer and these are far removed from the catastrophic local breakdowns that can be observed in many cases of transition to turbulence. I will describe some experiments carried out in my own low turbulence wind tunnel that I believe throws fresh light on the problem and explain the role of Tollmien-Schlichting waves in the transition process.

3. Experimental setup

All the observations to be presented here were made in a very low turbulence windtunnel of 3 ft x 3 ft cross-section on the boundary layer of a flat plate. A buried ear-phone 200 mm downstream from the leading edge was used to provide various forms of flow excitation on the centre-line of the plate. Careful surveys of the flow showed the base flat plate boundary layer to conform closely to the theoretical Blasius solution. The extremely low free-stream turbulence of less than 0.01% enabled controlled excitations to be carried out without serious contamination. The velocity in the free-stream was 20 m/sec and mostly measurements were made at about 1 m downstream of the leading edge.

Careful measurements of the transition process, made in this windtunnel with hot-wire anemometers, revealed the following sequence of events when only natural excitation was used. Firstly, very weak waves appeared above the background noise of the instrumentation whenever conditions were appropriate for the amplification of instability waves, that is at about 1 m downstream of the leading edge. The ripples seen on the signal arose from waves that were somehow excited by the random fluctuations in the external stream and were then amplified as they travelled downstream. The random nature of this excitation coupled with the selective frequency amplification process created highly modulated wavetrains. As the detecting instrumentation was moved progressively downstream the amplification became more selective, and the spectrum narrower, but still containing a band of modes in both frequency and spanwise orientation. Nevertheless, the evolution of individual Fourier elements was found to follow predictions of linear stability theory.

At some stage downstream the naturally created modulated signal began to show local deformations that no longer appeared as a simple modulation riding on the steady base flow. These local irregularities developed very rapidly into very large amplitude "events". During this phase of the process the amplitudes of the almost periodic wavetrains were relatively small and would be unlikely to be seen in any flow visualizations, nor would they have appeared above the background in a noisy wind tunnel experiment. The surprising aspect of this process was that if a regular periodic wavetrain of similar amplitude were to be artificially introduced into the flow it did not show any significant nonlinear behaviour. In fact, the amplitude of the regular wavetrain had to be some seven or eight times as large as the naturally modulated wave to exhibit strongly nonlinear characteristics. The naturally occurring waves appeared to generate a random sequence of "events" at relatively low amplitudes. Once the signal had shown a significant local distortion a turbulent patch seemed to arise inevitably shortly afterwards. The evolution of these signal distortions into regions of high frequency turbulence was extremely rapid.

So although the classical long wave instability defined by solutions of the Orr-Sommerfeld equation does play some role in the transition process it also seems necessary to consider the influence of wave modulation.

I want, therefore, to describe some experimental observations that illustrate the effects arising from modulations of the basic wavetrain. Some of the results are not new and have been presented before, but no one has responded or taken any notice of them. There are also some new results from current work that is being done by a research student, F.N. Shaikh.

4. Experiments on modulated wavetrains

The disturbed flow fluctuations produced by controlled excitations are much easier to explore experimentally than randomly created wavetrains that arise from the free-stream turbulence. That is why so much experimental work on boundary layer transition has been carried out on the disturbances created by simple time-periodic two-dimensional wavemakers of the sort used by Schubauer & Skramstad. As an alternative to the somewhat artificial periodic two-dimensional wave system used so often, a pulsed point source was used in my own experiments. This form of artificial excitation generated a more interesting response that contained a band of oblique modes of different frequencies while still retaining the advantage that it was deterministic. At the time it was felt that the resulting wave packet created in the flow was closer to the naturally occurring modulated wavetrains than the very constrained two-dimensional periodic waves so often studied.

Of course, provided the amplitudes of the waves generated are small enough, the linear superposition of simple periodic wavetrains will synthesis a pulse induced wavepacket. Although the original experimental work was carried out in order to verify a linear asymptotic theory of such a packet, the measurements obtained showed some new and somewhat unexpected features.

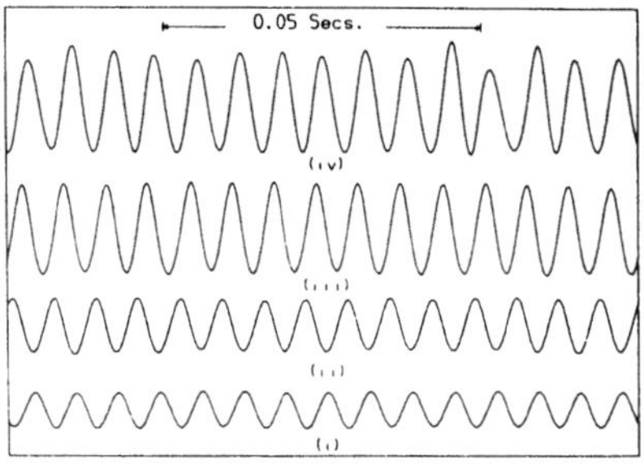

Fig. 1 Hot-wire signals from a sinusoidal excitation.

It turned out to be surprisingly hard to generate wavepackets that were large enough to measure easily without creating violent distortions in the hot-wire anemometer records. In fact, the packets that were recorded turned out to contain a certain amount of distortion of the type referred to, and these were found to have a pronounced spanwise character. Although this behaviour had all the indications of being caused by some nonlinear interaction, the measured peak-to-peak amplitudes in the packet were well within the bounds of linear behaviour for regular periodic wavetrains. This pattern of behaviour is illustrated in figures 1 and 2. In figure 1 the response to a periodic excitation at a point on a flat plate is shown at some location downstream from the source. The amplitude of the driver has been progressively increased from (i) to (iv). The response records did not show any marked nonlinear character until the maximum level (iv) had been reached, at which stage the signal shows some irregular subharmonic character. Figure 2 shows the hot-wire anemometer signal responses recorded when the buried earphone was excited by short-duration pulses of various amplitudes. In this experiment the flow conditions, the position of the hot-wire probe, the amplifier gains etc., were

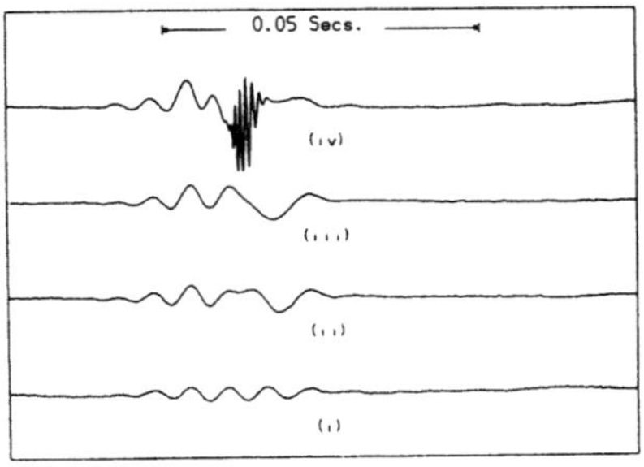

Fig. 2 Hot-wire signals from a pulsed excitation.

identical to those of the previous case of periodic excitation. Nevertheless it was only at the lowest excitation level that the pulse drive created the expected roughly Gaussian-shaped wavepacket. It was impossible to drive the system any harder than the level used in (i) without producing signals with the gross distortions shown in subsequent runs (ii) - (iv). The peak-to-peak signal amplitudes at which this occurred were much lower than those of periodic waves that showed nonlinear activity. More detailed experiments, with recordings taken at a range of downstream locations and amplitudes of excitation showed quite clearly that these signal distortions arose through the nonlinear evolution of the motion within the wavepacket and were not, as generally thought, to arise from flow nonuniformities or because of any significant local flow distortions at the source location. The modulated nature of the evolving wavetrain seemed to be clearly all important.

When the pulsed excitation was driven even harder than that required to create a distorted wavepacket another feature appeared in the records. The topmost trace, (iv) on figure 2, contains a record showing a burst of high frequency oscillations riding on the primary signal. The primary wavepacket was found to be quite repeatable at low amplitudes, but as the level of excitation was increased, not only did the degree of distortion grow but the pattern differed slightly with each realisation. This became particularly noticeable when the high frequency secondary burst appeared, making measurements harder to carry out in that regime.

Figure 3 shows the high frequency secondary burst of oscillations of about 1 kHz riding on the primary packet of roughly 150 Hz. The filtered high frequency component, plotted out here on a stretched scale, appears also as a distinct wavepacket. Unfortunately the shape and frequency content of the secondary packet varied from one realisation to the next, making their study difficult. Figure 4 contains the power spectra of five separate realisations obtained at the same position and physical conditions, together with the mean spectrum from 50 separate bursts. Even though the individual secondary bursts varied in both energy and frequency content the mean of 50 provides a useful measure of the event

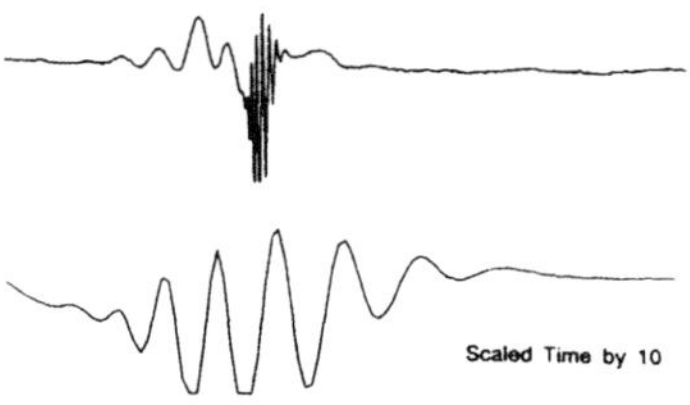

Fig. 3 High frequency component.

at that station. A series of mean spectra were obtained at closely spaced streamwise locations, and these are shown on figure 5. Although the sample size of 50 is not really adequate for showing the evolution of the spectral content of the burst the general trend is clear from the figure. The dominant frequency occurred at around 1 kHz and the energy content increased rapidly with downstream location as shown on figure 6. This figure shows that the secondary wave system grew exponentially until a distance of 104 cm from the leading edge has been reached, where the growth rate levelled off to a more moderate value.

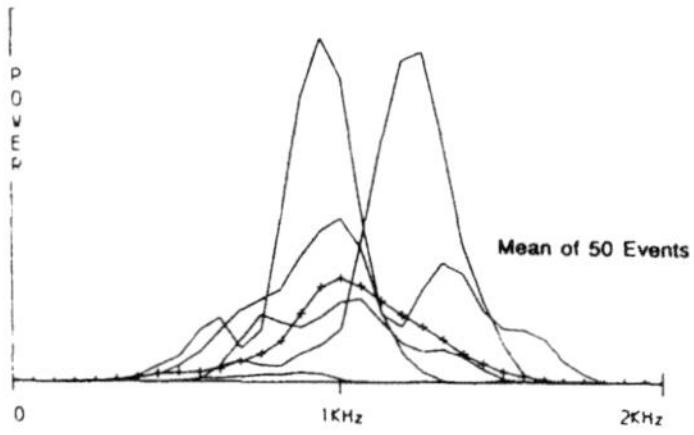

Fig. 4 Power spectra of secondary bursts.

The shapes of individual records of the secondary high frequency bursts from which the spectra in figures 5 and 6 were derived also contain some additional worthwhile information. A set of such records taken at 103 cm from the leading edge are shown on figure 7. Mostly the records are of clean bursts of waves of roughly Gaussian envelope form, whilst some show almost zero activity. Further downstream at 106 cm the character of the records has changed in that the envelopes are no longer of simple Gaussian form, but show modulations that are reminiscent of those seen in the original primary packet when the amplitudes were large. This signal characteristic together with the fact that the rate of energy growth is levelling off as shown on figure 6 strongly suggests that the secondary wavepacket burst is saturating and behaving like the primary one just prior to the formation of secondary waves. At this stage of development the irregular character of the secondary packets made experimental work difficult, but a number of recordings were taken even further downstream at the highest sampling rate possible with current equipment. Careful filtering of these records showed the occasional burst of very high frequency in the range of 4-5 kHz. One example of such a realisation is shown on figure 9.

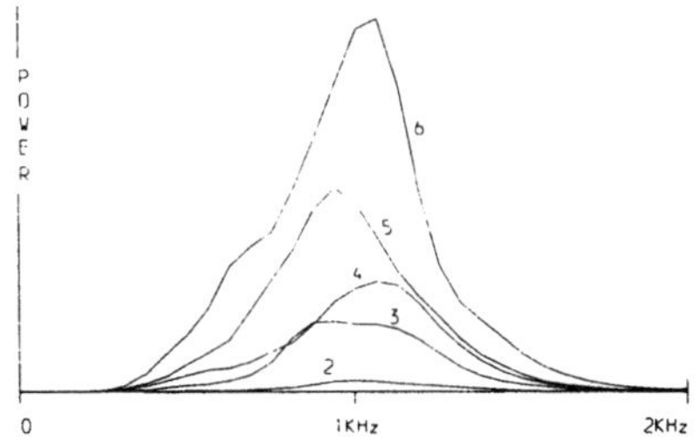

Fig. 5 Mean power spectra at various locations.

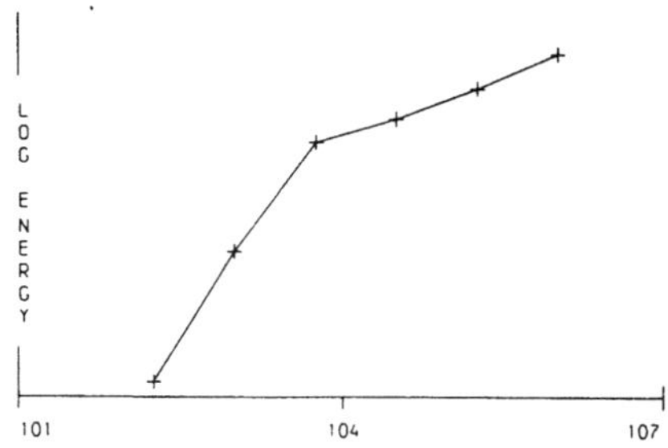

Fig. 6 Integrated power versus distance from L. E. CMS.

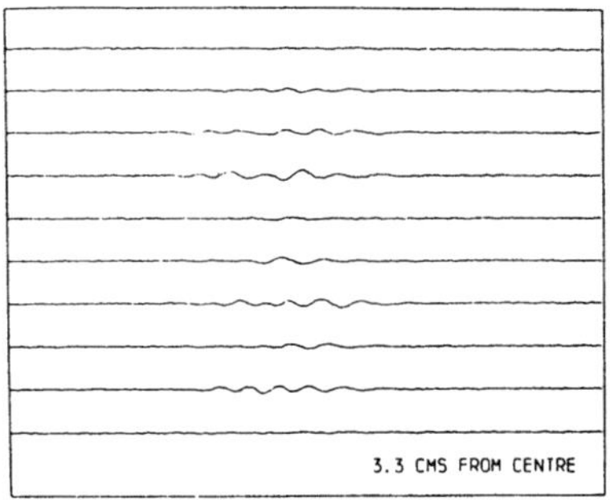

Fig. 7 Set of filtered secondary bursts at 103 CMS.

Although more details of the very high frequency tertiary packet could not be obtained, it was quite clear from the few results that were obtained that these waves amplified even more rapidly than the secondary packets and led to turbulent bursts a few mm further downstream.

The records described above refer to the pattern of behaviour with streamwise position measured at the outer edge of the boundary layer and at roughly 3.3 cm off the nominal

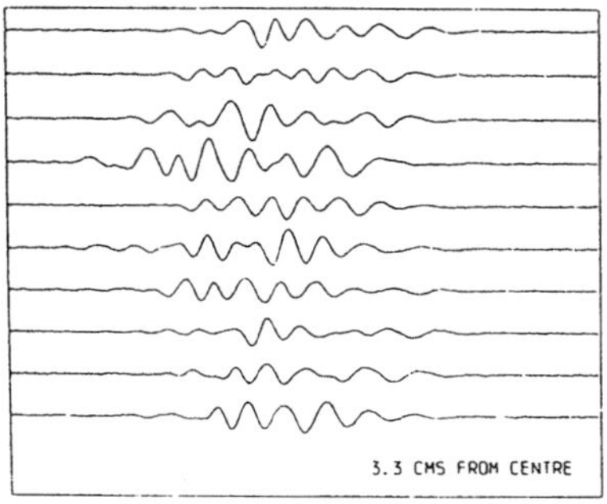

Fig. 8 Set of filtered secondary bursts at 106 CMS.

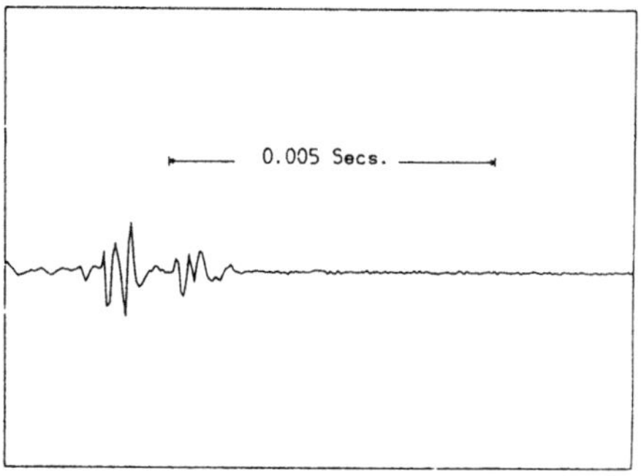

Fig. 9 Tertiary burst.

centre-line of the plate. This spanwise position was chosen because it showed the breakdown to higher frequencies earlier than on the centreline. More detailed measurements of these events were taken in the region within the boundary layer at a position of 110 cm from the leading edge. These measurements were taken in order to show something of the overall flow structure during the formation of a turbulent spot. Ensemble averaged hot-wire signals were recorded over 128 realisations at a number of closely spaced probe locations for the period of the packet passage. The physical region chosen covered only a small spanwise section of the packet where the primary wave amplitude was largest and where turbulence first appeared. The information is displayed as a set of contours of instantaneous streamwise velocity on a span/normal plane at various time instants on figure 10. Initially the contours consisted of horizontal lines representing the undisturbed Blasius boundary layer flow. At stage (1), some 0.1 secs after the packet was initiated by the pulse, the contours showed only some slight variations from the horizontal, indicating a thickening of the boundary layer over a region of 5 mm or so coupled with the beginnings of the formation of steep shear layers close to

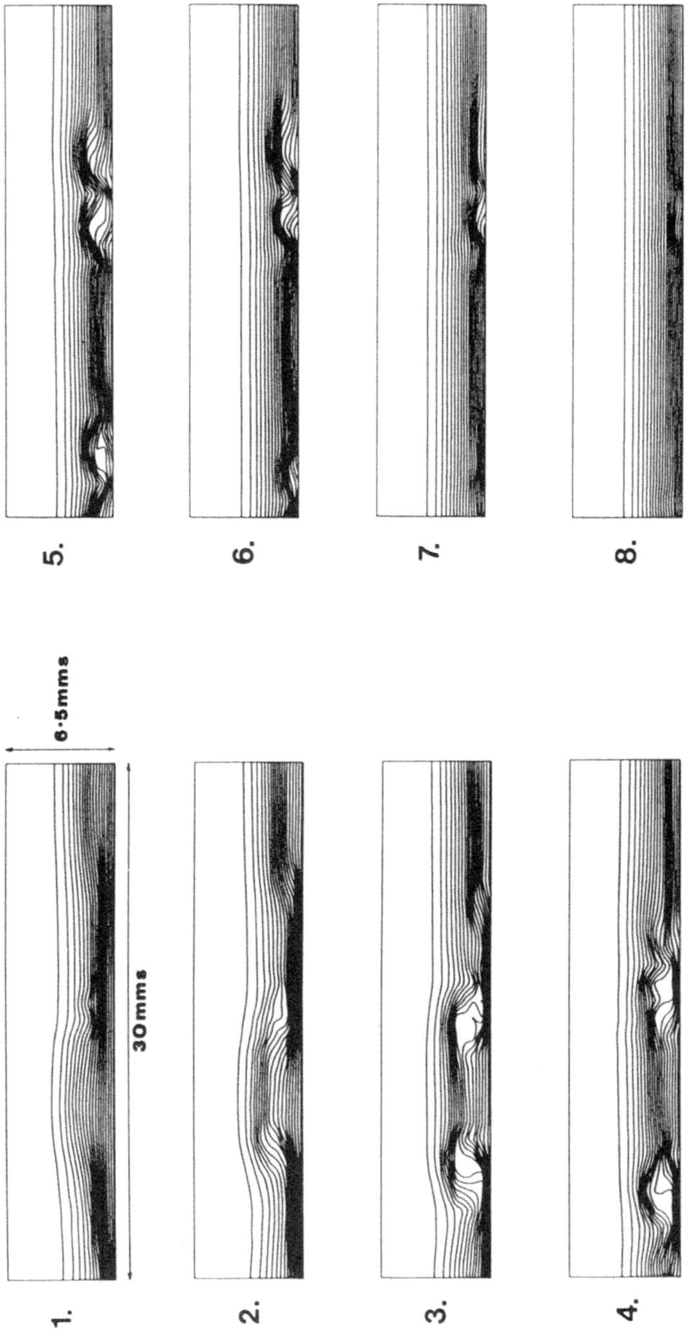

Fig. 10 Development of streamwise velocity contours at various time instants.

the wall on either side. The process continued to evolve and by stage (3) two "eyes" have formed with velocity profile histories shown in the sequence of profiles on figure 11. These profiles show a region of constant velocity bounded by steep shear zones. In fact the most intense shear layers form just over the "eyes". At later times the "eyes" move apart and the whole motion decays and the flow returns to the undisturbed form.

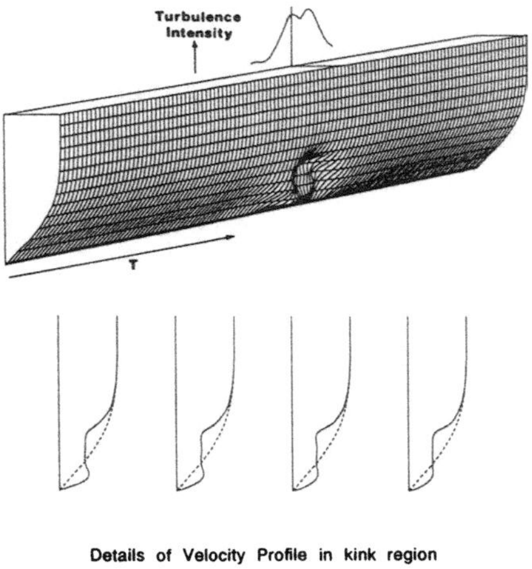

Fig. 11 Temporal development of velocity profile.

The data can also be displayed as a sequence of velocity profiles at a given spanwise location - as indicated by the arrow on the previous figure. The "kink" in the profile surface lasts roughly 0.05 secs, or one half of a Tollmien-Schlichting period. The signals recorded by the hot-wire when placed just at the edge of the boundary layer did not pick up any of the gross distortions measured within the layer but did indicate the passage of a high frequency patch or spot of turbulence just at the time period when the profile contained the very intense shear layer. The strength of the r.m.s. turbulence intensity detected is also indicated on the figure. It should be noted that there is no wave motion apparent in the above display. In fact the hot-wire did pick up strong wave activity, but when the velocity records were scaled to include the very large deviations in the region of the kink, the waves could no longer be resolved. Flow visualization of such an event would also fail to show the existence of any Tollmien-Schlichting waves.

An artificial excitation that is even more representative of natural excitation can be produced by feeding the buried earphone with a deterministic random noise sequence. The hot-wire anemometer records obtained downstream then consist of streams of modulated waves quite similar to those seen in naturally excited experiments. But by using a deterministic sequence the process could be repeated as often as required. This enabled clean ensemble averaged signals to be obtained at various measuring stations somewhat like those obtained for the short duration pulse, but with a richer range of possible nonlinear interactions. One example of the above type of excitation is shown on figure 12. Hot-wire records were taken directly downstream from the source point at 1 cm intervals in the range 0.8 to 1.2 m from the leading edge. At the upstream station the signal was just like filtered white noise, as might have been expected. In the region where the motion was governed by a linear system, the signal appeared as a modulated narrow band process, but as soon as nonlinear terms became important the whole character changed. The signal became irregular and no longer had the smooth appearance of band filtered noise, These irregularities developed far downstream over a distance of only a few centimeters and then individual realizations showed short bursts of turbulence at isolated parts of the record. The ensemble of the square of the signal was also formed so that the standard deviation about the mean of the ensemble could be computed and displayed. The magnitude of this component is shown for the last four streamwise locations. The plots show that over a small region of time the deviation was large. Detailed examinations from the individual records showed irregular very high frequency turbulence at that time instant. The details of this turbulent component were uncorrelated with one another and hence produced the large deviations shown on the last few records as the intensity of the turbulence rapidly grew downstream.

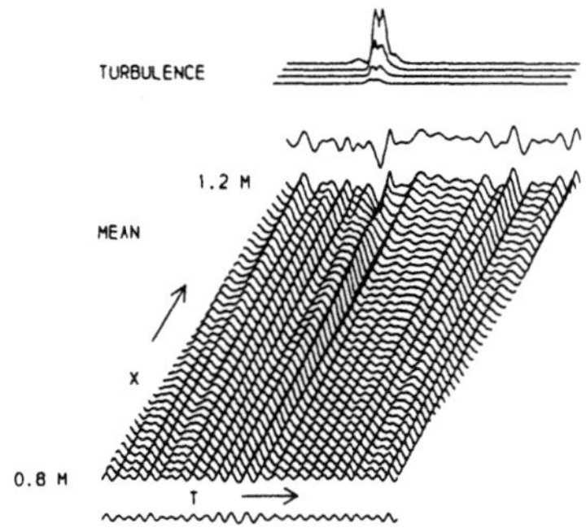

Fig. 12 Signal development from a random excitation.

Recently more detailed experimental measurements have been made on another more carefully constructed pseudo-random sequence of specified spectral composition. In the new work signals were obtained not only just outside the boundary layer, as in the earlier work, but some recordings were taken within the boundary layer, where quite violent motions occurred similar to those seen in the wave packet case. Figure 13 contains a sequence of four frames showing the response outside the boundary layer at various downstream locations to the same excitation signal. The upper part of each frame shows the hot-wire record while below the wavelet transform is shown as a contour plot. The wavelet transform is essentially formed by sweeping a wavepacket-like transform

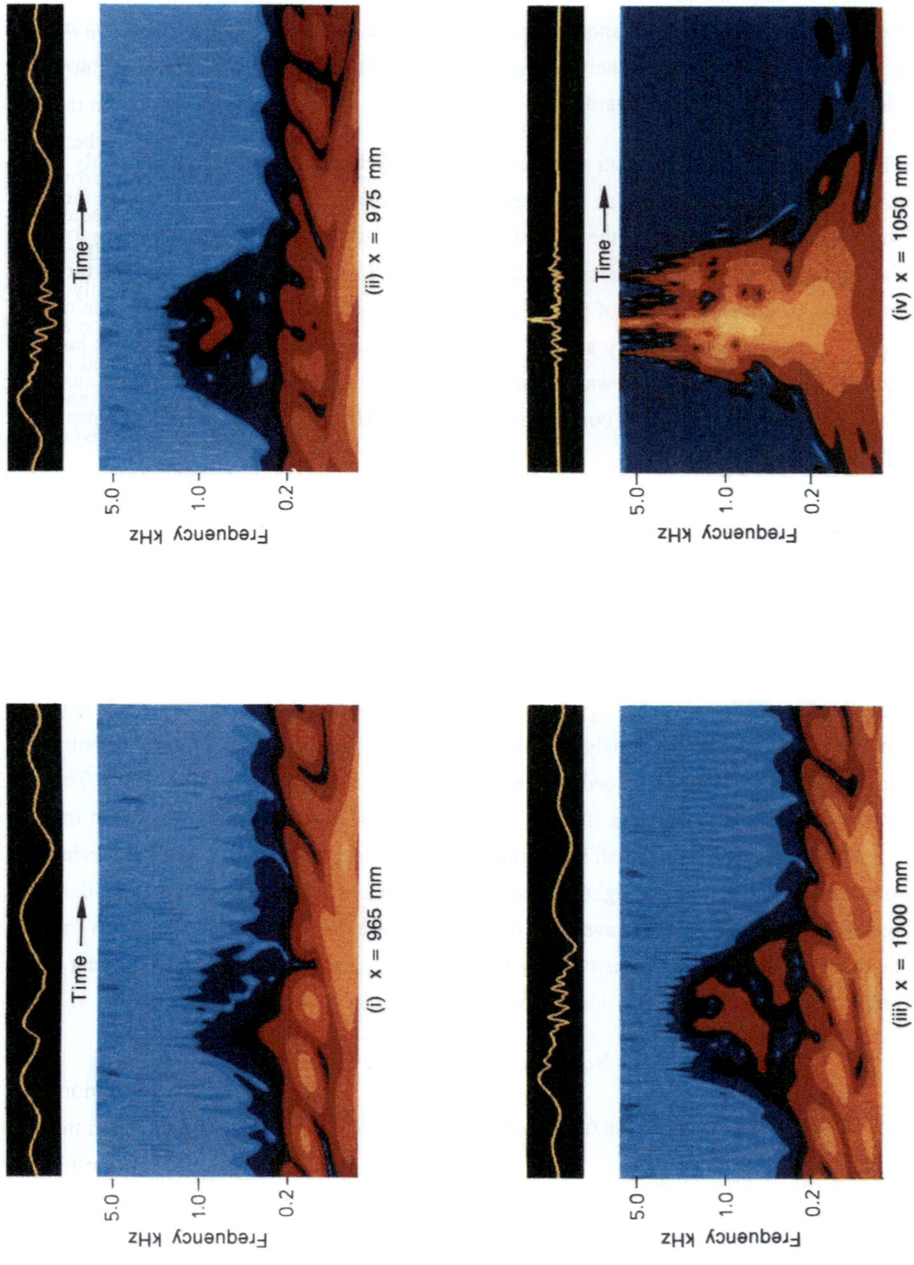

Fig. 13 Wavelet transforms

operator through the record. The shape of the wavelet is conserved so that the higher frequencies are explored in shorter time periods. The wavelet transform of a record produces a response that is, therefore, a function of both time and frequency. In these plots frequency increases upwards.

The first frame contains smooth modulated Tollmien-Schlichting waves, with possibly a hint of subharmonic activity as indicated in the transform display. The second frame, some 10 mm downstream of the first, shows some clearly discernible ripples riding on the Tollmien-Schlichting waves. These also show up quite clearly on the wavelet transform as a concentration of power at about 1 kHz. In frame three the amplitude of ripples increased significantly and by frame four this high frequency component has exploded into turbulence. The wavelet transform shows that at the final stage of evolution there is a considerable energy concentration around 1 kHz and there is also another peak at about 5 kHz.

5. Discussion

The experiments described mimic certain aspects of the naturally excited transition process. The modulated waves that result from either an isolated pulse excitation or a random sequence seem to be very important in the nonlinear phase of wave growth. The linearised equations of motion ignore the quadratic and higher interaction terms, whereas simple periodic waves, or slowly modulated wavetrains, only are affected to cubic order in amplitude (Nonlinear Schroedinger equation - see Stewartson & Stuart (1971)). The rapidly modulated wavetrains studied here show powerful quadratic effects. When the amplitudes are sufficiently small it is permissible to consider the quadratic nonlinear terms as a forcing function on the right hand side. This quadratic stress field is very dependent on the modulation of the wavetrain and of the phase of waves with respect to the modulation. This effect is illustrated by looking at a typical product of the component u and v of the perturbation. Consider a modulated wavetrain streamfunction

$$\psi(x, y, t) = \text{Real part } \{\ A\,(x, t)\ \phi(y)\ e^{i(\alpha x - \omega t)}\ \},$$

where A(x, t) is the modulation of the wavetrain $f\,(y)\ e^{i(\alpha x - \omega t)}\ \}$.

$$u = \frac{\partial \psi}{\partial y} = \text{Real part } \{\ A\,(x, t)\ \phi'(y)\ e^{i(\alpha x - \omega t)}\ \}$$

$$\text{and } v = -\frac{\partial \psi}{\partial x} = -\text{ Real part } \{i\,\alpha\, A\,(x, t)\ \phi(y)\ e^{i(\alpha x - \omega t)}\ \}$$

$$+ \frac{\partial A(x,t)}{\partial x} \phi(y) e^{i(\alpha x - \omega t)} \} .$$

The term arising from the product of u and the first term of v is the usual Reynolds stress that arises for periodic waves, the contribution from the wavetrain modulation that then appears through the second term is, however, very much larger for the types of wave being considered here. Of course the quadratic stress field will in general be highly three-dimensional and this will cause other effects. I believe, but I have no solid evidence for this view, that it is the magnitude of the stream-wise gradient of the stress rather than its magnitude that causes the flow to respond. It is this dipole-like structure that forces the hairpin vortex to form. The measurements of the streamwise velocity perturbation certainly indicate that there is a vortex aligned along the span that is associated with the steep shear layer. The vortex head is only some 10 mm long and there must be trailing vortices of equal strength on either side. The "eyes" in the cross-section are, I believe, the cores of the trailing vortex legs.

My interpretation of the sequence of events leading to the formation of a turbulent spot are as follows. Random excitation by broadband free-stream turbulence initiates weak Tollmien-Schlichting waves. These evolve downstream into a band limited wavetrain and therefore one that is highly modulated through the processes of selective amplification. Provided the level of the oscillation remains below some threshold the evolutionary process remains essentially linear, but if exceeded, another violent instability process is initiated. The dipole stress causes the baseflow vorticity to roll up to form a narrow hairpin vortex loop. By self induction this will develop rapidly and behave as if there is a finite-time singularity moderated by viscosity. The shear layer forms very rapidly indeed once the critical threshold has been exceeded. The resulting shear layer is very unstable to disturbances of appropriate frequencies and the rate of growth of the secondary waves, are also scaled by the shear layer, somewhat greater than those of the primary wavetrain. The secondary wave evolves into a nonlinear wave packet that must inevitably modify the local flow by generating finer steep shear layers. This process of spawning finer shear layers that support higher and higher frequency instabilities must roll off as viscosity inhibits the highest frequency instabilities. The fractal character of the process leads inevitably and very rapidly to a turbulent spot.

It seems that the shear layer, once initiated, always saturates with roughly the same gradient and therefore supports secondary waves of the same frequency, namely some 5 or 6 times those of the primary mode. This applies to successively higher modes. It is therefore not really surprising that the identical pattern of frequency cascade has been found in a modulated wavetrain and in a wavepacket.

Further downstream the small turbulent spots will spread out, more will form, and eventually a fully developed turbulent boundary layer flow will result.

6. Conclusions

Modulated waves play a very significant role in the transition process. Fully developed turbulence may well be modelled by random arrays of horseshoe vortices of various scales - some ideas of the sort were proposed by Theodorsen some 40 years ago. Disturbances other than purely wavelike ones must be considered, especially when nonlinearity is important. A sequence of hairpin vortices of finer and finer scale arise as a consequence of the instabilities of the preceding shear-layer/vortex-loop structure.

References

Reynolds, O. (1883). Phil. Trans. Roy. Soc. **174** p.935.

Schlichting. H. (1933). Nachr. Ges. Wiss. Göttingen, Math. Phys. Kl. p.182.

Schubauer, G.B. and Skramstad, H.K. (1947). NACA Rep. 909.

Stewartson, K. and Stuart, J.T. (1971). J. Fluid Mech. **48** pp. 529-545.

Taylor, G.I. (1938). Proceedings of 5th International Congress of Applied Mechanics, Cambridge, MA. pp 294-310 (1939).

Professor M. Gaster
Department of Engineering
University of Cambridge
Trumpington Street
Cambridge CB2 1PZ
U.K.
MG111@UK.AC.CAM.PHX

Role of Algebraic Instability In Transition and Turbulence

M.T. LANDAHL

Department of Aeronautics and Astronautics, Massachusetts Institute of Technology, Cambridge, MA, USA

I. INTRODUCTION

One of the major challenges in turbulence research has been to explain how the turbulence is created and to describe and analyze the space-time behavior of the turbulent eddies. For wall-bounded shear flows the pioneering experiments by Kline and his group (Kline et al. 1967, Kim et al., 1971) and by Corino and Brodkey (1969) brought out the special significance of the strongly intermittent processes in the near-wall region during turbulent bursting and identified alternating low- and high-speed flow regions,"streaks", to be the predominant flow structures there. The oscillations and lift-up of a low-speed streak were seen to initiate a rapid outflow (ejection) of fluid from the wall region and a subsequent breakup of the streak.

Many later investigators employing sophisticated experiments as well as numerical simulations have provided extensive data on the boundary layer turbulence structure. Particularly useful for providing a detailed picture of the instantaneous flow picture have been the numerical Navier-Stokes simulations such as those of Moin and Kim (1982), see Robinson (1991).

In view of the dominating effects of the bursting on the near-wall turbulence dynamics, and its controlling effect on the momentum transport in this region, it is essential to understand how such a strong and intermittent phenomenon can arise. Classical hydrodynamic instability theory gives only a limited guidance, at least in its linear version, since the mean velocity profile for a flat-plate boundary layer is stable to wave trains of infinitesimal amplitude. However, this theory starts from the examination of infinite sinusoidal wave trains which, as will be seen in the analysis of initial disturbances presented here, may exclude an important class of disturbances.

A theoretical understanding of the mechanisms at work in the wall region is crucial for making further advances in the field of boundary layer turbulence. The construction of efficient and accurate turbulence models for the prediction of mean velocity profiles requires that the dynamics of the fluctuating field is represented in a correct manner. In order to devise engineering methods through which the turbulence properties (such as the skin friction) could be manipulated and controlled, it is necessary to understand the cause-and-effect relationships in this flow region. Because turbulent flows are unsteady, viscous and highly three-dimensional flows in which nonlinearity plays an essential role, a complete analytical treatment has so far been found too difficult. However, considerable progress has been made in the area of flow instability and transition to turbulence, which shows many of the same difficulties.

We will demonstrate here how one may arrive at a qualitative understanding of the near-wall turbulence dynamics, and its relation to flow instability, with the aid of a simple model, in which the role of the interaction of the eddy with the strong mean shear is emphasized as the fundamental process. The crucial assumption in this model is that the evolution of an isolated, localized and three-dimensional eddy may be analyzed on basis of three different and separate time scales, namely on a shear interaction time scale, a viscous one, and a nonlinear one. For the near-wall region with its very large mean shear, the shear interaction time scale would be much smaller than the viscous and nonlinear ones. This allows for a simplified treatment of the viscous and nonlinear stages in which the asymptotic large-time solution for the shear interaction stage solution, which may be obtained from linear inviscid theory, provides the initial conditions for the later stages. The theoretical model is able to explain many of the observed features of the wall-layer turbulence and suggests mechanisms whereby the wall layer vorticity may become concentrated into discrete vortices. The proposed formulation may also be useful in the study of three-dimensional flow instability.

II. MODEL FORMULATION

The evolution of the velocity field

$$U_i(x_j,t) = \bar{U}(y)\,\delta_{i1} + \bar{W}(y)\,\delta_{i3} + u_i(x_j,t), \quad \langle u_i\rangle=0 \tag{1}$$

due to a weak three-dimensional eddy in a mean two-dimensional parallel shear flow $\bar{U}_1(x_j,t) = \bar{U}(y)\,\delta_{i1} + \bar{W}(y)\,\delta_{i3}$ is governed by the following equations (written in standard tensor notation with $x=x_1$ in the streamwise direction and the wall located at $x_2=y=0$):

$$\frac{\bar{D}u_i}{Dt} + \bar{U}'(y)u_2\,\delta_{1i} + \bar{W}'(y)u_2\,\delta_{3i} = -\frac{1}{\rho}\left[\frac{\partial p}{\partial x_i} - \frac{\partial}{\partial x_j}\left(\mu\frac{\partial u_i}{\partial x_j} + \tau_{ij}\right)\right] \tag{2}$$

$$\frac{\partial u_i}{\partial x_i} = 0 \tag{3}$$

where

$$\frac{\bar{D}}{Dt} = \frac{\partial}{\partial t} + \bar{U}\frac{\partial}{\partial x_1} + \bar{W}\frac{\partial}{\partial x_3} \tag{4}$$

$$\tau_{ij} = -\rho(u_iu_j - \langle u_iu_j\rangle) \tag{5}$$

and p is the perturbation pressure, angled brackets (< >) denoting average. The equations (2)-(5) have been obtained by the subtraction of the equations for the mean flow from the full set of equations. The solutions are to be subjected to the initial conditions that $u_1=u=u_o$, $u_2=v=v_o$ and $u_3=w=w_o$ at t=0, and that the velocities vanish at y=0 and y=∞. Also, the disturbance field is assumed to be horizontally localized so that the disturbance velocities, and the perturbation pressure, vanish at large horizontal distances. In view of the observed strong spatial intermittency in the wall region it is meaningful to consider the eddy in isolation so that the interaction with other eddies may be ignored.

By eliminating the pressure one finds the following equation for $u_2=v$, in which index notation for partial derivatives is used:

$$\bar{D}(\nabla^2 v)/Dt - v_x\bar{U}'' - v_z\bar{W}'' - \nu\nabla^4 v = q \tag{6}$$

where

$$q = \nabla^2 T_2 - \frac{\partial^2 T_i}{\partial x_i \partial x_2} = \nabla^2[(uv)_x + (vw)_z] - \frac{\partial}{\partial y}[(u^2)_{xx} + 2(uw)_{xz} + (w^2)_{zz}] \tag{7}$$

and

$$T_i = \frac{1}{\rho}\frac{\partial \tau_{ij}}{\partial x_j} \tag{8}$$

The evolution of the weak eddy may be thought to occur on three different time scales, namely

1) shear interaction time $t_s = 1/\bar{U}'_w$, ($\mu\bar{U}'_w$ = wall shear stress)
2) viscous interaction time t_v (to be defined later)
3) nonlinear interaction time $t_n = L/u_0$, L being a typical streamwise length scale of the initial disturbance and u_0 the scale of the initial streamwise disturbance velocity.

For the conditions in the wall region of a turbulent boundary layer with its large mean shear, t_s is clearly the smallest one. Therefore, in the spirit of matched asymptotic expansions, the asymptotic limit $t/t_s \to \infty$ of the shear interaction solution may be used to determine the inner (in t) limit for both the viscous and nonlinear stages, i.e., provide the initial conditions for the calculation of these later stages of eddy evolution.

III SHEAR INTERACTION STAGE

For the shear interaction stage the viscous and nonlinear effects have not yet had time to develop, and the flow is therefore governed by the inviscid and linear versions of equations (1)-(4), namely

$$\bar{D}u/Dt + \bar{U}'v = -(1/\rho)\partial p/\partial x \tag{9}$$

$$\bar{D}v/Dt = -(1/\rho)\partial p/\partial y \tag{10}$$

$$\bar{D}w/Dt + \bar{W}'v = -(1/\rho)\partial p/\partial z \tag{11}$$

$$\partial u/\partial x + \partial v/\partial y + \partial w/\partial z = 0 \tag{12}$$

where $\bar{D}/Dt = \partial/\partial t + \bar{U}\partial/\partial x + \bar{W}\partial/\partial z$, $\bar{U}' = d\bar{U}/dy$, $\bar{W}' = d\bar{W}/dy$, and the initial conditions and boundary conditions are as stated above. Taking the horizontal divergence of (9), (11) and using (12) one obtains

$$\nabla_h^2 p/\rho = -\bar{D}v_y/Dt + \bar{U}'v_x + \bar{W}'v_z,\ \nabla_h^2 = \partial^2/\partial x^2 + \partial^2/\partial z^2 \tag{13}$$

An integration of (8) over time following a fluid element yields

$$u = u_0(\xi,y,\zeta) - \bar{U}'l - P_x/\rho \tag{14}$$

where $\xi = x - \bar{U}t$, $\zeta = z - \bar{W}t$, l is the liftup of the fluid element, determined from

$$l = \int_0^t v\bar{D}t \equiv \int_0^t v(\xi + \bar{U}t_1, y, \zeta + \bar{W}t_1)dt_1 \tag{15}$$

and where

$$P = \int_0^t p\bar{D}t \tag{16}$$

Similarly for the spanwise velocity component w

$$w = w_0(\xi,y,z) - P_z/\rho \tag{17}$$

Integration of (12) over time and taking the initial conditions to be consistent with continuity gives for $\mathbf{P}$

$$\nabla_h^2 P/\rho = -v_y - u_{0\xi}(\xi,y,z) - w_{0z}(\xi,y,z) + \bar{U}'l_x + \bar{W}'l_z \tag{18}$$

where it has been assumed that $l=0$ for $t=0$.

Elimination of the pressure from (9) - (11) gives the Rayleigh equation for the instability of an inviscid two-dimensional parallel flow

$$\bar{D}\phi/Dt = \bar{U}''v_x + \bar{W}''v_z \tag{19}$$

where $\phi = \nabla^2 v$. By integrating this over time one finds

$$\phi = \phi_0(\xi,y,\zeta,t) + \bar{U}''l_x + \bar{W}''l_z \tag{20}$$

index zero denoting initial values, which may be handled by standard methods for the Poisson equation to yield the solution for v. For the near-wall region, in which y-derivatives are much greater than x- and z-derivatives (the eddy has a boundary layer character), the following approximation holds (Landahl, 1990) :

$$v = \int_0^{\infty} C(y,y_1)\phi(x,y_1,z,t)dy_1 = \int_0^{y} (y-y_1)\phi(x,y_1,z,t)dy_1 - y\int_0^{\infty} \phi(x,y_1,z,t)dy_1 \quad (21)$$

where $C(y,y_1) = (1/2)[|y-y_1|-y-y_1]$. Substitution of this into (20) gives, after an interchange of the time integrations, the following expression for the liftup **l**:

$$l = \int_0^{\infty} C(y,y_1) \frac{F_o(\xi_1,y_1,\zeta_1)-F_o(\xi,y,\zeta)}{\bar{U}(y)-\bar{U}(y_1)} dy_1 \quad (22)$$

where

$$F_o(x,y,z,t) = \int_{-\infty}^{x} [\phi_o(x_1,y,z) + \bar{W}'' l_z]dx_1 + \bar{U}'' l,\ \xi_1 = x-\bar{U}(y_1)t,\ \zeta_1 = x-\bar{W}(y_1)t \quad (23)$$

Since **l** appears on both sides of (22) it is necessary to use iteration, or time stepping (Landahl, 1990), to calculate its value.

We consider first the case of a one-dimensional parallel shear flow, with $\bar{W}=0$. For this it follows from (22), (23) that **l**, and from (14) u as well, will tend to a finite value as $t/t_s \to \infty$ and become a function of x alone, not of x and t, separately. From (21) and (23) it follows, upon substitution of ξ_1 as an integration variable instead of y_1, that then v will decay for large times as 1/t. Hence from (18) one sees that **P**, as well, will become a function of x, only. From (17) and (18) it thus follows that, in the limit of $t/t_s \to \infty$, both u and w will tend to functions of the Lagrangian variable x, only (and of y and z), but not of x and t, separately. In the analysis that follows we will denote these limiting solutions for u and w by u_∞ and w_∞, respectively.

The most important term in the expression (14) for the streamwise perturbation component is that given by the first term, which is proportional to the fluid element liftup **l**, and is the fundamental one appearing in Prandtl's (1925) mixing-length theory. The integral formulation (22), (23) is of course equivalent to the Rayleigh equation (19) which may provide exponentially growing solutions for a disturbance wave train for a shear flow having an inflection point in the mean flow component in the direction of the wave train. The advantage with the integral formulation is that it allows a possibility for a more direct study of how a disturbance of finite streamwise dimensions evolves into an unstable wave packet without the need for the normal-mode expansion of the initial disturbance (which would also need incorporation of the contribution from the continuous spectrum).

The above formulation it is particularly useful for the study of the behavior of the zero streamwise wave number component. This can be obtained most simply by intergrating the the disturbance equations over the streamwise variable (Landahl, 1980). Let overbar denote integrals over x, i.e.,

$$\bar{u} = \int_{-\infty}^{\infty} u dx \quad (24)$$

and similarly for the other components. Integration of (19) (considering first $\bar{W}=0$) gives

$$\bar{v}(y,z,t) = \bar{v}_o(y,z) \tag{25}$$

index o denoting initial values. Similarly one finds

$$\bar{u} = -t\bar{U}'\bar{v}_o(y,z) + \bar{u}_o(y,z) \tag{26}$$

$$\bar{l} = t\bar{v}_o(y,z) \tag{27}$$

the latter obtained from $\bar{D}\,l/Dt = \partial\bar{v}/\partial t$ and the initial condition $l_o = 0$.

This linear time dependence of the integrated quantities (i.e., their wave number zero streamwise component) is the consequence of algebraic instability (Landahl, 1980). Through this, which could arise for any parallel shear flow whether inflectional or not, the total streamwise perturbation momentum, but not u, will grow linearly with time. From (27) it is seen that this growth occurs because the streamwise dimension of the disturbed region will grow linearly with time, leading to the formation of low- or high-speed regions highly extended in the streamwise direction, i.e., low- or high-speed streaks. The calculations for the near-wall region carried out on basis of the model described (Landahl, 1990) show results in good qualitative agreement with the experiments (Kline et al. 1967, Kim et al., 1971) and numerical simulations (Moin and Kim, 1982, Alfredsson et al., 1991).

When a spanwise mean velocity component $\bar{W}$ is added, the algebraic instability takes on a different character. Integration of (19) over x gives for $\bar{v}$ the Rayleigh-type equation (with $\bar{W}$ replacing $\bar{U}$)

$$(\partial/\partial t + \bar{W}\partial/\partial z)\nabla^2\bar{v} - \bar{W}''\bar{v}_z = 0 \tag{28}$$

showing that v may experience exponential growth for spanwise travelling wave trains if $\bar{W}(y)$ has an inflection point. Integration of (22), (23) also shows that $\bar{l}$ will grow at least as fast as t^2 for small t if $\bar{v}_0$ has any spanwise variation.

From (22), (23) one also finds that $l \sim y$ as $y \to 0$, hence this Prandtl term automatically satisfies the noslip condition at the wall. The contribution from the pressure gradient, given by

$$\Delta u_p = -p_x/\rho, \tag{29}$$

will, following (18), also vanish at the wall for large times, if the initial velocity distribution is chosen so as to satisfy the noslip condition. A non-vanishing Δu_p at the wall may occur if the initial conditions are set by inviscid or nonlinear processes, or by disturbances originating upstream.

The shear interaction solution for large t/t_s gives an intermediate stage of eddy development which, in addition to producing the streaky structures (Landahl, 1990), reproduces many of the other interesting qualitative flow features seen in the experiments (Kline, et al., 1967, Kim et al., 1971) and numerical simulations (Moin and Kim, 1982, Johansson et al., 1991). For the streamwise perturbation vorticity $\gamma_1=\partial w/\partial y-\partial v/\partial z$ one finds (since $v\to 0$) that, in this long-time limit

$$\gamma_1 = -t\bar{U}'w_{\infty\xi} + w_{\infty y} \tag{30}$$

subscripts denoting partial derivatives. Similarly for the spanwise component $\gamma_3 = \partial v/\partial x-\partial u/\partial y$,

$$\gamma_3 = \partial v/\partial x-\partial u/\partial y = t\bar{U}'u_{\infty\xi} - u_{\infty y} \tag{31}$$

For large times the first term in both (30) and (31) will dominate. Thus, both the streamwise and the spanwise components will intensify for large times. The mechanism responsible for this is the stretching (in case of γ_3) and tilting (in case of γ_1) of the mean spanwise vorticity $-\bar{U}'$ by the gradients of the spanwise perturbation velocity component, w, which in the long-time limit becomes independent of time in a frame of reference moving with the fluid element, thus leading to a vorticity intensification that grows linearly with time. The enhancement of spanwise vorticity through this mechanism, together with the rotation by the mean flow, will lead to the formation of a spanwise thin shear layer which is highly tilted in the streamwise direction, as has indeed been found from conditional sampling of experimental and numerical results (Johansson et al., 1991). Similarly, the intensification of streamwise vorticity, as shown by the shear interaction solution, may lead to the formation of concentrated vortices oriented primarily in the streamwise direction. Of course, viscosity will eventually limit the shear intensification due to the interaction of the disturbance with the mean shear, as will be seen in the analysis below.

In the viscous and nonlinear stages we are to consider $t/t_v=O(1)$ and $t/t_n=O(1)$, respectively. Therefore, since $t_s/t_v<<1$ and $t_s/t_n<<1$, in the analysis of these later stages, we may, in the spirit of matched asymptotic expansions, consider the limiting results for $t/t_s\to\infty$ as the "outer" limits of the "inner" shear interaction solution (in time), which provides the "inner" limits of the "outer" viscous and nonlinear solutions. Thus, the limiting shear interaction solution is used to supply the initial conditions for the viscous and nonlinear stages of the eddy evolution. This asymptotic viewpoint makes it possible to simplify the problem of the eddy evolution considerably, since it allows separate treatment of the three stages. We will first consider the viscous stage.

IV. VISCOUS STAGE

On the basic assumption that the three fundamental time scales t_s, t_v and t_n are well separated we shall, in treating the viscous stage, consider t_v/t_n to be small, so that for $t/t_v=O(1)$ the nonlinear effects have not yet had time to develop. Therefore, the viscous stage can be treated by linear theory. For the streamwise component we have

$$\bar{D}u/Dt + \bar{U}'v = -(1/\rho)\partial p/\partial x + \nu\nabla^2 u \tag{32}$$

For $t/t_s \to \infty$, v and p both tend to zero, hence we have the following simplified equation to solve:

$$\bar{D}u/Dt = \nu\nabla^2 u \tag{33}$$

which for zero viscosity yields the result $u = u_\infty(\xi,y,z)$, in consistency with the linear inviscid solution for the shear interaction stage. For small viscosity one might therefore consider a solution of the form

$$u = f(\xi,y,z,t) \tag{34}$$

For the Laplacian of u one then has

$$\nabla^2 u = f_{\xi\xi} + f_{yy} + f_{zz} - 2\bar{U}'t\, f_{y\xi} + (\bar{U}'t)^2 f_{\xi\xi} \tag{35}$$

For large times the last term will dominate, hence the equation for u may be approximated by

$$\partial f/\partial t = \nu(\bar{U}'t)^2 f_{\xi\xi} \tag{36}$$

which may be reduced to the ordinary diffusion equation with the aid of the substitution

$$\tau = \nu(\bar{U}')^2 t^3/3, \tag{37}$$

which yields

$$\partial f/\partial\tau = f_{\xi\xi} \tag{38}$$

The solution of this, subject to the initial condition $f(\xi,y,z,0)=u_\infty(\xi,y,z)$, is given by

$$f = \frac{1}{\sqrt{4\tau\pi}}\int_0^\tau u_\infty(\xi_1,y,z)\exp[-\frac{(\xi-\xi_1)^2}{4\tau}]d\xi_1 \tag{39}$$

From τ one finds the viscous time scale to be

$$t_v = [L^2/\nu(\bar{U}')^2]^{1/3} \tag{40}$$

The solution (39) describes the effect of viscous diffusion on the convected disturbances and shows that the linear intensification of the shear with time obtained from (30) or (31),

$$\partial u/\partial y = -\bar{U}'t\, f_\xi + f_y \tag{41}$$

exhibited by the first term on the right-hand side will not continue to grow forever but will eventually, for $t/t_v = O(1)$, be quenched by the viscous diffusion. For the inner limit solution u_∞ in the form of a Gaussian hat one finds (Landahl, 1990) that the shear will grow to a maximum at $t/t_v=(3/2)^{1/3}$ and then decay.

The above analysis bears a definite resemblance to that employed in linear stability theory for the "inner viscous region" at the so-called critical layer, at which the instability wave moves with a phase velocity equal to the mean velocity. In the stability theory there appears also an "outer" viscous layer located at the wall and governed by a different scaling. In the present problem one must also consider the possibility that a viscous correction near the wall is required because of the nonslip condition there. In the expression (13) for u, the contribution from the term involving the liftup does not need any such correction to lowest order, because **l** is proportional to y and thus gives zero velocity there (however, a first-order viscous wall layer correction to $\partial u/\partial y$ may be needed). On the other hand, the contribution due to the induced streamwise pressure gradient, given by (22) may (however unlikely) need a viscous correction. One can construct this by considering the solution of

$$\frac{\partial u}{\partial t} = \nu \frac{\partial^2 u}{\partial y^2} \tag{42}$$

with the initial condition $u = -\Delta u_p = \mathbf{P}_x$ at $t = 0$. This is the well known problem of determining the viscous flow due to a plane boundary suddenly moved in a fluid at rest, and which has the solution

$$\Delta u_{vwall} = \mathbf{P}_x \operatorname{erf}[y/\sqrt{4\nu t}] \tag{43}$$

Similarly, the viscous wall correction to the spanwise component of the velocity perturbation induced by pressure gradients is given by

$$\Delta w_{vwall} = \mathbf{P}_z \operatorname{erf}[y/\sqrt{4\nu t}] \tag{44}$$

V. NONLINEAR STAGE

We turn now to the nonlinear stage of eddy evolution. We assume that the nonlinear time scale is substantially shorter than the viscous one so that the nonlinear development can be handled by inviscid dynamics. If the converse were true, namely that the nonlinear time scale were much larger than the viscous one, the viscosity would have made the disturbances decay before nonlinearity had time to set in, so that the nonlinear effects would not be interesting. The nonlinear inviscid case is also the primary one to consider for arriving at an understanding of how a turbulent eddy may be regenerated, since in the absence of nonlinearity the eddy motion would eventually decay. For the case of $t_n/t_v=O(1)$ the effects of viscosity and nonlinearity must be treated together, which requires a more difficult analysis not considered here.

From the behavior of the shear interaction solution for large t/t_s one sees that the pressure induced by the disturbance, equation (18), tends to zero, and the horizontal pressure gradients therefore may be neglected in the analysis of that stage of flow evolution. When the horizontal pressure gradients are neglected the nonlinear flow evolution may be handled most conveniently in a Lagrangian frame of reference, following Russell and Landahl (1984). Thus, the velocity components are taken to be functions of the material coordinates ξ,η,ζ and time, where ξ,η,ζ are the values of x,y,z at the time t=0 (now being considered as the "outer" nonlinear time, $t/t_n=O(1)$ for which the asymptotic large t/t_s behavior has been established). For the horizontal Eulerian coordinates one finds, when the horizontal pressure gradients are neglected, that

$$x = \xi + t[\bar{U}(\eta)+u_\infty(\xi,\eta,\zeta)]$$

$$z = \zeta + t\, w_\infty(\xi,\eta,\zeta) \tag{46}$$

From continuity it follows that ,in the transformation from the Lagrangian to the Eulerian system, the Jacobian of the transformation must satisfy

$$J \equiv \frac{\partial(x,y,z)}{\partial(\xi,\eta,\zeta)} = 1 \tag{47}$$

which gives

$$\frac{\partial\eta}{\partial y} = x_\xi\, z_\zeta - x_\zeta\, z_\xi \tag{48}$$

or, with the use of (45) and (46)

$$y - \eta \equiv l = \int_0^\eta \frac{d\eta}{(1+tu_{\infty\xi})(1+tw_{\infty\zeta})-t^2 u_{\infty\zeta} w_{\infty\xi}} - \eta \tag{49}$$

It follows from (48,49) that a singularity in y is possible for finite $t = t_{ns}$ for points for which

$$(1+t_{ns}\, u_{\infty\xi})(1+t_{ns} w_{\infty\zeta})-t_{ns}{}^2 u_{\infty\zeta} w_{\infty\xi} = 0 \tag{50}$$

which would occur at a critical time $t=t_{ns}$, of order t_n , required to satisfy (50) in some point in the disturbance flow field. The kinematical interpretation of the condition (50) is that the fluid elements converge horizontally to a point at that position making an infinite vertical velocity necessary at that point because of continuity. Of course, when the singularity is approached, pressure gradients and viscous stresses will no longer be negligible. How these effects may modify the behavior near the singular point is considered in recent papers by Cowley et al. (1990) and Van Dommelen and Cowley (1990), who discusses the use of this theoretical approach for the study of unsteady boundary layer separation. They find that their effects on the time and position for the appearance of the singularity are generally small. Thus, the direct action of adverse pressure gradients to produce boundary layer separation, which has been proposed in the past as a mechanism involved in sublayer bursting, does not seem to be an important factor in this process.

Particle paths in a turbulent boundary layer near such a kinematic singularity computed from (48) (Landahl, 1988) show remarkable similarity to those observed during sublayer bursting in the visualization experiments by Kim et al.(1971).

A second kind of strong nonlinear development is that associated with local small-scale instability of the shear layers produced by the stretching and tilting of the mean shear by the disturbance. A characteristic time scale for this instability is given by the inverse of the induced shear in the layer of enhanced shear and its characteristic length scale is set by the shear layer thickness. The instability time scale may be smaller or greater than the nonlinear time scale t_n, which essentially measures the time required for the distortion of the initial flow field to reach the the critical time t_s

given by (50). How a local small-scale instability can create large-scale vortical motion was discussed in Landahl (1990). He showed that the Reynolds stresses generated by a patch of small-scale instability will induce moments of momenta into the large-scale flow causing vortical motion in the large-scale flow to appear with components in both the spanwise or streamwise directions depending on the instability wave direction. In particular, a local cross-flow (spanwise) instability wave with only a small offset angle to the mean flow direction will generate streak formation.

VI. TURBULENCE MODELLING

The model presented above could also be used for modelling of the turbulent momentum transfer terms in the near-wall region. The behavior with y near y=0 of these is particularly important for the application of standard turbulence models such as the k-ε model or algebraic eddy viscosity models in which Van Driest type damping factors must be introduced for the near-wall behavior to obtain good results throughout the boundary layer. Especially for three-dimensional boundary layers, the results are found to be quite sensitive to the form of damping factors used (Lindberg, 1991). Here we will give a brief sketch of how the formulation for the evolution of a localized disturbance could be used to construct a near-wall turbulence model.

The fundamental starting relation is that given by (6) in which the nonlinear terms on the right-hand side are regarded as given source terms, which could be expected to be locally intense and intermittent in regions of strong nonlinear activity, such as where shear-layer instability or nonlinear singular behavior as described by (49), (50) may take place. The evolution of the disturbances initiated by each such event could then be followed with the aid of the model. Assuming the time scale separations as stated above one would expect that the main dynamical behavior would be that described by the shear interaction stage, i.e., that the disturbance evolves during this initial stage primarily in a linear inviscid manner. The net effect of the intermittent sources initiated at different instances and positions could then be determined by Duhamel superposition and the desired averaged quantities such as $\langle u^2 \rangle$ and $\langle uv \rangle$ obtained in the classical Wiener manner.

The Reynolds shear stress is most easily expressed in terms of the lift-up variable l. From (14), and the definition $v=\bar{D}l/Dt$ one has, neglecting the effect of the streamwise pressure gradient on u,

$$\langle uv \rangle = -\bar{U}'\langle l\bar{D}l/Dt \rangle \tag{51}$$

which, when averaged over the averaged lifetime t_m over the eddy, gives

$$\langle uv \rangle = -(1/2)\bar{U}'\langle l^2 \rangle / t_m \tag{52}$$

For the mean-square streamwise velocity fluctuations one finds in a similar manner, again when the contribution to u from the streamwise pressure gradient is ignored

$$\langle u^2 \rangle = [\bar{U}']^2 \langle l^2 \rangle \tag{53}$$

Since, according to the linear inviscid theory, l grows linearly with time, the streaks will overlap for large times, therefore $\langle l^2 \rangle \sim t$ (cf random walk theory). Thus, the use of the shear interaction stage solution requires for consistency that one takes into account a finite lifetime of the streak, set either by viscous or nonlinear effects, or both. An estimate of the average lifetime may be obtained from the division of (53) by (52), which gives

$$t_m = - (1/2\bar{U}')\langle u^2\rangle/\langle uv\rangle \quad (54)$$

A preliminary calculation using the data from the numerical simulations of the flow in a channel by Moin and Kim (1982) indicates that t_m is fairly constant with a value of 20 - 30 in the inviscid wall region outside the viscous wall region, $y^+ < 5$ The viscous time scale (39) yields a value for the lifetime of about 20 for a typical streamwise length scale of about 100 in wall units which appears to be fairly consistent with (54).

VII. CONCLUSIONS

Despite the extreme idealizations on which the proposed model is based, it is able to point to plausible dynamical mechanisms for the evolution of the turbulent eddies in the wall layer seen in the experiments and in the numerical simulations. Central to the model is the early shear interaction evolution stage of the eddy, which, because of the assumed distinct separation of the shear interaction time scale from the viscous and nonlinear ones, is described by a linear inviscid theory. The long-time asymptotic shear interaction solution can, in the spirit of the method of matched asymptotic expansions, be used to provide initial conditions for the viscous and nonlinear stages. The solutions for these later stages show both how viscosity can limit the vorticity intensification due to stretching and tilting of the mean vorticity, and how nonlinearity may act to produce a singularity in the nonlinear stage, manifesting itself as a highly localized vertical jet reminiscent of the ejection motion observed in the early visualization experiments (Kline, et al., 1967, Corino & Brodkey, 1969, Kim, et al., 1971)

The long-time shear interaction limit by itself provides an intermediate stage solution showing many interesting properties. It gives that the horizontal perturbation velocity components become constant in a frame of reference moving downstream with the mean velocity. Through the action of stretching and tilting of the mean vorticity, which in the absence of viscosity and nonlinearity will continue forever in this Lagrangian frame of reference, both the streamwise and spanwise vorticity components continue to intensify forever so as to become proportional to time in the long-time limit. This simple mechanism may provide an explanation for the intermittent appearance of intense shear layers, and streamwise vortices, in the wall layer. The shear interaction solution also demonstrates how streamwise streaks (not to be confused with the streamwise vortices) may form through the liftup of fluid in regions continuously growing in the streamwise direction due to algebraic instability (Landahl, 1990).

Instability of shear layers formed in this manner, and the strong ejection due to the nonlinear effects from horizontal convergence of fluid elements, can act as sources for a new initial disturbance setting off a new eddy and thus make it regenerate (Landahl, 1975). Eddy regeneration is an essential part of the maintenance of wall turbulence. Mechanisms of such kinds may also be operative in the sequence of events leading to the transition of a laminar flow to turbulence. Whether horizontal convergence of fluid elements as described by (50), or local shear layer instability is the dominating mechanism in this regeneration process is still not clear. For fully developed turbulence in a free shear flow, the mechanisms exhibited by the shear interaction solution may also be important, for example in understanding rapid distortion theory.

The new formulation is expected to be particularly useful in the instability problem for the study of how an initial disturbance in a boundary layer evolves over long times to produce unstable wave packets, streaks, and secondary instabilities. Some initial efforts of numerical calculations along such lines appear quite promising.

Professor Keith Moffatt kindly pointed out to me that some of the results presented in this paper, such as the viscous interaction time scale, could also be found from an analysis based on a spectral treatment, as shown in his 1965 paper.

REFERENCES

Corino, E.R. & Brodkey, R.S. 1969 A visual investigation of the wall region in turbulent flow. J. Fluid Mech. **37**, 1.

Cowley, S.J., L. L. Van Dommelen, L.L. & S.T. Lam, S.T. 1990 On the use of Lagrangian variables in descriptions of unsteady boundary layer separation. Transactions of the Royal Society, Series A **333**, 343.

Kim, H.T., Kline, S.J. & Reynolds, W.C. 1971 The production of turbulence near a smooth wall in a turbulent boundary layer. J. Fluid Mech. **50**, 133.

Kline, S.J., Reynolds, W.C., Schraub, F.A. & Runstadler, P.W. 1967 The structure of turbulent boundary layers. J. Fluid Mech. **30**, 741.

Landahl, M.T., 1975 Wave breakdown and turbulence. SIAM J. Applied Math. 28, 775.

Landahl, M.T. 1980 A note on algebraic instability of inviscid parallel shear flows. J. Fluid Mech. 98, 243.

Landahl, M.T. 1988 Linear and non-linear mechanisms in boundary layer turbulence. International J. for Num. Methods in Fluids 8, 1183.

Landahl, M.T. 1990 On sublayer streaks. J. Fluid Mech. 212, 243.

Lindberg, P.-A. 1991 Near-wall turbulence models for 3D boundary layers IMHEF REPORT T-91-9, EPFL, Lausanne.

Moffatt, H.K. 1965 The interaction of turbulence with strong wind shear. In Proceedings of the URSI-IUGG International Colloquium on Atmospheric Turbulence and Radio Wave Propagation, eds. Yaglom A.M. & Tatarsky, V.I., Nauka, Moscow.

Moin, P. & Kim, J. 1982 Numerical investigation of turbulent channel flow. J. Fluid Mech. 118, 341.

Prandtl, L. 1925 Bericht uber Untersuchungen zur ausgebildeten Turbulenz. Z. angew. Math. Mech. 5, 136.

Robinson, S.K 1991 Coherent motions in the turbulent boundary layer. Ann. Rev. Fluid Mech. **23**, 601.

Russell, J.M. & Landahl, M.T. 1984 The evolution of a flat eddy near a wall in an inviscid shear flow. Phys. Fluids **27**, 557.

Van Dommelen, L.L. & Cowley, S.J. 1990 On the use of Lagrangian description of unsteady boundary-layer separation. J. Fluid Mech. **210**, 593.

Active Control of Skin Friction and Separation in Certain Class of Wall-bounded Flows

I. Wygnanski

Abstract. Experimental investigations of large coherent structures in turbulent shear flows bypassed the boundary layer in the presence of a strong, adverse-pressure-gradient and the wall-jet. Both are wall bounded flows having one characteristic in common: their mean velocity profile is inviscidly unstable to two dimensional perturbations. Thus the identification of the large coherent structures with the predominant instability modes might be extended to this class of flows and could be quantitatively analyzed. The apparent similarity between the wall-jet and a combination of a free jet and a boundary layer is explored was and the relevance of the solid surface to the evolution of the large coherent structures was assessed. From this point of view, the boundary layer in a strong adverse pressure gradient, is regarded as a wake evolving in the vicinity of a solid surface. In both flows the significance- of the outer region is accentuated while the no-slip conditions at the solid surface are maintained.

The response of these flows to external, two dimensional excitation is currently being investigated in an attempt to further the understanding of the interactions between the inner and the outer structures in a turbulent boundary layer. While doing so, two technologically important effects were discovered. The mean flow in the wall-wake remained attached in spite of the strong adverse pressure gradient which caused separation in the absence of the excitation while the skin friction in the wall-jet was reduced as a consequence of the excitation.

Introduction

Investigations of large coherent structures in turbulent shear flows concentrated either on wall bounded flows like a boundary layer or a channel (Willmarth, 1975a, 1975b) or on free shear flows like the mixing layer, the wake and the jet (Ho & Huerre, 1984, Wygnanski et al, 1986; Weisbrot & Wygnanski, 1988). In the latter category of flows, the large coherent structures were clearly identified as the predominant instability modes and were quantitatively analyzed in this context. In fact, one may formally show that the coherent eddies existing in a two-dimensional, turbulent, mean flow are, to a first approximation, governed by the Orr-Sommerfeld equation provided one decomposes the equations of motion into coherent and incoherent motion (Hussain, 1983), one neglects the second order products of both motions and one restricts the divergence of the mean flow. The triple decomposition, therefore, enables one to link the hydrodynamic stability analysis with quantitative definitions of coherent structures.

However, the large coherent structures in wall bounded flows are much more complex and more difficult to identify then in free shear flows. Many forms of such structures were observed visually (Kline et al, 1967; Falco, 1977) but no consensus was reached as to their origin and their precise association with the enhancement of the skin friction or with the rate of growth of the boundary layer. Since Hussain and Reynolds (1970, 1972) failed to excite growing two-dimensional modes in a turbulent channel flow, it was tentatively concluded that plane instability modes are not related to the large coherent structures observed in channels and boundary layers. However, instabilities akin to the ones responsible for the generation of hair-pin vortices in a laminar boundary layer (Craik, 1985), may still be identified with the creation of large coherent structures in fully turbulent wall bounded flows. In fact "sub-layer streaks" were recently modelled by Landahl (1990), by using an instability concept associated with three dimensional intermittent disturbances growing algebraically in the direction of streaming.

Free shear flows are inviscidly unstable, while boundary layers in the absence of an adverse pressure gradient are not (Fig. 1). There is enough evidence (Katz et al, 1989) to suggest that a boundary layer on the verge of separation resembles a free mixing layer and it responds to external stimuli, like a mixing layer. The receptivity of this boundary layer is attributed to an inviscid instability which may dwarf any other form of instability and generate large, predominantly two-dimensional, coherent structures. The wall-jet is inviscidly unstable (Fig. 1) in its outer region and may thus possess large coherent structures characteristic of a plane turbulent jet. The wall-jet might be the ideal flow configuration for resolving the intricate interactions between the outer and the inner structures in a turbulent boundary layer, because it offers a larger degree of flexibility and controllability of flow parameters than a boundary layer does, regardless of pressure gradient. This flexibility stems from the fact that, the vorticity in the outer layer depends on the added momentum flux which for a given free stream and jet efflux velocity depends also on the dimension of the nozzle. Thus the susceptibility of the wall-jet to external perturbations might also depend on the ratio between the wave-length of the perturbation, the width of the nozzle and the upstream boundary layer thickness. Consequently the structure of the wall-jet can be progressively altered and the importance of the outer vertical layer on the wall region, can be evaluated.

The boundary layer in a strong adverse pressure gradient, may be regarded as a wake adjacent to a solid surface and might be subjected to the same kind of stability analysis. In both flows the significance of the outer region is accentuated while the no-slip conditions at the solid surface are maintained. The vorticity near the surface of the wall-jet has an opposite direction to the vorticity in its outer layer and may detract from its strength. On the other hand, the vorticity near the surface of the wall-wake may reinforce the vorticity in its outer region. The superposition of an external stream on the wall-jet weakens the vorticity in the outer region until it resembles a turbulent boundary layer in the limit. Although one has less flexibility with the wall-wake than with the wall-jet in an external stream, the understanding of the former flow is technologically very important. In particular, the control of the large scale eddies in a turbulent boundary layer having a vanishingly small wall stress (Stratford, 1959) might lead to a delay in boundary layer separation which is omnipresent in most flows around solid bodies. Preliminary observations associated with the wall-jet and wall-wake are presently discussed, stressing some features common to both flows. Many more measurements might be required before the concept is proven but the results obtained to date, are encouraging.

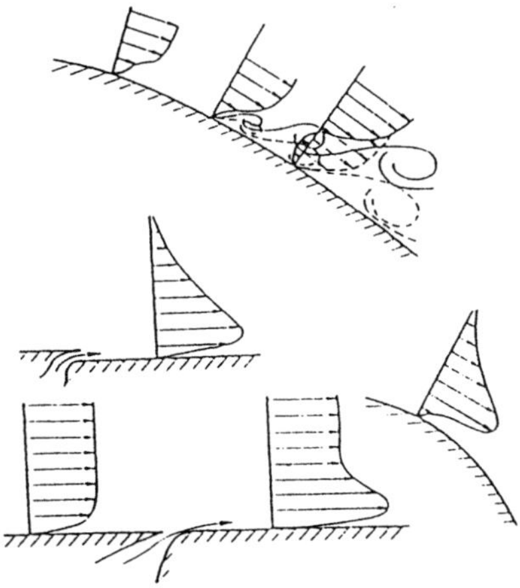

Fig. 1

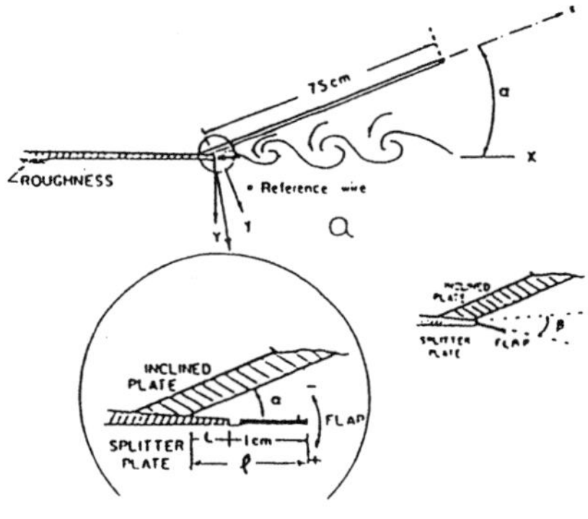

Fig.2 A schematic of the wedge flow apparatus.

The Mean Flow in the Forced Wall-wake

The flow over a solid convex wedge was investigated (Fig. 2). The wedge consisted of two long flat plates which were hinged together to provide a discontinuity in the slope of the surface and thus in pressure gradient. An 18° or a 22° wedge angles, which could not normally have been negotiated by the upstream boundary layer without separation, were used in the experiment. The different wedge angles depended on the detailed geometrical conditions at the discontinuity. A thin metal flap spanned the entire test section at the hinge location and, in the absence of the forced oscillations, the flow separated at its trailing edge. The surface of the splitter plate was roughened upstream in order to ensure that the boundary layer approaching the discontinuity was thick, turbulent and two-dimensional. The velocity profiles measured are typical for a turbulent boundary layer in this range of Reynolds numbers (i.e., the shape factor $H \equiv \delta^* / \theta = 1.6$ for $Re_{\delta^*} = 800$). The maximum turbulent intensity, u'/U is approximately 7% 50 mm upstream of the hinge. The displacement amplitudes of the flap did not exceed +l mm. The imposed streamwise component of velocity oscillations were measured by a hot-wire located in the potential flow near the flap. The velocity perturbation was recorded together with the forcing signal, and the data were ensemble averaged in order to remove the free stream turbulence of the wind tunnel and other sources of random disturbances from contaminating the results

The flow generated downstream of separation resembles the classical, unforced mixing layer observed in the absence of pressure gradient. The vorticity thickness, δ_v, obtained by fitting a straight line to the mean velocity profile at the point of inflection (Brown & Roshko, 1974) increases linearly with X, as it does in the classical mixing layer. The slope of the vorticity thickness $d\delta_v/dX = 0.23$ is in good agreement with previously made measurements in which the boundary layer on the splitter plate was tripped (Weisbrot & Wygnanski, 1988). The pressure recovery coefficient $C_p=(P-P_{init})/1/2\rho U_{init}^2$ was less than 0.1 over the distance of 60 cm because the flow was fully separated (Fig. 3).

Activating the flap at frequencies corresponding to Strouhal numbers based on initial momentum thickness $[St=f\theta/U]_{init}$, which were smaller than 0.01 and reference amplitudes $u'/U_{init}>0.25\%$, resulted in a reattachment of the flow to the surface and a pressure recovery of $C_p \cong 0.5$ over the corresponding distance considered before (Fig. 3). The flow was partially reattached at reference amplitudes as low as 0.1%. One may conclude, therefore, that the externally introduced oscillations were sufficiently amplified to allow the boundary layer to overcome the adverse pressure gradient imposed by the divergence of the walls and the existence of the flap at the discontinuity.

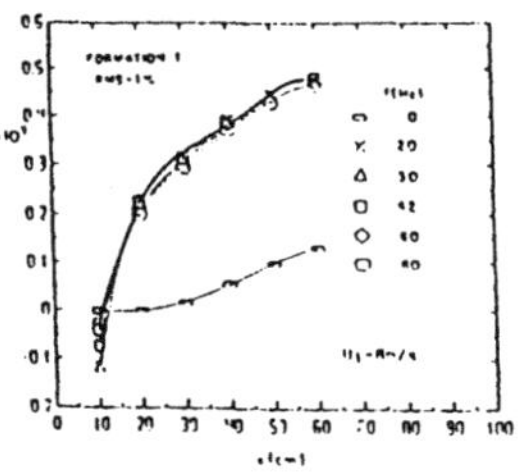

Fig. 3 The pressure distribution.

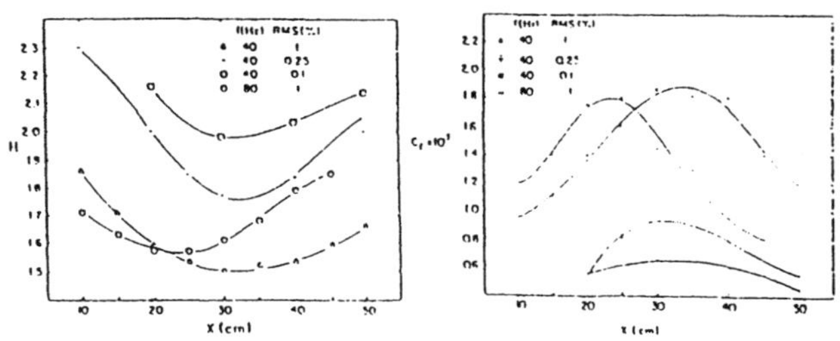

Fig. 4 The variation of H and C_f with X.

The dependence of some important boundary layer parameters, like the shape factor H, and the skin friction coefficient C_f, on X is plotted in Fig. 4 for some of the frequencies tested and for the 3 amplitudes at a single frequency f = 40 Hz. The momentum thickness increases almost linearly with X. The shape factor decreases initially with X and then gradually increases towards the the trailing edge of the plate, while the skin friction coefficient behaves in an opposite fashion. The X location at which H is smallest corresponds to the location at which C_f is largest, as might have been expected from the momentum integral equation provided the pressure increases monotonically with X. The location at which H has a minimum appears to depend primarily on the frequency of forcing and not on the amplitude, while the minimum value of H attained depends mostly on the amplitude of the imposed oscillations. The value of fX/U at which H attains its minimum is approximately 2, suggesting that the wave-length of the imposed oscillations determines the preferred reattachment distance.

Fig. 5 Mean velocity and turbulent intensity profiles meansured at X = 10 cms.

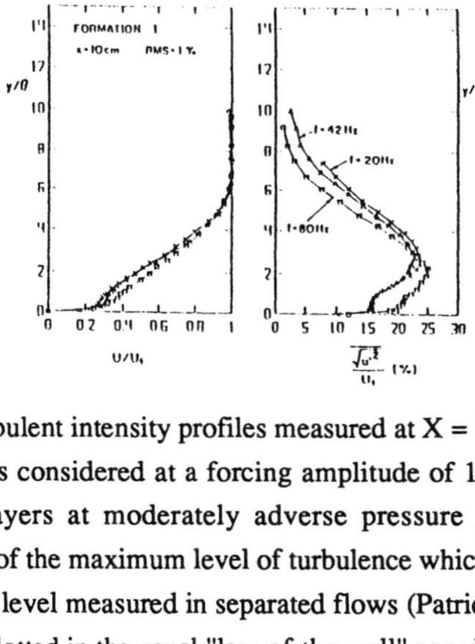

Representative mean velocity and turbulent intensity profiles measured at X = 10 cm are plotted in Fig. 5 for three frequencies considered at a forcing amplitude of 1%. These profiles are typical of boundary layers at moderately adverse pressure gradients (Bradsgaw, 1967) with the exception of the maximum level of turbulence which exceeds 20% and is more characteristic of the level measured in separated flows (Patrick, 1986). When the mean velocity profiles are plotted in the usual "law-of-the-wall" coordinates the universal logarithmic distribution remains unchanged regardless of the forcing frequency or amplitude:

$$U^+ = 5.6 \log Y^+ + 5 \qquad (1)$$

The maximum value of Y^+ at which the equation applies depends not only on the pressure gradient but also on the amplitude and the frequency of forcing. It can be best correlated with the dependence of H on these parameters. The general representation of the "law-of-the-wake" also applies in all cases considered, i.e.,

$$W = C \sin^2\left[\pi \frac{Y}{2\delta}\right] \qquad (2)$$

but the constant C depends on the frequency and amplitude of the forcing (Fig. 6). It seems that in a "wall-wake" there is a large region in which the turbulent stress is approximately constant in contradistinction to the measurements done in a wall-jet at moderate Reynolds numbers (Wygnanski et al, 1991)

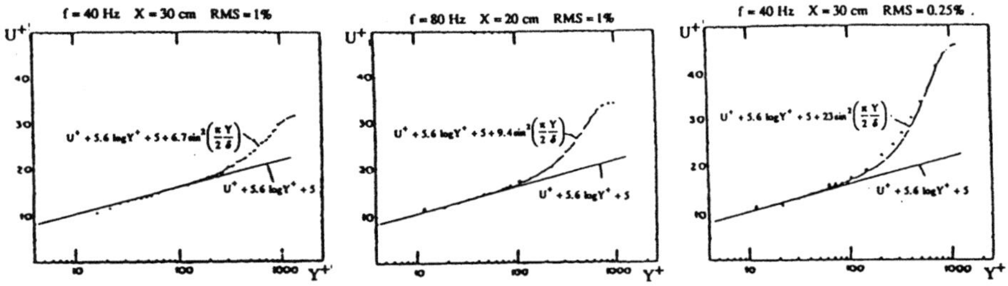

Fig. 6 Mean velocity profiles plotted in wall coordinates.

The Transformation of the Wall Jet in Streaming Flow to a Boundary Layer

The experimental research was carried out in a low turbulence wind tunnel. The wall-jet originated from a two dimensional nozzle flush mounted into the floor of the tunnel in order to minimize its effect on the evolution of the boundary layer. The slot-width was altered by moving the entire front plate in the X-direction and inserting appropriate filling strips in the tunnel floor. The jet turns rapidly in the direction of streaming and the effects of its deflection could not be detected beyond 10 slot widths in most test conditions. The velocity was mostly measured in the developed region of the wall-jet starting from 20 slot-widths downstream of the nozzle and extending sometimes beyond 200 slot-widths. The Reynolds number at the nozzle, Re_j, was altered by changing the efflux velocity, U_j, and the width of the slot, b. The ratio between the free stream velocity and the jet erflux velocity was also altered by varying the velocity in the tunnel, U_∞ and by varying U_j. Typical streamwise evolution of mean velocity profiles are shown in Fig. 7 where the top set corresponds to an initial velocity ratio $U_\infty / U_j = 0.085$, the central set to 0.59 and the bottom to 0.93. The effect of the upstream boundary layer is progressively stronger from top to bottom. These three sets of data were chosen because they may represent different categories of instability. The velocity profiles in the top data-set contain only one inflection point in the outer region. The outer region in this case resembles a free jet at all

X locations. There are two inflection points in the outer region of the wall-jet near the slot of the second set, underlining the importance of the upstream boundary layer in this case. The relatively strong U_j overcomes this velocity defect whereupon downstream of the fourth station the mean-velocity profile no longer contains two inflection points. The effect of the upstream boundary layer is much stronger in the bottom set which resembles the development of a wake in the vicinity of a solid surface. Since the ratio of the free stream velocity to the jet velocity approaches unity near the nozzle the flow appears to relax directly to an ordinary boundary layer without passing through the intermediate stage of a wall-jet embedded in an external stream. The wall-jet in streaming flow may therefore represent the boundary layer in the limit.

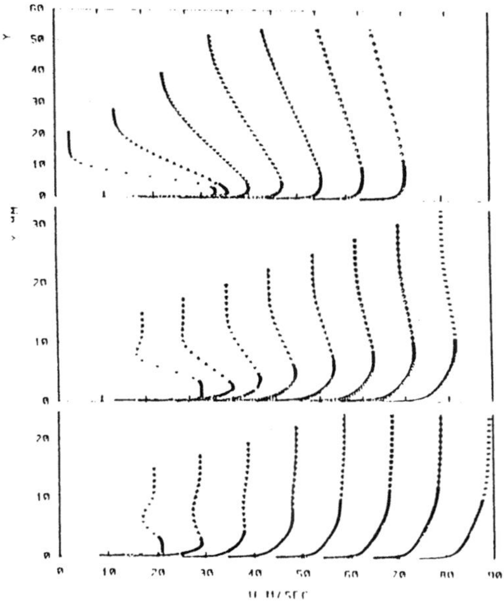

Fig. 7 The evolution of a wall-jet in an external stream.

The attainment of similar mean velocity profiles is not essential in analyzing the evolution of the large coherent structures but it is very convenient because it enables one to reduce significantly the amount of computations and measurements required to prove the concepts outlined. It is well known that a complete self preservation (i.e. self similarity of the velocity distribution and of the turbulent stresses) of a turbulent wall-jet may only be possible in especially tailored pressure gradient (Irwin, 1973). Nevertheless many investigators managed to collapse some of the data, the mean velocity profiles in particular, on self similar plots. Using a Gallilean transformation one should consider $(U_{max} - U_\infty)$ to be a dominant quantity governing the development of the flow. When local value of $[(U - U_\infty))/(U_{max} - U_\infty)]$ was expressed as a function $f(\eta)$ {where $\eta = Y/Y_{m/2}$

represents the dimensionless distance from the wall while $Y_{m/2}$ is the distance measured from the surface where the quantity $[(U-U_{\infty}))/(U_{max}-U_{\infty})] = 1/2\}$ most of the mean velocity profiles measured in each data-set, collapsed onto a single curve (Fig. 8). The collapse was much better when $U_{\infty}/U_j \leq 0.4$ than for higher ratios of U_{∞}/U_j. The shape of the mean velocity profile was affected by the ratio U_{∞}/U_j while a change in Re_j did not result in a visible alteration of the normalized shape of the velocity distribution. The rate of spread of the wall-jet is greatly affected by the velocity ratio and by Re_j (Fig.9). It varies almost linearly in the direction of streaming as it did in the absence of external flow.

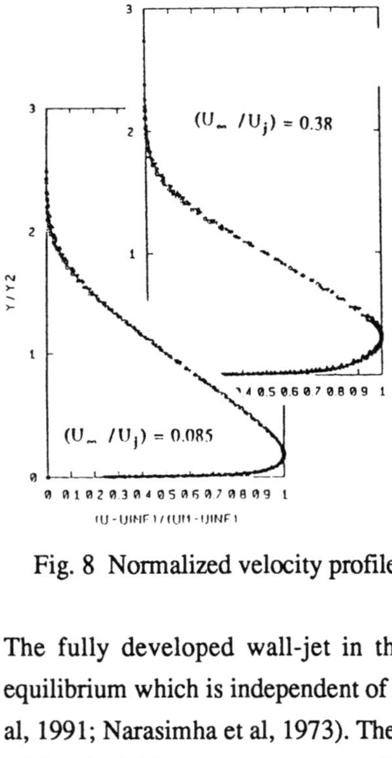

Fig. 8 Normalized velocity profiles.

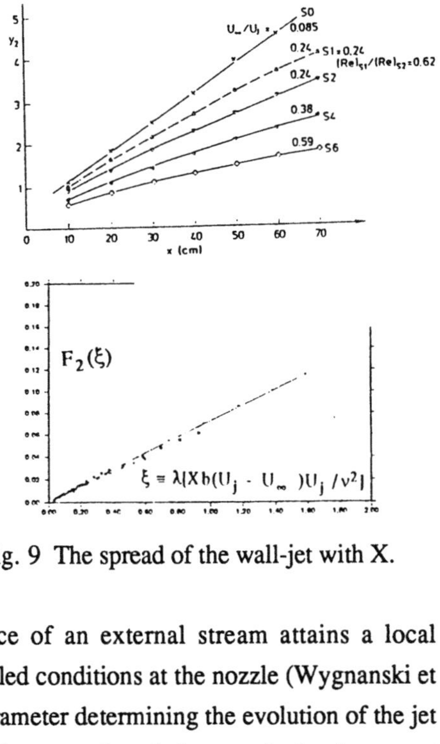

Fig. 9 The spread of the wall-jet with X.

The fully developed wall-jet in the absence of an external stream attains a local equilibrium which is independent of the detailed conditions at the nozzle (Wygnanski et al, 1991; Narasimha et al, 1973). The sole parameter determining the evolution of the jet is it's initial, kinematic momentum flux, J. This approach scaled correctly the three most important parameters in the wall-jet: the width of the flow $Y_{m/2}$ the decay or the maximum velocity in the jet, U_m and the local wall stress. τ_w

$$\frac{U_m \nu}{j} = F_1(\xi) \quad ; \quad \frac{Y_{m/2} J}{\nu^2} = F_2(\xi) \quad ; \quad \tau_w / \rho\left[\frac{\nu}{J}\right] = F_3(\xi) \qquad (3)$$

where $\xi = [XJ/\nu^2]$ while $J = U_j^2 b$ and U_j is the jet velocity at the nozzle exit. The width or the slot is b; ρ and ν are the density and kinematic viscosity or the fluid respectively. X is a streamwise distance measured from the nozzle.

It made these parameters independent of Re_j and enabled an independent evaluation of τ_w. The question of scaling took different proportions than originally anticipated because it appears that the flow was not scaled correctly in the past. The addition of a free stream complicates the problem because one regards the maximum velocity in the wall-jet as an excess velocity over the free stream, thus:

$$F_1(\xi) = (U_m - U_\infty)\nu / J \tag{4}$$

One has also to assume that the paramount parameter governing the flow is the excess momentum flux: $\int_0^\infty (u - U_\infty) u \, dY$ which for a uniform jet velocity in the absence of an upstream boundary layer degenerates to $b(U_j - U_\infty)U_j$ leading to two possible definitions of the streamwise dimensionless distance ξ:

$$\xi = Xb(U_j - U_\infty)U_j / \nu^2 \qquad \text{or} \qquad \xi = \lambda Xb(U_j - U_\infty)U_j / \nu^2 \tag{5}$$

where $\lambda = (U_j - U_\infty)/(U_j + U_\infty)$.

When the upstream boundary layer is thick, the simplified definition of excess momentum flux might no longer be adequate because one has to account for the momentum loss in the upstream boundary layer.

The utility of the definition of ξ becomes apparent when one substitutes the observed, self similar, normalized form of the velocity profile (Fig. 8), into the momentum integral equation

which makes $\tau_w/\rho\ (\nu/J)^2$ a function of ξ, λ and of $U_\infty/(U_m - U_\infty)$ while being independent of Re_j. Therefore, when $U_\infty/(U_m - U_\infty)$ is assumed to be negligible, the similarity conditions of the wall-jet in absence of an external stream are satisfied. When $U_m \rightarrow U_\infty$ a different set of conditions for self similarity prevail. This explains some additional difficulties in analyzing this flow.

The Forced Turbulent Wall-jet without an External Stream

This flow was examined first because of it is well known as being self similar with respect to the selected parameters. External excitation of free shear flows had a strong effect on a variety of time averaged quantities, like the mean velocity distribution, the rate of spread of the flow, the turbulent intensity and the Reynolds stress. It is thus natural to start the present discussion of the experimental results by exploring parametrically the effects of forcing on the same quantities in the wall-jet.

A variety of forced flows were considered in the present experiment, they differed in jet momentum; slot width; frequency and amplitude of the imposed oscillations; and finally the manner in which forcing was introduced to the flow. A change in each one of the above parameters required concomitant traverses across the flow at various X locations. A sample of the data is shown in Fig. 10. The velocity profile plotted on the left (Fig. 10a) exhibits self similarity with respect to $Y_{m/2}$ and U_m irrespective of the X location while the profile plotted on the right (Fig. 10b) proves the existence of self similarity irrespective of the changes in all other parameters considered. Both profiles are compared to the dimensionless, unforced, velocity profile (Wygnanski et al, 1991) and represented by solid curves. One may conclude that two dimensional external excitation does not affect the normalized form of the velocity distribution within the range of parameters considered. Replotting the velocities adjacent to the wall on an expanded scale (Fig. 10c), and comparing them to unforced velocity measurements at otherwise identical conditions, indicate small but consistent differences between the two sets of data. These differences will be discussed later in some detail.

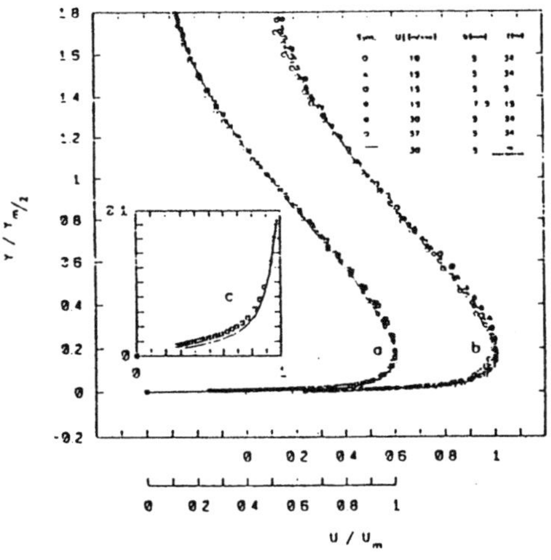

Fig. 10 Velocity profiles using self-similar coordinates.

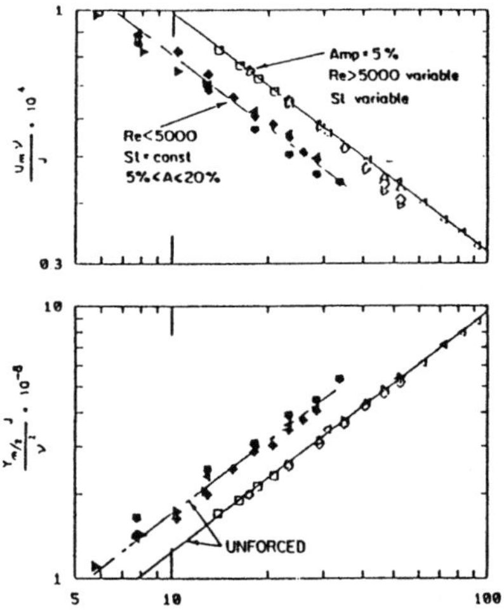

Fig. 11 The evolution of F_1 and F_2 with ξ.

We were surprised to discover that neither the rate of spread nor the decay of the velocity scale were affected by the forcing. The solid lines drawn in Fig. 11 represent the unforced data accumulated for $Re_j \geq 7500$ while the broken lines correspond to $Re_j \leq 5000$. The symbols plotted represent data corresponding to various frequencies of excitation at nozzle Strouhal numbers $St_j = f \cdot b/U_j$ varying between $3 \leq St_j \cdot 10^3 \sim 17$ (corresponding to $St = fn^3/J^2$ of $0.4 \leq St \cdot 10^{15} \leq 444$) and amplitudes, based on the streamwise velocity perturbation near the nozzle, ranging from 2% to 20%. Although some consistent differences depending on the level of forcing and on the Strouhal number can be detected they do not alter our initial conclusion about the lack of sensitivity of these mean-flow parameters to the external excitation. The cumulative effect of the threshold in Re_j occurring around $Re_j \cong 5000$ appears to be much more significant then that of the two-dimensional excitation. Dimensional analysis of the independent parameters in the externally excited wall-jet suggests that variables other than ξ, might affect the mean velocity distribution. Whenever the external excitation is harmonic, the broadening or the distortion of mean flow will depend on the square of the local amplitude of the velocity perturbations. In some cases the finite amplitude might be introduced directly by the forcing mechanism, while in others, the flow itself might act as an amplifier. We shall now examine the global effect of the excitation on the loss of momentum resulting from skin friction drag. The velocities plotted in Figs. 10c suggest that forcing the wall-jet might have an effect on the drag. Since the local skin friction was most conveniently determined from the slope of the mean velocity profile near the wall (Wygnanski et al, 1991), the procedure was repeated presently in-spite of the fact that it is not considered in the literature as being very reliable (Launder & Rodi, 1981). The velocity measurements plotted in Fig. 12 are shown in dimensional form because this representation enables the reader to assess the quality and the amount of data acquired in the viscous sub-layer. Sometimes more than 10 data points were included in the linear fit made to the velocity profile near the wall. There is no doubt that even a relatively low level of forcing (the data plotted correspond to an initial amplitude of 2%) results in a reduction of the wall shear stress, τ_w. Heat loss from the hot-wire to the wall was assumed to be negligible (Wills, 1962) since the flow was turbulent and most of the data was taken at distances from the wall ranging from 50 to 80 wire diameters while the Re based on the wire diameter varied from 0.7 to 4.

Plotting $\tau_w/\rho\ (\nu/J)^2$ versus ξ, and comparing the results with the skin friction measured in the absence of forcing indicates clearly the differences in τ_w in spite of the logarithmic .scale chosen (Fig. 13). At no instance did τ_w increase above its nominal unforced value. Reductions in the skin friction of approximately 10% were prevalent at most frequencies corresponding to initial excitation amplitudes which were lower or equal to 5%. However, reductions in τ_w of approximately 40% were also recorded by forcing at much higher amplitudes corresponding to 10% or 20% of the efflux velocity at the nozzle. Alternately, forcing at a preselected frequency which is amplified by the flow achieved the same drag reduction at a much lower input amplitude (Fig. 13).

In attempting to sort out the independent contribution of frequency, amplitude and Reynolds number on the local τ_w, one may assume that the dimensionless $\tau_w/\rho\ (\nu/J)^2$ depends on the distance from the nozzle XJ/ν^2 and on the local amplitude of the coherent perturbations present in the flow regardless of their origin. We computed the phase-locked, ensemble-averaged velocity signals and from them deduced the local r.m.s. levels of the coherent motion. These quantities were then integrated across the flow to provide a measure of the intensity of the coherent motion at a given X station. It transpired that a local coherent motion having an average amplitude of 0.3% may be responsible for a reduction in τ_w of approximately 15% while an additional increase in local amplitude by a factor of 10 resulted in an incremental reduction in $\tau_w/\rho(\nu/J)^2$ of an additional 15% only. This suggests that the reduction in the local skin friction is not linearly dependent on the strength of the coherent motion.

A comparison was made between the mean velocity distributions of the forced and the unforced wall-jets in the vicinity of the surface by using the conventional "wall coordinates". It seems that external excitation modifies the velocity distribution in what appears to be the logarithmic region. It was shown that the slope, A, of the logarithmic profile ($U^+ = A \log Y^+ + B$) appears to be a universal constant and is equal to 5.5 while the additive constant B was strongly dependent on Re_j ($4 < B < 10$). This conclusion applies to the forced wall-jet as well. It also appears that the logarithmic fit to the forced data may only be applied at larger values of Y^+ when compared to the unexcited flow.

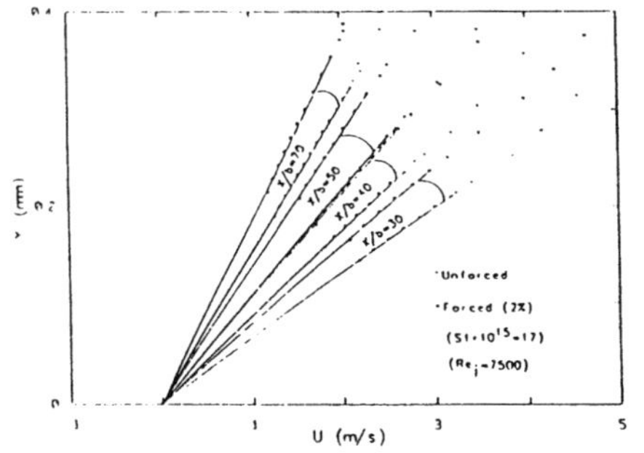

Fig. 12 The effect forcing on $\partial U/\partial Y$ near the wall.

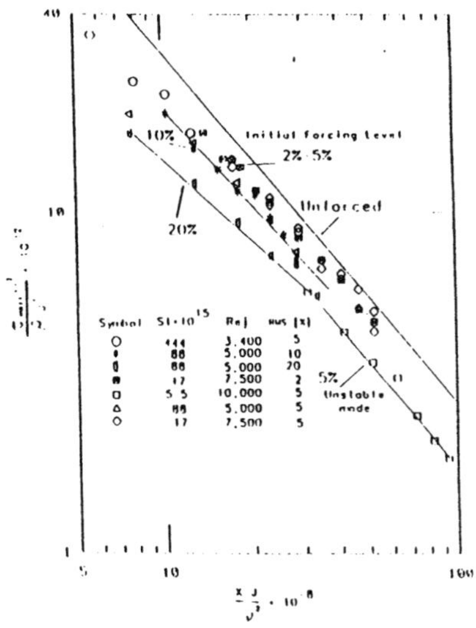

Fig. 13 The evolution of the dimensionless skin friction with ξ.

Some of the underlying assumptions used in deriving the logarithmic velocity profile do not apply to the wall-jet (i.e. that neither the logarithmic region is far removed from the location at which $Y=Y_m$ nor is the stress constant within that region). Consequently the existence of the logarithmic region appears to be somewhat fortuitous. The introduction of forcing did not only affect τ_w but also reduced the extent of the constant stress layer (Fig 14). The total and the viscous stress distributions in the inner part of the wall-jet are plotted at the top of this figure in order to assess the effects of excitation on these quantities. The wall-stress, which is entirely viscous at the surface, is reduced by the two dimensional forcing. However for $Y^+ > 10$, the effect of excitation on the viscous stress vanishes since forcing seizes to have an effect on the mean velocity profiles in this region, and thus on $\nu(\partial U/\partial Y)$. At $Y^+ > 30$ where the contribution of the viscous stress to the total stress becomes vanishingly small, the latter decreases rapidly with increasing distance from the wall and changes sign between $60 < Y^+ < 100$ depending on the forcing and on Re.

The lateral distribution of turbulent energy production $-(u'v')\partial U/\partial Y$ with increasing distance from the surface, also plotted on Fig. 14, indicates that external excitation reduces the Reynolds stress and with it the turbulent production. The integrated turbulent production in the inner region of the wall-jet between the surface and the location at which $Y = Y_m$ defined by:

$$\text{PRODUCTION} = \frac{1000}{U_m^3}\int_0^{Y_m} -\langle u'v'\rangle \frac{\partial U}{\partial Y}\, dY \qquad (6)$$

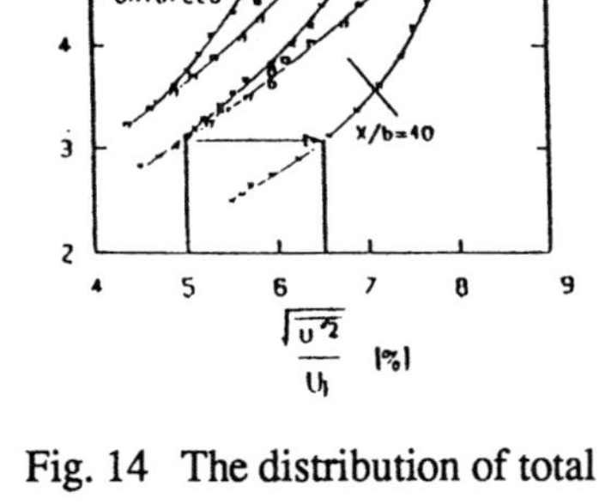

Fig. 14 The distribution of total and viscous stress and turbulent production near the surface.

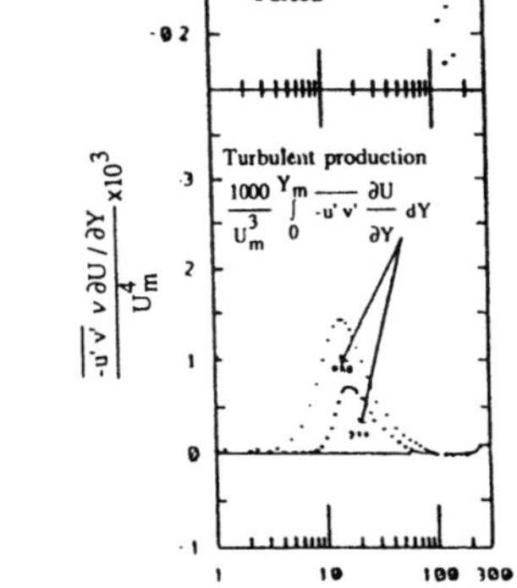

Fig. 15 The distribution of u' near the surface

is reduced by an approximate factor of 3 when the flow was forced at St $10^{15} = 5.5$ at X/b = 70 (i.e. forcing reduced the value of the production integral from 0.668 to 0.214).

This is a significant effect which presumably alters the entire turbulent energy balance in this flow and not just the scales of the large coherent eddies. Since the mean velocity gradient vanishes at $Y^+ \cong 150$, the Y locations at which $u'v' = 0$ and at which $\partial U/\partial Y = 0$ do not coincide (Kruka & Eskinazi, 1964) leading to a region of a weak but negative turbulent production just below the location at which $Y=Y_m$.

Forcing the wall-jet reduces the turbulent intensity in the vicinity of the surface and this reduction is strongly affected by St while all other parameters are maintained constant. The reduction in u'/U_j exceeds in some places 25% as may be deduced from figure 15 (see arrow comparing one data point at X/b=40 and St=5.5 which provides an example). Since the production of turbulence increases the level of u', the reduction in u' due to forcing might be a direct result of the reduced production level near the wall.

The Evolution of Harmonic Perturbations in the Wall-jet and Wall-wake

The propagation and amplification of small amplitude wavy disturbances, in slightly divergent-free shear flows was discussed by Crighton and Gaster (1976) and others. The application of this analysis to the wall-jet is, in principle the same, with the exception of the "no slip" boundary conditions at the wall and the need to use the viscous form of the Orr-Sommerfeld equation.

This problem too was analyzed by Gaster (1974) in conjunction with the stability of the divergent boundary layer. The stability of the non-divergent, laminar wall-jet was considered by Tsuji et al (1977) We shall therefore, not repeat these derivations and only dwell on some specific aspects stemming from the application of this model to the turbulent wall-jet.

Harmonic excitation of both flows enabled us to digitize the measured velocities at any streamwise location together with the forcing signal used to deflect the flap located at the apex of the wedge in the wall-wake case or near the nozzle in the wall-jet. This signal, plotted at the bottom of each figure (Fig. 16) provided the necessary phase reference with respect to which other measurements taken anywhere in the flow could be referred and ensemble-averaged. Since there is a possibility of a sub harmonic resonance the long time series were subdivided into segments which contain two waves of the forcing frequency. Samples of such phase-locked and ensemble averaged traces of the streamwise component of velocity are plotted above the forcing signal. The periodicity of the velocity signal is self evident but its harmonic distortion varies with Y. The least distorted signals appear at the outer edge of the flow where the velocity is not contaminated by turbulence. One may observe a slow phase advance in the maximum velocity recorded in the wall-jet with increasing distance from the wall provided $Y/Y_{m/2} < 1$. At this Y location, which approximately coincides with the inflection point of the mean velocity profile, a sudden phase shift occurs which is followed by a gradual phase delay with increasing Y. The data presented for the wall-wake accentuates the sudden phase shift in the outer flow. One may Fourier decompose these signals and plot the transverse distributions of amplitude and phase of the signals at the frequency of forcing. The transverse distribution thus obtained, is compared with the amplitude distributions calculated from the linear stability model. Both measured and calculated amplitude distributions are normalized by their local maxima which occurs fairly close to the wall.

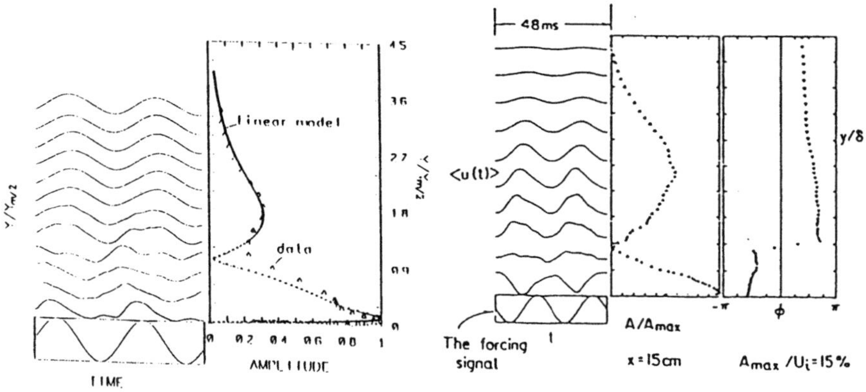

Fig. 16 The amplitude distribution in a forced wall-jet - comparison with linear model, and the amplitude and phase distribution in a wall-wake.

When similar comparison was done in highly unstable, unbounded turbulent shear flows like the mixing layer, the interaction between the coherent motion and the incoherent turbulence was not accounted for. Since these flows were dominated by an inviscid instability the inviscid Orr-Sommerfeld equation was solved, and the role of the Reynolds number did not have to be considered (Weisbrot & Wygnanski, 1988). The primary instability in these flows is affected by viscosity in spite of the fact that they are also inviscidly unstable. The no-slip condition at the surface provides significant viscous stresses and enhances the dissipation which may cause the harmonic motion to decay. The incoherent turbulent fluctuations may increase this dissipation and the non linear interactions between the coherent and incoherent fluctuations may provide a cascade mechanism through which the effects of viscosity are enhanced. All these interactions, which are not accounted for in the model, might be lumped together into an equivalent viscous term by introducing an eddy viscosity (Tam & Chen, 1979; Liu, 1971; Marasli et al, 1989). Since the Reynolds number appears as a parameter in the Orr-Sommerfeld equation a choice of a fictitious Re introduces an indeterminacy not present in the inviscid calculations. For example, the experimental results plotted in figure 16 agree reasonably well with theoretical model provided the eddy viscosity was 10 limes larger than the viscosity of the fluid.

The disturbances at the fundamental frequency in a wall-wake may also be viewed as an array of spanwise vortices extending across the entire boundary layer (Fig. 16). The 180° shift in phase which depicts the center of the eddies is convected along the surface at approximately one-half of the free stream velocity while moving away from the surface with increasing X.

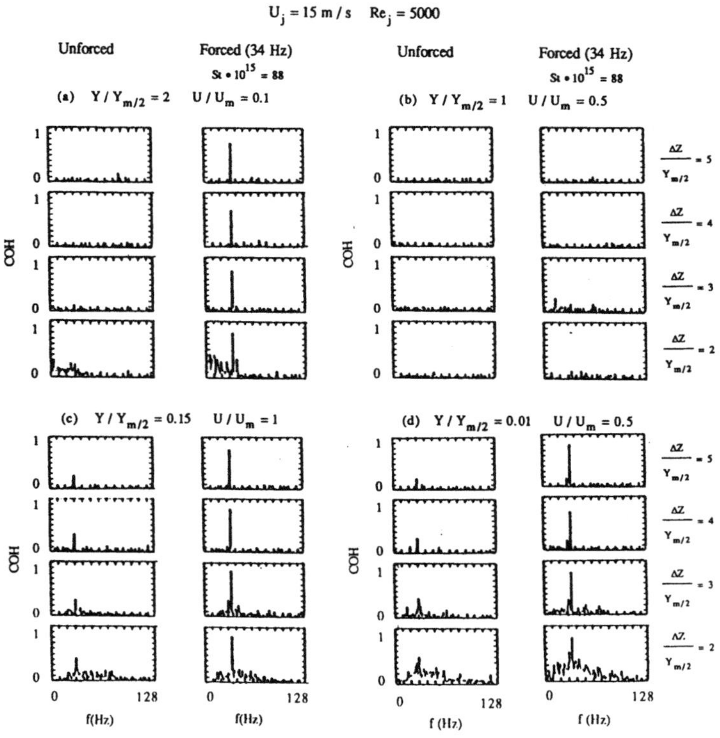

Fig. 17 Spanwise coherence in a wall-jet.

We shall now explore the changes that took place in the large scale structures in response to the external excitation. We shall particularly look for effects occurring near the solid surface in order to ascertain the reasons for the reduction in τ_w. The improvement in the two-dimensionality of the large scale structures is one such effect which had been observed in both the mixing layer and in the wake (Wygnanski et al, 1986; Wygnanski et al, 1979). The coherence spectra of the velocity fluctuations, sensed by two probes separated in the spanwise direction provides a convenient measure of the. degree of two-dimensionality attributable to each scale (or frequency) of the motion. Two point correlations of the broad-frequency signal also yield a measure of two-dimensionality which is biased somewhat towards the most energetic eddies. The data plotted in figure 17 represents coherence at spanwise separation distances ranging from 10 to 130 mm corresponding typically to more than 5 characteristic widths of the flow. The measurements were done at a variety of Y locations in the presence and in the absence of external excitation. The frequency chosen to be shown for the wall-jet was 34 Hz corresponding to $St=fv^3/J^2=88 \cdot 10^{-15}$. The results plotted in figure 17 correspond to four Y locations: starting at the outer edge of the wall- jet ($Y/Y_{m/2} \cong 2$), proceeding to the vicinity of the inflection point of the mean velocity profile ($Y/Y_{m/2} \cong 1$), going down to the location at which $U \cong U_m$ and ending in the vicinity of the wall. The spanwise coherence in the outer part of the unforced jet is negligible provided $\Delta Z/Y_{m/2} > 2$ (Fig. 17a). For $\Delta Z/Y_{m/2} \leq 2$ the coherence in the unforced flow exceeds the value of 0.15 over a broad range of low frequencies and Y locations, suggesting that the large scale structures occurring naturally in the flow are somewhat coherent over this span. There is also a distinct peak in the coherence level around 30 Hz. Within the inner part of the unforced wall-jet which is even noticeable at $\Delta Z/Y_{m/2} = 5$. If this St is associated with the natural coherence of the large eddies, then the forcing frequency used in the experiment is very close to being the frequency which undergoes naturally the highest level of amplification at this particular X location.

External excitation increased the coherence to 0.75 at the outer edge of the wall-jet (Fig. 17a) but failed to have any effect near $Y/Y_{m/2} = 1$ (Fig. 17b). This effect may be explained with the aid of figure 16 which is representative of the coherent amplitude distribution of the streamwise velocity fluctuations in the forced wall-jet. The maximum amplitude of the coherent motion in the outer part of the jet occurs around $Y/Y_{m/2} = 1.8$ while the minimum occurs at $Y/Y_{m/2} = 1$ and therefore one should not expect any improvement in the coherence in the vicinity of this Y location. The improvement in the coherence observed in the inner region of the wall-jet (at $Y/Y_{m/2} = 0.15$ & 0.01) can also be associated with the relatively large amplitude of the local oscillations. It is interesting to note that the spanwise coherence in the forced wall-jet did not diminish by increasing ΔZ and it was actually higher near the surface than in the outer edge of the jet.

Similar measurements are also done in the wall-wake (Fig. 18) where the maximum coherence in the unforced flow is approximately 0.3 beyond a separation distance of ΔZ = 30mm which is commensurate with the local boundary layer thickness. External excitation increased this value to 0.8 over most of the boundary layer with the exception of the height above the wall where the local velocity $U/U_1 \cong 0.7$. This height corresponds to the average location of the centers of the passing eddies as they were detected by the 180° phase shift in the phase locked amplitudes. Three dimensional plots of the spanwise correlation coefficient taken at various heights above the surface are shown in figure 19 for the forced and unforced flows. The contour plots representing the natural flow exhibit some periodic behavior near the surface of the wall-jet but none near the surface of the wall-wake because of the massive separation which occurs under these conditions. Forcing, not only increased the spanwise extent of the highly correlated region but made it much more periodic in time. The primary eddies appear to be two-dimensional and regular.

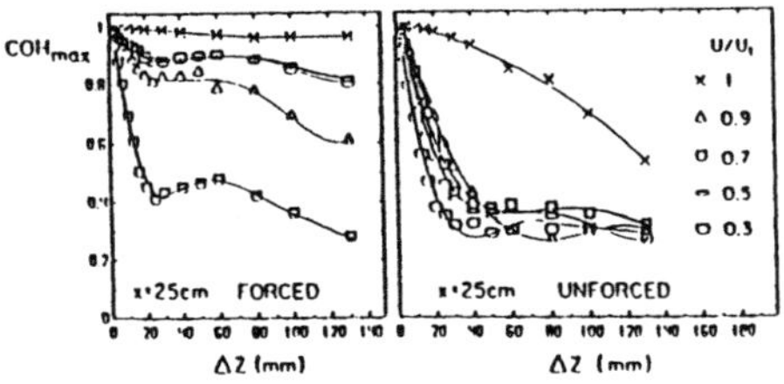

Fig. 18 Maximum spanwise coherence in a wall-wake.

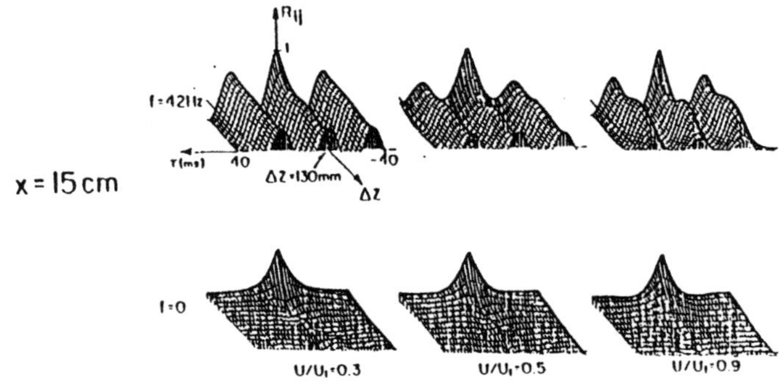

Fig. 19 Correlation measurements in the T-Z plane of a wall-wake.

The unexpected reduction in the mean-skin friction prompted us to examine the phase locked distribution of this quantity over two periods of the forcing frequency. The data was based on the phase locked analysis of $[\partial\langle U\rangle/\partial Y]_w$ measured at various streamwise locations. The oscillations in the skin friction corresponded to the oscillations imposed on the flow. It is interesting to note that the maximum skin friction observed during any phase of the forced oscillation at any streamwise location was approximately equal to the mean skin friction of the unforced flow.

Conclusions

Two dimensional excitation of the wall-jet in the absence of an external stream has no appreciable effect on the rate of spread of the jet nor on the decay of its maximum velocity. Careful examination of the flow near the surface (i.e. for $0 < Y^+ < 100$) reveals some profound differences which manifest themselves in reducing the skin friction. Local reductions of 30% in the wall stress, as a consequence of such an excitation, were not uncommon. The skin friction drag which is the only contributor to the loss of momentum in this flow was also reduced by a comparable amount. The production of turbulent energy near the surface was reduced resulting in lowering the intensities of the streamwise component of the velocity fluctuations. These effects, which were observed in the fully developed region of the wall-jet (i.e. at X/b > 30), are insensitive to the method of forcing but they are sensitive to the frequency and the amplitude of the excitation.

External excitation enhanced the two-dimensionality and the periodicity of the coherent motion. This enhancement is clearly visible near the surface and near the free interface of the turbulent flow. The large coherent structures in this flow might be identified with the most amplified, primary instability modes, of the mean velocity profile. Detailed stability analysis confirms this proposition though not at the same level of accuracy as it did in many free-shear flows.

The preliminary observations in the wall-wake lead to similar conclusions and represent also a successful attempt to control and delay the separation of a turbulent boundary layer. The introduction of harmonic, two-dimensional oscillations results in a reattachment of the flow and changes the proportions between the "wake" and the "wall" functions whose linear combination represents the streamwise velocity distribution in a turbulent boundary layer. It does not, however, alter the universal form of these functions.

Acknowledgement

The presentation was based on a M Sc. thesis of Mr. B. Nishri at Tel Aviv University and on the work done in collaboration with Prof. M. Zhou, Dr. Y. Katz and Mr. E. Chorev at the University of Arizona. Both projects were supported in part by grants from AFOSR and monitored by Dr. J. McMichael.

References

Bradshaw, P. 1967, J. Fluid Mech., **29**: 625.

Brown, G. L. and Roshko, A. 1974, J. Fluid Mech., **64**: 775.

Craik, Alex, D. D., Wave Interactions and Fluid Flows, 1985, Cambridge University Press

Crighton, D.G. and Gaster, M., 1976, J. Fluid Mech., **77**: 397

Falco, R. E., 1977, Phys. of fluids Suppl. **20**: 124

Gaster, M., 1974, J. Fluid Mech., **66**: 465.
Ho, C. M. and Huerre, P., 1984, Ann. Rev. Fluid Mech., **16**: 365.
Hussain, A. K. M. F. and Reynolds, W. C., 1970, J. Fluid Mech., **41**: 241
Hussain, A. K. M. F. and Reynolds, W. C., 1972, J. Fluid Mech., **54**: 241-261
Hussain, A.K.M.F., 1983, Physics of Fluids, **26**: 2816-2850
Irwin, H.P.A.H., 1973, J. Fluid Mech., **61**: 33
Kalz, Y., Nishri, B., and Wygnanski, I., 1989, AIAA Paper 89-0975.
Kline, S. J., Reynolds, W. C., Schraub, F. A., and Runstadler, P. W., 1967, J. Fluid Mech., **50**: 133.
Kruka, V. and Eskinazi, S., 1964, J. Fluid Mech., **20**: 555.
Landhal, M.T., 1990, J. Fluid Mech., **212**: 593-614
Launder, B. E. and Rodi, W., 1981, Prog. Aerospace Sci., **19**: 81.
Liu, J. T. C., 1971, Phys. Fluids, **14**: 2251.
Marasli, B., Champagne, F.H., and Wygnanski, I., 1989, J. Fluid Mech., **198**: 255-273
Narasimha, R., Narayan, K. Y., and Parthasarathy, S. P., 1973, Aeronaurical J., **77**: 335.
Patrick, W. P. 1986, United-Technologies Res. Center Report R85-915555.
Stratford, B.S., 1959, J. Fluid Mech., **5**: 1-35.
Tam, C.K.W. and Chen, K.C., 1979, J. Fluid Mech., **92**: 303
Tsuji, Y., Morikawa, Y., Nagatani, T. and Sakou, M., 1977, Aeronautical Quarterly, **XXVIII**: 235
Weisbrot, I., and Wygnanski, I., 1988, J. Fluid Mech., **195**: 137-159
Willmarth, W. W., 1975a, Annual Review Fluid Mech., **7**: 13.
Willmarth, W. W., 1975b, Structure of turbulence in boundary layers, in: "Advances in Applied Mechanics", Academic Press, New York.
Wllls, J. A. B., 1962, J. Fluid Mech., **12**: 3.
Wygnanski, I., Champagne, F., and Marasli, B., 1986, J. Fluid Mech., **168**: 31
Wygnanski, I., Fiedler, H., Oster, D. and Dziomba, B., 1979, J. Fluid Mech., **93**: 325
Wygnanski, I., Katz, Y., and Horev, E., 1991, J. Fluid Mech., **234**: 669-690

Address of the author

I. Wygnanski

Department of Aerospace Engineering, University of Arizona, U. S. A.

and

Department of Fluid Dynamics & Heat Transfer, Tel Aviv University, Israel

Rapid Distortion Theory and the structure of turbulence

Julian Hunt Nicholas Kevlahan

Abstract. This is a review of how, when linear distortions are applied to turbulent velocity fields, certain changes to some or all components of the turbulence can be calculated using linear theory. Important examples of such distortions are mean and random straining motions, body forces, interactions with other flows (eg. waves). This theory is usually known as Rapid Distortion Theory (RDT) because it is valid for *all* kinds of rapidly changing turbulent flows (RCT), when the distortion is applied for a time (defined in a Lagrangian frame) T_D that is short compared to the 'turn-over' or decorrelation time scales T_L or $\tau_D(k)$ of the energy containing eddies or smaller eddies of scale k^{-1}, respectively. However, for certain kinds of distortion the theory is also applicable to slowly changing turbulence (SCT) where T_D is of the order or greater than T_L.

New insight about the structure of slowly changing turbulence has been derived from RDT by considering different strain rates, initial conditions and time scales. RDT calculations show that in shear flows, whatever the initial form of the energy spectrum $E(k)$ (provided it decreases with wavenumber k faster than k^{-2}) or of the anisotropy, for a long enough period of strain, $E(k)$ always tends to the limiting form where it is proportional to k^{-2} for the small scales. Other statistical properties, such as ratios of Reynolds stresses, are also insensitive to the initial conditions. By contrast RDT shows how turbulent flows without mean strain or with irrotational strain are significantly more sensitive to initial and external conditions. These conclusions are consistent with those drawn from Direct Numerical Simulations (DNS) and experiments at moderate Reynolds numbers.

The eddy structure of small scale turbulence can be studied by calculating the distortion of random Fourier components in a velocity field caused by different kinds of large scale straining motions. This method is used here in conjunction with an analysis of how the different types of straining motions in different regions of the small-scale turbulence are affected by the distortion. Interestingly, linear analysis shows that the vorticity vector tends to become aligned with the middle eigenvector of the rate of strain tensor, which is consistent with DNS for turbulent flows having a continuous spectrum.

At sufficiently high values of the Reynolds number, even in homogeneous and inhomogeneous distorted turbulence (mean straining flows, body forces and boundaries), the nonlinear processes act over a wide enough spectrum to ensure that at small scales the energy spectrum $E(k)$ has an approximately universal form (as proposed by Kolmogorov). However, at the same time different components of the spectrum have an anisotropic structure. This can be estimated by applying RDT to each wavenumber component of the spectrum and taking the time of distortion to be approximately equal to the turnover time $t(k)$ appropriate to the value of k.

1. Introduction

Turbulence is a recognisable state of nature but it has no rigid definition; it is rather like certain diseases which are defined by a collection of symptoms called a *syndrome*. In the case of turbulence these 'symptoms' include randomness with a finite probability density function, strong vorticity, a complex highly three-dimensional velocity field, motion over a large and continuous range of length scales, and greatly increased effective values of viscosity and diffusivity. Many 'chaotic' flows, such as particular kinds of thermal convection, have some, but not all, of these 'symptoms'. The transition to turbulence only occurs when the Reynolds number (which is the ratio of inertial to viscous forces) or, in some cases the imposed disturbance, exceeds certain thresholds.

The conclusion from much recent research is that turbulence, even on the smallest scales, does not have a general statistical structure that is quite independent of initial and boundary conditions. Only for certain aspects of turbulence and in certain classes of flow does it appear that the eddy structure and statistical properties are insensitive to external conditions and large-scale motion. For example, Kolmogorov's inertial range theory is a good first approximation for the two-point second moments of the small scales and spectra in all turbulent flows at sufficiently high Reynolds number, but the higher order moments certainly do not have a universal structure (Hunt & Vassilicos 1991, Frisch 1991). The intermediate or energy containing scales of turbulence and the Reynolds shear stresses do not generally have a universal structure, but in a wide class of shear flows similar eddy structures and statistical properties are found for these scales of motion (Townsend 1975, Hussain 1986). However, even in these flows, the largest scales that span the shear flow (such as wakes and jets) are quite specific to the particular mean flow profile and the history of the turbulence (Bevilaqua & Lykoudis 1978, Mungal & Hollingsworth; see review by Hunt 1992).

Flow visualisation and, more recently, computational studies have shown that within most turbulent velocity fields there are large eddy 'structures', within which the local distribution of velocity persists over time and space, and which have a repeatable flow pattern. These 'eddies' can be defined mathematically in a number of different ways, but most criteria for different types of eddies are based on a local strain rate or local variation of velocity (Wray & Hunt 1990, Adrian & Moin 1988, Mumford 1982). The structures of interest range in scale from very small scale vortical eddies that may be present in almost all high Reynolds number flows (eg. Hunt & Vassilicos 1991) to large scale 'coherent structures' that span the entire width of shear flows and have characteristics peculiar to those flows (Hussain 1985).

Both quantitative models and qualitative analyses have shown how the salient features of the dynamics and kinematics of the whole flow, both instantaneously and statistically, can be explained in terms of the *local* dynamics and kinematics of these eddies (eg. Falco 1991, Hunt 1992, Townsend 1956). This approach is a more physically based and scientifically more fruitful method for studying the essential nonlinear and multi-scale dynamics of turbulence than that of developing complex statistical models based on uncertain mathematical approximation and hypotheses that are difficult to justify by physical arguments.

The primary aims of this review are to show how analyses of the linearised equations of motion, in the framework of Rapid Distortion Theory (RDT), can be used to explain some of the significant kinematical and dynamical aspects of the statistical properties and eddy structures in turbulence. (The aspects explained in this paper are briefly described in the summary.) Recent reviews have been written by Cambon 1992 and Hunt 1992b.

RDT uses linearised equations to describe the changes to a given velocity field $\mathbf{u}_0(\mathbf{x}, t)$ when it is subject to a distortion. Usually the given or initial velocity field is defined in terms of Fourier or other appropriate series, where amplitudes and phases are random. Then the problem reduces to calculating the 'transfer function' M_{in} which determines how the amplitudes and phase of each component changes during the distortion. Once the transfer function is known (as in linear control theory) it is straight forward to calculate how the same distortion affects a wide range of initial turbulent velocity fields. For example some kinds of distortion are highly sensitive to the variation in initial conditions and others are not; the implications of this are discussed later.

Note that, although the methods of RDT are more general than those of linear hydrodynamic stability theory (HST), in fact these two types of calculation using the linearised solution of the equations of motion, are aimed at exploring quite different phenomena. Most HST calculations focus on solutions that grow exponentially in space and/or time, ie. solutions of the form $\mathbf{u} \propto \exp(-i(\alpha x + \sigma t))$, where the imaginary parts of α and σ are positive. Experimentally, these are found to correspond to large-scale modes that span the whole mean flow; these are the modes that initially grow and determine the transition to turbulence. In a fully turbulent flow, the linearised modes of HST where Im $\sigma = 0$ (considered to be perturbations on the *mean flow* ignoring the fluctuations) correspond approximately to the largest scale eddy motion (Barcilon *et al.* 1974). RDT calculations may grow exponentially (as shown in §3), but usually they are applied to flows or parts of flows, or scales of eddy motion, where the disturbances do not grow exponentially fast (eg. Townsend 1976).

The idea of using linearised equations to describe how an initially prescribed velocity field is distorted by a shear or a straining flow began with Kelvin (1887). (His solution can be generalised to the nonlinear case Craik & Criminale 1985.) The same mathematical formalism was first applied to distortions to turbulence by Taylor (1935) and was subsequently developed by Batchelor (1953), Batchelor & Proudman (1954) and Townsend (1976). The original application of RDT was to calculate second order statistics such as changes in the intensities, correlations and (longitudinal one-dimensional) energy spectra resulting from the homogeneous distortion of homogeneous isotropic turbulence. The predictions of RDT were found to be satisfactory when compared to experimental results from a wind tunnel with a sudden contraction (eg. Tucker & Reynolds 1968).

RDT has since been extended and applied to other flows where the turbulence is undergoing inhomogeneous distortion, eg. around bluff bodies (Hunt et al. 1991), large scale flow in distorted ducts and non-isentropic compressible flow (Goldstein 1978), flow over hills and waves (Carruthers & Hunt 1990), and flow in internal combustion engines (Reynolds 1980).

A less obvious extension of RDT has been its use as a way of analysing the structure of turbulence. In this case RDT is used to model individual realisations, instead of statistics averaged over many realizations. Recently Lee, Kim & Moin (1990) have shown that an RDT shear calculation for slowly changing turbulence is able to produce the characteristic long, streaky structures observed in experiments on the turbulent boundary layer.

Another way in which RDT may be used to investigate the structure of turbulence is to divide the Fourier transform of the velocity into separate large-scale and small-scale parts and to model the interaction of the two scales as a rapid distortion of the small scales by the large scales (Kida & Hunt 1989). By dividing the flow into a number of structure types and applying a variety of large scale distortions (corresponding to specific large scale structures) it is possible to see how various types of interaction affect the structure of the small scales and hence the properties of the turbulence. Dynamical quantities, such as vorticity production, can also be measured as a function of distortion to determine if the simple dynamics of this model can produce properties observed in experiments and Direct Numerical simulation (eg. the concentration of high intensity vorticity into thin filaments). The results of such an investigation will be presented in this paper.

The two-scale interaction model described above satisfies the usual condition of RDT, namely that it is valid for short times, because it considers local regions of the flow where it is assumed that each local small-scale region will only be subject to distortion by a given large-scale structure for a short time. One might also imagine trying to use

RDT to analyse an entire flow over long times. The nonlinearity and independence from initial conditions which characterise turbulent flow make such a linear analysis seem unjustified. There is, however, the possibility that for some initial conditions and some types of strain the linear RDT solution reaches a steady 'eigensolution' that is independent of the initial conditions, or a 'statistical eigensolution' which is only weakly dependent on initial conditions, after averaging.

An example of such an 'eigenflow' was given by Hunt & Carruthers (1990). They showed that for shear distortion of flows with energy spectra initially decreasing faster than $O(k^{-2})$ the energy spectrum reaches a limiting form of $E(\chi) \propto \chi^{-2}$, where χ is the local (distorted) wavenumber. This means, for example, that a flow with initial energy spectrum $E(k) \propto \exp(-k^2)$ will change to a power law spectrum. This is exactly what is observed in Direct Numerical Simulations where the initial sharply peaked spectrum broadens to a steady power law spectrum after the turbulence has experienced significant straining by a shear flow.

Only turbulent flows (or components of such flows) that are independent of initial and boundary conditions can reasonably considered to be demonstrations of a 'universal nature of turbulence' (such as the small scale equilibrium theory of Kolmogorov 1941). Hence, the above example of eigensolutions in turbulent flows is important because of the evidence it provides as to what sort of turbulent flows may have universal characteristics. In this case, the mean shear imposes a certain statistical property and structure. This explains why many models of turbulence structure which neglect the effects of initial conditions are quite satisfactory for shear flows, but not for other kinds of distorted flows.

Although RDT is not itself a theory of turbulence, it can be used in conjunction with turbulence theories as guide to the modelling of dynamical terms. For example, in the closure theory EDQNM (Eddy Damped Quasi Normal Markovian) compressible RDT has recently been used to give the form of the pressure-dilatation term (Durbin & Zeman 1992). These models may also be tested by comparing their results for rapid distortions with those predicted by RDT. Numerical experiments often produce surprising results, but cannot, of course, explain them. The other role of RDT in relation to models and simulations of turbulence is to help explain their results, eg. the alignment of vorticity and eigenvectors of the rate of strain which is presented in this paper.

2. Formulation of RDT for homogeneous flows

In this section we derive the equations for incompressible RDT and the criteria for their validity, starting from the Navier-Stokes equation. For a treatment of compressible

RDT see, for example, Goldstein (1978). We focus here on homogeneous flows where the integral scale of the turbulence, L_x is much less than the length scale H over which the mean flow or the statistical properties of the turbulence vary. We consider the turbulence evolving in *time*; the results describe the distortion of turbulence along the mean streamlines in a statistically steady flow.

2.1 Derivation of RDT equations. Consider the turbulent velocity, $\mathbf{u}^*(\mathbf{x},t)$, pressure, $p^*(\mathbf{x},t)$, and vorticity, $\boldsymbol{\omega}^*(\mathbf{x},t)$. Let $\mathbf{u}^*(\mathbf{x},t)$, $p^*(\mathbf{x},t)$, and $\boldsymbol{\omega}^*(\mathbf{x},t)$ be divided into mean and fluctuating components: $\mathbf{u}^*(\mathbf{x},t) = \mathbf{U}(\mathbf{x},t)+\mathbf{u}(\mathbf{x},t)$, $p^*(\mathbf{x},t) = P(\mathbf{x},t)+p(\mathbf{x},t)$, and $\boldsymbol{\omega}^*(\mathbf{x},t) = \boldsymbol{\Omega}(\mathbf{x},t)+\boldsymbol{\omega}(\mathbf{x},t)$, and substitute into the Navier-Stokes equations for incompressible flow,

$$\begin{aligned}\frac{\partial u_i}{\partial t} + U_l\frac{\partial u_i}{\partial x_l} + u_l\alpha_{il} + u_l\frac{\partial u_i}{\partial x_l} &= -\frac{\partial p}{\partial x_i} + \nu\frac{\partial^2 u_i}{\partial {x_l}^2}, &(1)\\ \frac{\partial u_j}{\partial x_j} &= 0, &(2)\\ \omega_i &= \epsilon_{ijk}\frac{\partial u_k}{\partial x_j}, &(3)\end{aligned}$$

where

$$-\frac{\partial^2 p}{\partial {x_l}^2} = 2\alpha_{lm}\frac{\partial u_m}{\partial x_l} + \frac{\partial u_l}{\partial x_m}\frac{\partial u_m}{\partial x_l}, \tag{4}$$

and where α_{ij} is the deformation tensor, $\partial U_i/\partial x_j$. If the input turbulence is homogeneous one may replace $u_i(\mathbf{x},t)$ by its Fourier transform,

$$\hat{u}_i(\boldsymbol{\chi},t) = \int u_i(\mathbf{x},t)\exp[-i\boldsymbol{\chi}\cdot\mathbf{x}]\,d\mathbf{x}, \tag{5}$$

and obtain equations for the evolution of the Fourier transform of the velocity,

$$\begin{aligned}\frac{d}{dt}\hat{u}_i(\boldsymbol{\chi},t) = \; & -\chi^2\hat{u}_i(\boldsymbol{\chi},t) - \alpha_{il}\hat{u}_l(\boldsymbol{\chi},t) + 2\tfrac{\chi_i\chi_l}{\chi^2}\alpha_{lm}\hat{u}_m(\boldsymbol{\chi},t)\\ & -i\,\chi_i\mathbf{P}_{lm}(\boldsymbol{\chi},t)\textstyle\int \hat{u}_l(\boldsymbol{\chi}',t)\hat{u}_m(\boldsymbol{\chi}-\boldsymbol{\chi}',t)\,d\boldsymbol{\chi}',\end{aligned} \tag{6}$$

where $\mathbf{P}_{lm}(\boldsymbol{\chi},t)$ is the projection operator (projects a vector into the plane perpendicular to $\boldsymbol{\chi}$),

$$\mathbf{P}_{lm}(\boldsymbol{\chi},t) = \delta_{lm} - \frac{\chi_l\chi_m}{\chi^2}. \tag{7}$$

In order that equation (6) is homogeneous, it is necessary that the wave number $\boldsymbol{\chi}$ changes according to the equation,

$$\frac{d}{dt}\chi_i = -\chi_j\alpha_{ji}. \tag{8}$$

This is equivalent to an equation for the conservation of wavefronts and shows that the motion of allowed wavenumbers in Fourier space is solenoidal. The Fourier transform of the vorticity, $\hat{\omega}(\chi, t)$, is related to $u_i(\chi, t)$ by the equation,

$$\hat{\omega}_i(\chi, t) = -i\,\epsilon_{ijk}\chi_j\hat{u}_k(\chi, t). \tag{9}$$

Note that in cases where the input velocity is not homogeneous the Fourier representation cannot be used, and so the velocity must be represented as a sum of a set of orthogonal basis functions calculated specifically for the initial flow. These functions are usually determined by the method of proper orthogonal decomposition (Lumley 1965). If the turbulence is homogeneous in only one or two directions the Fourier transform may be applied in those directions.

The first three terms in equation (6) are linear in the fluctuating component and represent changes in $\hat{u}_i(\chi, t)$ due to viscous stresses and the interaction between the fluctuating and mean components. The last term on the right hand side of equation (6) represents changes due to interactions between fluctuating components of different wavenumbers. The rapid distortion approximation consists in assuming that this nonlinear term is negligible during the course of the distortion. Thus, the RDT approximation is equivalent to assuming that the fluctuating components do not interact with themselves and are only affected by the mean strain during the distortion (note that equation (8) for the evolution of the wavenumber is the same in both the linear and nonlinear cases). Expressed in terms of energy exchange, the RDT approximation takes into account transfer of energy from the mean flow to the fluctuating components, and redistribution of energy in χ-space, but not the nonlinear exchange of energy between Fourier amplitudes of different wavenumber. Usually the viscous term is ignored since it merely adds an exponential decay factor to the solutions.

If the initial values of the Fourier transform of the velocity field are known one can calculate the distorted Fourier transforms of velocity and vorticity using RDT and then transform back to physical space to find the distorted velocity and vorticity fields. It is also possible to find directly a number of useful spectral quantities such as the energy spectrum tensor, $\Phi_{ij}(\chi, t)$,

$$\Phi_{ij}(\chi, t) = \frac{1}{2}\overline{(\hat{u}_i^*(\chi, t)\hat{u}_j(\chi, t) + \hat{u}_i(\chi, t)\hat{u}_j^*(\chi, t))}. \tag{10}$$

Note that if the turbulence is initially isotropic the initial energy spectrum tensor may be calculated from the initial energy spectrum and initial wavenumbers,

$$\Phi_{ij}(\mathbf{k}, t) = \frac{1}{4\pi k^4}(\delta_{ij}k^2 - k_ik_j)E(k, 0). \tag{11}$$

Where $\chi(t = 0) = \mathbf{k}$. The energy spectrum tensor is the three-dimensional Fourier transform of the correlation function, $R_{ij}(\mathbf{x}, \mathbf{x} + \mathbf{r}; t)$,

$$R_{ij}(\mathbf{x}, \mathbf{x} + \mathbf{r}; t) = \overline{u_i(\mathbf{x}, t)u_j(\mathbf{x} + \mathbf{r}, t)}, \tag{12}$$

It is usually convenient to consider the integrated spectrum function, $E_{ij}(\chi, t)$,

$$E_{ij}(\chi,t) = \frac{1}{2}\int_{|\boldsymbol{\chi}|=\chi} \Phi_{ij}(\boldsymbol{\chi},t)\, dS(\boldsymbol{\chi}), \tag{13}$$

where the integral is over a sphere of radius χ in wavenumber space, so that, $E_{ij}(\chi,t)$ gives the contribution to $\overline{u_i u_j}$ from length scales of magnitude χ^{-1}. Thus, for example, E_{11} gives the energy of the x-component of velocity as a function of time, and $E_{ii}(\chi,t)$ (or $E(\chi,t)$) gives the total energy contribution from length scale χ^{-1}. $E(\chi,t)$ may be integrated once more to give the total energy of the distorted flow, $\overline{u_i^2(t)}$,

$$\frac{1}{2}\overline{u_i^2(t)} = \int_0^\infty E(\chi,t)\, d\chi. \tag{14}$$

Similar quantities may be defined and calculated for vorticity (Batchelor 1953, chap. 3).

Simple solutions to the RDT equations exist for pure rotational distortion, pure irrotational straining and pure shear (Townsend 1976). For example, Townsend's analytical solutions for simple shear (magnitudes of plane irrotational straining and rotational distortion are equal) of inviscid turbulence, $U_1 = \alpha x_3$, are:

$$\hat{u}_1(\mathbf{k},t) = \frac{k^2}{k_1^2+k_2^2} Q_1 \hat{u}_3(\mathbf{k},0) + \hat{u}_1(\mathbf{k},0) \tag{15}$$

$$\hat{u}_2(\mathbf{k},t) = \frac{k_1 k_2}{k_1^2+k_2^2} Q_2 \hat{u}_3(\mathbf{k},0) + \hat{u}_2(\mathbf{k},0) \tag{16}$$

$$\hat{u}_3(\mathbf{k},t) = \frac{k^2}{k^2 - 2\beta k_1 k_3 + \beta^2 k_1^2} \hat{u}_3(\mathbf{k},0) \tag{17}$$

where $\mathbf{k} = (k_1, k_2, k_3)$ is the initial wavevector, $\boldsymbol{\chi} = (k_1, k_2, k_3 - \beta k_1)$ is the distorted wavevector, and,

$$Q_1 = -\frac{k_2^2}{k_1(k_1^2+k_2^2)^{1/2}} \arctan\left(\frac{\beta k_1 (k_1^2+k_2^2)^{1/2}}{k^2 - \beta k_1 k_3}\right) + \frac{\beta k_1^2 (k^2 - 2k_3^2 + \beta k_1 k_3)}{k^2(k^2 - 2\beta k_1 k_3 + \beta^2 k_1^2)} \tag{18}$$

$$Q_2 = \frac{k^2}{k_1(k_1^2+k_2^2)^{1/2}} \arctan\left(\frac{\beta k_1 (k_1^2+k_2^2)^{1/2}}{k^2 - \beta k_1 k_3}\right) + \frac{\beta (k^2 - 2k_3^2 + \beta k_1 k_3)}{k^2 - 2\beta k_1 k_3 + \beta^2 k_1^2} \tag{19}$$

$$\beta = \alpha t \tag{20}$$

Note that for homogeneous three-dimensional distortions consisting of arbitrary proportions of irrotational and rotational straining over finite times even the linearised RDT equations cannot be solved in closed form for arbitrary initial conditions (Craik & Criminale 1986), and so one must resort to asymptotic methods.

Thus, we have seen that RDT provides a straightforward method for calculating changes in turbulent velocity and vorticity fields and their second order moments given the initial velocity (or vorticity) field and the distorting field, specified in terms of a deformation tensor. The next question to be addressed is under which conditions the RDT linearisation is justified.

2.2 Criteria for the RDT linearisation. As suggested by its name, the usual criterion for the application of RDT is that the distortion takes place so rapidly that the inertial forces of the distorted turbulence have no effect, ie.,

$$T_D \frac{u(l)}{l} \ll 1, \tag{21}$$

where l is the length-scale of the distorted turbulence and T_D is the time-scale of the distortion. Another criterion is that the magnitude of the nonlinear turbulent terms be small compared with the distorting terms,

$$\frac{u^2(l)/l}{u(l)U/L_D} \ll 1, \tag{22}$$

where L_D is the length-scale of the distorting velocity field. The nonlinear terms will lead to significant transfer of vorticity into directions *away* from that of the maximum applied straining (linear analysis says that vorticity will only be transferred to the direction of maximum applied straining).

Thus, equation (22) must be modified to take into account the nonlinear rotation of vorticity if RDT is to be valid for all components of velocity and vorticity:

$$\frac{u^2(l)/l}{\theta(T_D)u(l)U/L_D} \ll 1 \tag{23}$$

where $\theta(T_D) = \exp((\lambda_{\min} - \lambda_{\max})T_D)$ and $\lambda_{\min}$ and $\lambda_{\max}$ are the moduli of the minimum and maximum values of the principal strains of the deformation tensor. This means that if the strain is strong and isotropic the nonlinear terms may be neglected for all time (Batchelor 1955). The value of $\theta(T_D)$ will depend on the typical eddy structures making up the turbulent flow. Combining criteria (21) and (23) we have the general conditions on velocity, lengths and times for applying the RDT linearisation:

$$\frac{u(l)}{l} \ll \max\left(\frac{U}{L_D}\theta(T_D), \frac{1}{T_D}\right) \tag{24}$$

These conditions may be summarised as follows: RDT is always valid as long as either the strain rate is large enough (taking into account nonlinear rotation effects), or the period of the distortion, T_D, is short enough. Note that this does *not* imply that the time-scale of the large eddies must be less than the time-scale of the small eddies.

In certain circumstances, considered below, RDT (despite its name) may actually be applied over *long* periods of time (ie. it may be used to analyse slowly changing turbulence) if the nonlinear term remains small compared to the linear terms.

The criteria derived above for the application of RDT to a turbulent velocity field all arise from an analysis of the turbulent Navier-Stokes equations in physical space, before the Fourier transform has been taken. These criteria are based on the relating acceleration to magnitudes of velocity by assuming that the motions are similar to oscillatory wave motion. They do not take into account the actual spatial structure of the turbulence.

An alternative new way of looking at the conditions under which RDT is valid is to *retain* the nonlinear terms and examine their magnitude in Fourier space. The nonlinear term in equation (6) is the nonlinear one and is seen to be responsible for the nonlinear exchange of energy between wavenumbers in spectral space, and is essentially the nonlinear transfer term, $T(\chi)$ in the energy evolution equation for a distorted turbulent flow,

$$\begin{aligned}\frac{\partial E(\chi)}{\partial t} &= -2\nu\chi^2 E(\chi) - 2\alpha_{ij}\int_{|\boldsymbol{\chi}|=\chi} \Phi_{ij}(\boldsymbol{\chi})\,dS(\boldsymbol{\chi}) \\ &+\alpha_{ij}\frac{\partial}{\partial\chi}\int_{|\boldsymbol{\chi}|=\chi}\frac{\chi_i\chi_j}{\chi}\Phi_{ll}(\boldsymbol{\chi})\,dS(\boldsymbol{\chi}) - T(\chi).\end{aligned} \tag{25}$$

The first term on the right hand side is the decay of energy due to viscosity, the second is the transfer of energy from the distorting flow to the distorted flow, and the third is the redistribution of energy in Fourier space caused by the deformation of the distorted flow and $T(\chi)$ is the transfer of energy from all wavenumbers of magnitude less than χ to all wavenumbers of magnitude greater than χ.

It is clear from inspection of the equations in Fourier space that the neglect of the nonlinear terms required by RDT is equivalent to assuming that during the distortion no energy is being transferred between length-scales except due to deformation of the velocity field. This means that an alternate RDT criterion based on the organisation of the distorted velocity field may be given: RDT may be applied for long times and for any level of strain as long as the transfers of energy between length-scales due to the distortion are much larger than the nonlinear transfer of energy, ie.,

$$\overline{|T(\chi)|} \ll \min\left(\overline{2\alpha_{ij}\int_{|\boldsymbol{\chi}|=\chi}\Phi_{ij}(\boldsymbol{\chi})\,dS(\boldsymbol{\chi})},\ \overline{\alpha_{ij}\frac{\partial}{\partial\chi}\int_{|\boldsymbol{\chi}|=\chi}\frac{\chi_i\chi_j}{\chi}\Phi_{ll}(\boldsymbol{\chi})\,dS(\boldsymbol{\chi})}\right) \tag{26}$$

Different turbulent flows will have different forms for $T(\chi)$ depending mainly on the type and distribution of structures they contain. Indeed for some special flows (eg. the symmetric 'error function' eddy, $u_\theta(r) = r\exp(-\frac{1}{2}\alpha^2 r^2)$, or the potential spiral vortex, $\mathbf{u}(r,\theta) = -(cr^{\frac{1}{\alpha}-1})\hat{\mathbf{r}} - (c/r)\hat{\boldsymbol{\theta}})$ it can be shown that $T(\chi)$ is identically zero. In that case the nonlinear inertial terms are initially zero and the RDT form of the equations is exact. However, in most subseqent distortions these terms will cease to be zero and the RDT solution will only be approximate. Thus the criteria for the validity of RDT should be based on the *organisation* of the velocity field, as well as its magnitude. Incidentally, this approach helps us understand the RDT approximation in terms of energy dynamics instead of inertial forces and time-scales. The organisational and energetic basis of this criterion makes it extremely useful when RDT is used as a way of exploring the structures of turbulence.

2.3 Initial and boundary conditions in RDT. An aspect of RDT that distinguishes it from models of turbulence such as EDQNM (Eddy Damped Quasi-Normal Markovian) is that the initial conditions may be non-statistical. This would suggest that RDT is unsuitable for investigating characteristic structures of nonlinear turbulence, which because of its nonlinearity quickly becomes uncorrelated with its initial conditions and often reaches a 'universal' form. Models such as EDQNM are best suited for use with turbulent flows having a universal statistical form (eg. shear flows). An RDT analysis should be useful for other, non-universal, flows. RDT should also be valuable in analysing flow which vary slowly with time and are weakly dependent on initial conditions—these are characteristic 'eigensolutions' or 'statistical eigensolutions'. In such cases the form of the energy spectrum (though not its *magnitude*) reaches a statistically steady state if RDT is be applied over a long time. (This does not necessarily imply that it is *valid* over a long time though experimentally it is found that the predicted form of $E(k)$ is correct.) This is another example of an energy and organisationally based criterion for the application of RDT. An example of such an 'eigenflow' will be given in §3.2 below. Since the various types of turbulent flow show different degrees of correlation with initial conditions, the models of turbulence are not all equally effective at predicting their flow properties. Table 1 below compares the effectiveness of various models for different flows (defined by time scales) and indicates which flows can be treated using RDT. Table 1 shows that RDT gives better results than the 1-point eddy viscosity and second order Reynolds stress transport models for certain quantities in many types of flows, and is comparable to the more complex 2-point EDQNM closure model. This is perhaps surprising considering that RDT is a linear theory and was originally intended for flows with $T_D/T_L \ll 1$.

In RDT and in DNS the boundary conditions can be applied before any statistical averaging is performed. This contrasts with the Reynolds stress models where the boundary conditions can only be specified in terms of the moments or other statistical quantities. This involves some serious assumptions near the boundary (Hunt 1992a). For the RDT or DNS computations on a rigid boundary layer of thickness l_δ for an eddy of length-scale l at high Reynolds number $u(l)l/\nu \gg 1$ the boundary conditions are,

$$n < l_\delta \qquad \mathbf{u} = 0 \qquad \text{on } n = 0$$

Property of flow	Observations or numerical simulations	Time scales	1-P models		2-P models	
			EVM	2RST	RDT	EDQNM
(i) $\frac{\overline{u_i u_j}}{K}, \frac{p\partial u_i/\partial x_j}{KS}$		$T_L S \leq 1$	Y (Sh), N N (NSh), N	Y (G)	Y (P)	Y (G)
		$T_L S \gg 1$	N	Y (P)	Y (G)	Y (G)
(ii) Correct sensitivity of property (i) to the *initial* $E(k)$	None	$T_D/T_L \ll 1$	Y	Y	Y	Y
	Some	$T_D/TL \sim 1$	N	N	Y	Y
	None (Sh)	$T_D/T_L \gg 1$	Y (Sh)	Y (Sh)	Y	Y
	Some (NSh)	$T_L S \geq 1$	N (NSh)	N (NSh)	Y	Y
(iii) Correct sensitivity of property (i) to the initial anisotropy b_{ij}, c_{ij}	Some	$T_D/T_L \ll 1$	N	Y (approx.)	Y	Y
	Weak	$T_D/T_L \sim 1$	N	Y	Y (?)	Y
	None (Sh)	$T_D/T_L \gg 1$	Y (Sh)	Y (Sh)	N (Sh)	N
	Some (NSh)		N (NSh)	Y (NSh)	Y (NSh)	Y
(iv) Correlation of $u_i(\mathbf{X})$ with initial (or boundary) turbulence	Strong	$T_D/T_L < 1$	N	N	Y (G)	Y (?)
	Weak	$T_D/T_L \geq 1$			Y (P)	Y

Table 1: Comparative merits of 1-point (1-PM) and 2-point (2-PM) models for different flow properties and turbulent flows (defined by time-scales) in homogeneous strains with shear (Sh) and without shear (NSh) away from rigid boundaries or other interfaces. T_L is the time-scale of the turbulence, T_D is the imposed time-scale of the distortion, and S is the applied strain rate, and K is the kinetic energy per unit mass. EVM = Eddy Viscosity Models; 2RST = Second Order Reynolds Stress Transport; RDT = Rapid Distortion Theory; EDQNM = Eddy Damped Quasi-Normal Markovian. Y = Yes, can be computed or is correct in principle; N = No, (G) = Good Model in practice; (P) = Poor Model in practice. (From Hunt 1992).

$$n > l_\delta \quad \mathbf{u} \cdot \mathbf{n} = 0 \quad \text{on } n = 0, \tag{27}$$

where n is the distance from the wall, and l_δ is the viscous layer generated on the wall by the eddy moving over wall. (For typical laboratory scale turbulence where $ul/\nu \sim \omega^2$, $l_\delta \sim (l_\delta \nu / u(l))^{1/2}$.). Typically, if $n \gg 2l$ the wall boundary condition is irrelevant for an eddy of scale l. This is why in shear boundary layers where $L_x \sim (1/2)n$, the direct effects of the wall on eddies is small, whereas in shear free boundary layers $L_x \sim n$, the direct effects are large.

3. Turbulent structures and statistical models of turbulence

3.1 RDT and other methods of calculation. In this section we discuss how RDT has been used in the investigation of turbulent structures and present some new results on structural and dynamical changes in turbulence undergoing arbitrary two-dimensional distortions. This distortion is considered as a model of some of the dynamics of the interaction between the large-scales and small-scales of turbulence. We also examine the phenomenon of turbulent 'eigenflows' which were discovered by Hunt & Carruthers (1990) on the basis of an RDT analysis of turbulent shear flows. This result suggests why similar forms of spectra are found in turbulent shear flows at values of the Reynolds number (Re) too small to produce the universal structure predicted by Kolmogorov's nonlinear cascade model. At these values of Re RDT and experiments show that no universal form exists for unsheared turbulence, which explains the success of statistical turbulence models in predicting moments and a wide range of spectra in shear flows, but not in unsheared flows.

RDT was originally developed to predict changes in the statistics of rapidly distorted turbulent flows, its application to the analysis of structures in turbulence is new. However, it is also suited to the analysis of eddy structure of turbulence, which is not possible with purely statistical models such as eddy viscosity or K–ϵ. The only requirements for the application of RDT are that one of the sets of criteria for the application of RDT be satisfied in order that the linearisation be valid. Even if the criteria are not met an RDT analysis may still be valuable since, by comparing the dynamics produced by the linear RDT theory with corresponding results from experiment or Direct Numerical Simulation (DNS), the degree of dependence of particular dynamical quantities (such as vorticity production) on nonlinear processes may be gauged.

In order not to prejudice the result the exploration of the eddy structure of turbulence using RDT is to begin by considering an isotropic, homogeneous turbulent velocity field (either from experiment, Direct Numerical Simulation, or randomly generated with prescribed energy spectrum) and then applying to it a distortion typical of the flow to be examined (eg. apply a strong shear to investigate turbulent boundary layers). After the distortion has been applied the flow is analysed to look for any changes in the number and distribution of various structure types.

3.2 RDT and statistical turbulent 'eigenflows'. In this section we briefly review the results of Hunt & Carruthers (1990) who used Townsend's (1976) RDT analysis of sheared turbulence to look for universal features in the spectra of the small-scale turbulence in shear flow.

Consider a uniform shear $\mathbf{U} = (\alpha x_2, 0, 0)$ rapidly distorting an initially homogeneous and isotropic turbulent velocity field $\mathbf{u}$. The Fourier components of $\mathbf{u}$ will evolve as given by equations (15), (16), (17). The spectra will be expressed in terms of the local wavenumber, $\boldsymbol{\chi}$, (which is what is actually measured),

$$\chi^2 = k_1^2 + (k_2 - k_1\beta)^2 + k_3^2, \tag{28}$$

where $\beta = \alpha t$. Since the input velocity field is assumed to be isotropic the components of $E_{ii}(\chi)$ may be calculated from equation (11),

$$\begin{aligned} E_{11}(\chi,t) &= \int_{|\boldsymbol{\chi}|=\chi} \frac{E(k,0)}{4\pi k^2}\left\{\frac{k^2}{(k_1^2+k_3^2)}(a+b)\left((a+b)-\frac{2k_1k_2}{k^2}\right)\right. \\ &\quad \left. + \frac{(k_2^2+k_3^2)}{k^2}\right\} dA(\mathbf{k}), && (29) \\ E_{22}(\chi,t) &= \int_{|\boldsymbol{\chi}|=\chi} \frac{E(k,0)(k_1^2+k_3^2)}{4\pi\chi^4}\, dA(\mathbf{k}), && (30) \\ E_{33}(\chi,t) &= \int_{|\boldsymbol{\chi}|=\chi} \frac{E(k,0)}{4\pi k^2}\left\{\frac{k_1^2k_3^2}{k^2(k_1^2+k_3^2)}\left(\frac{-k^2}{k_3^2}a+\frac{k^2}{k_1^2}b\right)^2\right. \\ &\quad \left. - \frac{2k_1k_2k_3^2}{k^2(k_1^2+k_2^2)}\left(\frac{-k^2}{k_3^2}a+\frac{k^2}{k_1^2}b\right)+\frac{(k_2^2+k_3^2)}{k^2}\right\} dA(\mathbf{k}), && (31) \end{aligned}$$

where,

$$\begin{aligned} a &= \frac{-k_3^2}{k_1(k_1^2+k_3^2)}\left\{\arctan\left(\frac{k_2}{(k_1^2+k_3^2)^{1/2}}\right) - \arctan\left(\frac{k_2-\beta k_1}{(k_1^2+k_3^2)^{1/2}}\right)\right\}, && (32) \\ b &= \beta k_1^2(k^2 - 2k_2^2 + \beta k_1 k_3)/k^2\chi^2, && (33) \\ k^2 &= \chi_1^2 + (\chi_2 + \beta\chi_1)^2 + \chi_3^2. && (34) \end{aligned}$$

We consider initial spectra of Gaussian and power law and forms,

$$\begin{aligned} E(k,0) &= u_0^2(kL)^N \exp(-k^2L^2) \quad \text{where } N \text{ is an integer and } N \geq 0, && (35) \\ E(k,0) &= \frac{u_0^2 L}{(1+k^2L^2)^P}, \quad \text{where } 1 < P. && (36) \end{aligned}$$

The integrals for the $E_{ii}(\chi, t)$ can then be evaluated for $\beta \gg 1$ by asymptotic analysis. It is found that in the range $1 \ll \chi L \ll \beta$ E_{11} is much larger than the others, so $E(\chi, t) \approx E_{11}(\chi, t)$ and,

$$E(\chi, t) \propto \frac{1}{k^2} \int_{|\boldsymbol{\chi}|=\chi} \beta^2 E(kL, 0)\, dA(\mathbf{k}). \tag{37}$$

The major contribution to this integral comes from the region $|\hat{\theta}| \ll 1$ and $|\phi| \ll 1$, where $\hat{\theta} = \frac{1}{2}\pi + (1 - \theta)/\beta$. So for $\beta \gg 1$ the integral (37) becomes,

$$E(\chi, t) \propto \int_{-\infty}^{\infty} \int_{-\infty}^{\infty} \frac{\beta E([\chi^2 L^2/\beta^2 + \chi^2 L^2(\hat{\theta}^2 + \phi^2)]^{1/2})}{\chi^2 L^2}\, d(\hat{\theta}\chi L)\, d(\phi\chi L). \tag{38}$$

Thus, if $1 \ll \chi L \ll \beta$ if $E(k, 0) = o(k^{-2})$ when $kL \ll 1$,

$$E(\chi, t) = \beta \left(\int_0^{\infty} \hat{k} E(\hat{k}, 0)\, d\hat{k} \right) / \chi^2 L^2. \tag{39}$$

Therefore for large enough strain as long as the initial energy spectrum decreases faster than k^{-2} the energy spectrum of sheared turbulence will tend to the limiting form of

$$E(\chi, t) \propto \beta \chi^{-2}. \tag{40}$$

Note that this result will hold for a broader class of initial spectra than considered above, any spectrum with a falloff faster than k^{-2} will suffice.

Figure 3 shows spectra from a number of experiments and simulations of shear flows. These results show a tendency towards a χ^{-2} region at large wavenumbers for large values shear. The result (39) means that also if the Reynolds number is moderate a shear flow will tend to maintain the $k^{-5/3}$ spectrum of very high Reynolds number turbulence, as has been observed in experiment (Monin & Yaglom 1971 and reviewed by Hunt & Vassilicos 1991).

This analysis has provided two very important pieces of information. First, it has shown that the form of the large wavenumber spectra in sheared turbulent flows (eg. turbulent boundary layers) may be determined by linear distortion by the mean shear, rather than by any complex nonlinear effects. Since k^{-2} is close to $k^{-5/3}$ the spectra of shear flows are often discussed in terms of the spectral predictions of Kolmogorov theory which is based on a nonlinear cascade of energy to small scales! Secondly, the fact that the spectrum tends to the χ^{-2} form gives an indication of the type of structure one would expect in turbulent shear flows.

The energy spectrum, $E(\chi, t)$, of a function is proportional to the square of the modulus of the complex coefficients, a_n, of the Fourier series of $u(\mathbf{x})$,

$$E(\chi, t) \propto |a_n(t)|^2 \quad \text{where} \quad u(\mathbf{x}) = \sum_n a_n(t) \exp(i\chi_n(t)x). \tag{41}$$

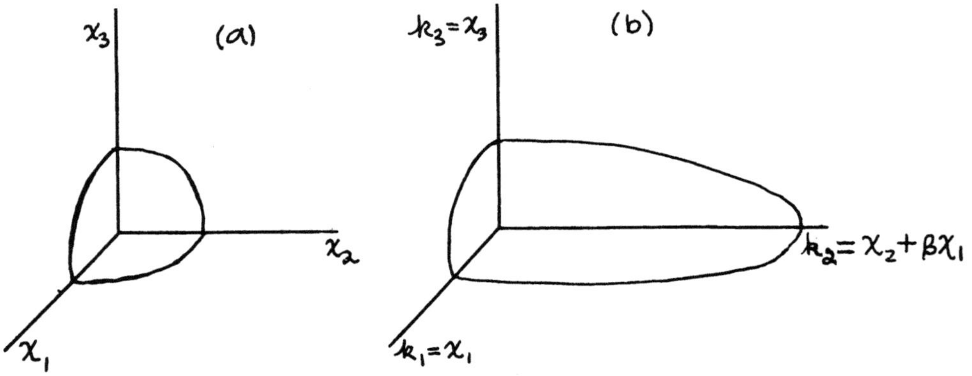

Figure 1: Schematic diagram showing how a spherical surface in the Fourier space of the local distorted wavenumber in a shear flow originates from a highly distorted region of wavenumber space at $t = 0$. (a) Spherical shell in Fourier space at $t = T$, (b) shape of region in Fourier space of initial wavevectors which are mapped to the spherical shell by the shear distortion.

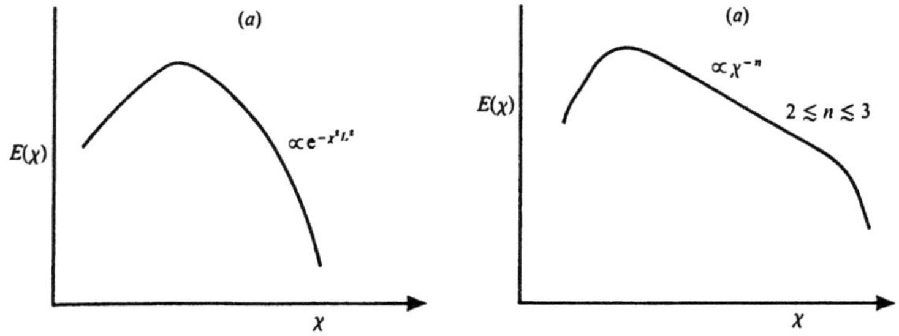

Figure 2: Schematic diagram of the change in energy spectra upon the application of strong shear, (a) before shear, (b) after shear.

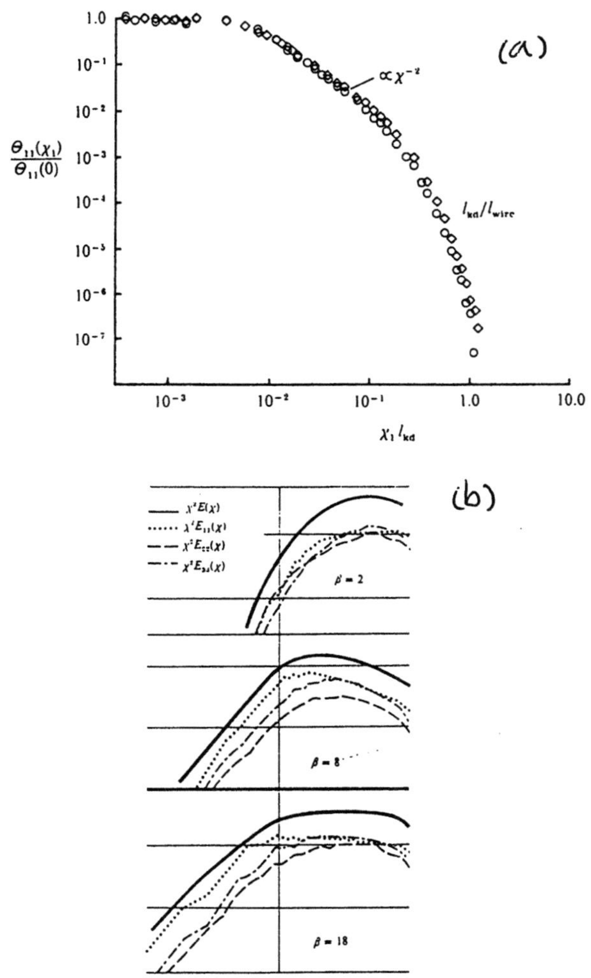

Figure 3: Changes in the energy spectra resulting from shear. (a) One dimensional spectrum measured by Champagne *et al.* (1970) in shear flow at two positions downstream. (b) $\chi^2 E(\chi, t)$ and $\chi^2 E_{ii}(\chi, t)$ in Direct Numerical Simulations of Rogallo (1981) of homogeneous turbulence in uniform shear ($Re_\lambda \approx 80$).

If $u(x)$ and all its derivatives are piecewise continuous then,

$$|a_n| = O(n^{-p}) \quad \text{as} \quad n \to \infty, \tag{42}$$

$$E(\chi, t) = O(\chi^{-2p}) \quad \text{as} \quad \chi \to \infty, \tag{43}$$

where $p-1$ is the lowest order continuous derivative (Courant & Hilbert 1953, p. 73). Thus, if the velocity is discontinuous $E(\chi,t) \propto \chi^{-2}$ as $\chi \to \infty$, but the converse is not necessarily true. All that can be said is that if, as in a shear flow $E(\chi,t) \propto \chi^{-2}$, there will be singularities in the velocity, but not necessarily simple discontinuities. These singularities could also be an accumulating oscillation where the frequency of oscillation becomes singular (Hunt & Vassilicos 1991). These singularities (actually sharp gradients when viscosity is included) in the velocity would be associated with elongated eddy structures, and just such 'stretched' eddies are seen in both DNS and experiments of shear flows.

3.3 Turbulent structure. RDT has been used to calculate the effect of uniform plane shear upon homogeneous, isotropic velocity field computed 'cheaply' using the method of Kinematic Simulation (KS) (Lee *et al.* 1990, Carruthers *et al.* 1991). In KS a turbulent-like 3-dimensional velocity field is generated from a truncated 3-dimensional Fourier series (usually about 38 terms is sufficient). The series is constructed so that the velocity field satisfies continuity. The magnitudes of the Fourier coefficients are chosen so that the velocity field has a prescribed energy spectrum, while the phases are chosen randomly from a set distributed isotropically on the unit sphere. KS produces accurate first and second order statistics (Fung *et al.* 1991) and allows quantities such as pressure and derivatives of velocity to be calculated exactly directly from the Fourier representation of the velocity. KS is also ideally suited for use with RDT since RDT calculates the evolution of Fourier coefficients. The velocity field produced using this combination of KS and RDT is shown in figure 4 below and compared with experimental, and Direct Numerical Simulation (DNS) results. Figure 4 shows good qualitative agreement between the RDT+KS results, the experiment and DNS. All three plots show extremely elongated eddy structures which lead to extreme gradients of velocity in the y-direction, consistent with the RDT analysis based on the asymptotic energy spectra of sheared turbulence. This is a case where the linear RDT analysis has captured all the essential dynamics, and has provided a simple structural explanation of the phenomenon.

3.4 Perturbations to small scale high Reynolds number turbulence caused by the effect of shear and stratification. In turbulent flows at very high Reynolds number, even if they are inhomogeneous and distorted, the small scales are approx-

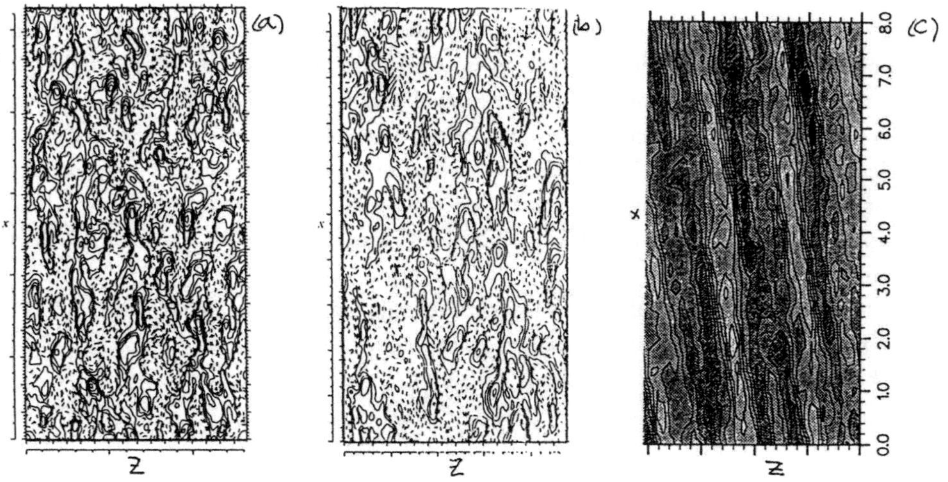

Figure 4: Comparison of the effect of uniform shear upon turbulent flow, diagrams show contours of streamwise velocity. (a) Experiment on turbulent channel flow (Lee, Kim & Moin 1990) $\beta = 8$, $\alpha = 35.1$, (b) Direct Numerical Simulation (Lee, Kim & Moin 1990) $\beta = 8$, $\alpha = 33.6$, (c) RDT + KS, $\beta = 10$, $\alpha = 50$, Von Karman energy spectrum, $E(k,0) = 1.196k^4/(0.558 + k^2)^{17/6}$.

imately isotropic, as hypothesised in the universal equilibrium theory of Kolmogorov and as observed in many oceanic and atmospheric experiments (reviewed by Hunt & Vassilicos 1991, Gibson 1991, Van Atta 1991). However, observations also show that there are systematic departures from isotropy that depend on the nature of the mean straining flow, the large scale motions and the effects of body forces. RDT can be applied to estimate these effects, because over a short enough time (defined later) the change caused by any mean strain or linear fluctuating body force on the fluctuating velocity field is larger than the changes caused by the nonlinear interactions of the turbulence (from the analysis in §).

The novel aspect of this analysis is that we analyse a statistically steady process by performing an unsteady calculation of the changes to the turbulence over a certain range of length scales of order k^{-1} that occur over a finite 'relaxation time scale' $\tau(k)$ whose magnitude depends on k. In high Reynolds number turbulence the relaxation time is determined by the nonlinear interactions between eddy motions on the same scale leading to $\tau(k) \sim \epsilon^{-1/3}k^{-2/3}$ whereas at low or moderate Reynolds numbers the large scale motions determine the relaxation time scale for all *scales*, $\tau(k) \sim L_x/u_0$. This approach is standard in statistical physics (for example in calculating the effect of polymers in fluids).

First consider the effect of mean shear on the high wavenumber spectrum. It follows from equations (15) to (20) that changes in the Fourier components when $t < \tau_k$ and when k is large enough that τ_k is small (i.e. $k \gg \epsilon^{-1/2}\alpha^{3/2}$) are given by

$$\begin{aligned} \hat{u}_1(\mathbf{k},t) &= \hat{u}_1(\mathbf{k},0) - \left(1 - \frac{k_1 k_2}{k^2}\right)\alpha t \hat{u}_3(\mathbf{k},0) && (44) \\ \hat{u}_2(\mathbf{k},t) &= \hat{u}_2(\mathbf{k},0) - \frac{k_1 k_3}{k^2}\alpha t \hat{u}_3(\mathbf{k},0) && (45) \\ \hat{u}_3(\mathbf{k},t) &= \hat{u}_3(\mathbf{k},0) - \frac{k_1 k_2}{k^2}\alpha t \hat{u}_3(\mathbf{k},0) && (46) \end{aligned}$$

Thence

$$\frac{1}{2}\overline{(\hat{u}_1^*\hat{u}_3 + \hat{u}_1\hat{u}_3^*)} = \Phi_{13}(\mathbf{k},t) = -\left(\frac{k_1k_3}{k^2} + \left(1 - \frac{k_1^2}{k^2}\right)\right)\alpha t \Phi_{33}(\mathbf{k}) \tag{47}$$

Assuming the turbulence is initially isotropic this leads to a change in Reynolds stress in a developing flow

$$-\overline{u_1u_3} = 2/5\alpha t\overline{u_3^2}(t=0) \tag{48}$$

The one-dimensional cospectrum $\Theta_{13}(\mathbf{k},t)$ can be obtained from (47) by substituting $\tau(k)$ to t and integrating over k-space in the region K_1 (eg. $k = 1/3K_1$ to $3K_1$) to obtain

$$\Theta_{13}(\mathbf{k},t) = C\alpha\epsilon^{2/3}K_1^{-7/3} \tag{49}$$

where C is of order 1.0. This result was first obtained experimentally, supported by physical and dimensional arguments by Wyngard & Coté (1972). It has been rederived in several ways using approximate statistical models such as EDQNM (Bertoglio 1979). However, this analysis can be extended to derive the corrections to Θ_{11}, Θ_{33}; namely

$$\Theta_{11} = \Theta_{11}\,(\text{isotropic})\left(1 - O(\alpha^2\tau^2)\right) \tag{50}$$

$$\Theta_{11} = \alpha_K \epsilon^{2/3} K_1^{-5/3} - O\left(\alpha^2 \epsilon K_1^{-3}\right) \tag{51}$$

$$\Theta_{33} = \frac{4}{3}\alpha_K \epsilon^{2/3} K_1^{-5/3} - O\left(\alpha^2 \epsilon K_1^{-3}\right) \tag{52}$$

Note that the correction to the normal stress is much smaller than that to the Reynolds shear stress cospectra, which is consistent with the experiments.

Derbyshire & Hunt (1992) have used the same approach to consider the combined effects of stable stratification and mean shear on the turbulence spectrum and thence explain why such large differences have been observed between the effects of stratification on low Reynolds number laboratory turbulence and on high Reynolds number atmospheric turbulence. From the study of Komori *et al.* (1983) and Hunt *et al.* (1988), the Reynolds shear stress and vertical variance in a developing shear flow with stable stratification determined by the buoyancy frequency N are given by:

$$-\overline{u_1 u_3} = \frac{2}{5}(\alpha t)\overline{u_3^2}(t=0) - \frac{2}{15}(\alpha t)\overline{u_3^2}(t=0)t^2N^2 \ldots, \tag{53}$$

$$\overline{u_3^2}(t) = \left(1 - \left(\frac{4}{35}\alpha^2 + \frac{4}{5}N^2t^2\right)\right)\overline{u_3^2}(t=0). \tag{54}$$

Using the same argument as in (50) the correction of the spectrum caused by stable stratification can be found by including both the RDT shear and buoyancy terms of (53),

$$\Theta_{13}(K_1) \sim -\Theta_{33}\alpha\epsilon^{-1/3}K_1^{-2/3}\left(1 - O(\epsilon^{-2/3}K_1^{-4/3}N^2)\right) \tag{55}$$

Now, from (54) the shear and buoyancy correction to the isotropic value of $\Theta_{33}(K_1)$ is,

$$\Theta_{33}(K_1) \sim \frac{4}{3}\alpha_K \epsilon^{2/3} K_1^{-5/3}\left[1 - O\left(\epsilon^{-2/3}K_1^{-4/3}\left(\alpha^2 + N^2\right)\right)\right], \tag{56}$$

which is of the same order as the correction to Θ_{13}. Here α_K is the Kolmogorov constant $\simeq 1.5$.

A different analysis is necessary for low Reynolds number turbulence whose strain spectrum ($\propto K_1^2\Theta_{13}(K_1)$) decreases with K_1 in this range. In this case the largest strain rate and therefore the characteristic time scale $\tau(k)$ is determined by the *large scales* and not the small scales, i.e. $\tau(K)$ ($\sim L/u_0$) is independent of K where $u = 0$ is the rms turbulence. Then from (53) as the Richardson number increases $|\Theta_{13}(K_1)|$ decreases over the whole range of K_1, ie.,

$$\Theta_{13}(K_1) \sim \Theta_{13}(K_1)(Ri = 0) - O\left[(L/u_0)^3\alpha N^2\right]\Theta_{33}(K_1) \tag{57}$$

Comparing (55) and (57) shows that since the corrections for high Reynolds number in (55) are small for $Ri \lesssim 0.2$, the shear stress ratio $(-\overline{u_1u_3})/u_3^2$ is expected to vary slowly for high Reynolds number turbulence over this range of Ri. The low Reynolds number correction in (54), however, is larger than the respective high Reynolds number correction and so the shear stress ratio decreases significantly in the typical 'low' Reynolds number turbulence found in the laboratory and computed with D.N.S. (eg. Van Atta 1991). To the extent that stable stratification does affect the high wavenumber spectrum in high Reynolds number turbulence, the greater correction to Θ_{13} in (55) than to Θ_{33} in (56) explains why the shear stress ratio decreases when $Ri \gtrsim 0.2$.

Thus we have seen that in moderately stratified turbulent flows at high Reynolds numbers (with $Ri \lesssim 0.25$) the *structure* of the turbulent velocity field (eg. the shear stress ratio) is controlled by shear, although the velocity profiles and statistics will vary. We have also seen that under certain conditions (i.e. $t\, dU/dz \sim 4$ and $Ri \sim 0.25$) the effects of shear and buoyancy effectively cancel and the shear stress $\overline{uw}$ becomes zero. Note that in this case the structure of density fluctuations are controlled primarily by density forces.

This is another example of how RDT analysis has given insight into the structure of turbulent flows using simple hypotheses, and analytical methods.

4. Structures and dynamics in distorted turbulence

In this section we model the interaction between the large-scales and small-scales of turbulence as a rapid distortion of the small-scales by the large-scales. The results presented include changes in dynamical quantities and a structural analysis.

Three-dimensional homogeneous, isotropic turbulence generated using KS is subjected to combinations of rotational and irrotational plane distortions ranging from pure rotation through pure shear to pure irrotational strain. The type of distortion experienced by an eddy is, in practice, related to the type of large-scale structure which is distorting the small-scales. For example, a large-scale eddy would induce a rotational distortion, whereas the regions between eddies would induce irrotational distortions and an intermediate position around an eddy would usually lead to a combination of these effects, i.e. a shear distortion. The picture of turbulence we put forward is one where the small-scale eddies are advected past the large-scale eddies thereby passing through regions producing a variety of distortions. The effect of these distortions was calculated using RDT. For arbitrary combinations of rotational and irrotational distortion the RDT equations cannot be solved analytically for general initial conditions, so the equations were solved numerically using an adjustable step-size method. A variety of dynamical quantities (eg. vorticity production, alignment of eigenvectors of rate of strain with vorticity, angle between velocity and vorticity) were measured before and after the distortions.

The purpose of this investigation is to determine which aspects of the dynamics of turbulence can be accounted for by the simple linear dynamics of RDT. Or, more specifically, what aspects of 'real' turbulence that are not seen in random or Gaussian turbulence may be seen when Gaussian turbulence (generated by KS) is subjected to a rapid distortion? An important case where the origin of a dynamical effect is unclear is the alignment of the middle eigenvector (associated with the middle eigenvalue) of the rate of strain tensor with the vorticity. This alignment is *not* seen in Gaussian or random turbulence (Shtilman *et al.* 1992). Ashurst *et al.* (1987) and Vincent & Meneguzzi (1991) have observed such an alignment in Direct Numerical Simulations and believed it to arise from poorly understood nonlinear effects.

The flow is also divided into different structure types using an algorithm we developed based on values of pressure, speed and the parameter Σ defined as,

$$\Sigma = \frac{S_{ij}^2 - \Omega_{ij}^2}{S_{ij}^2 + \Omega_{ij}^2}, \tag{58}$$

where S_{ij} is the rate of strain tensor, and Ω_{ij} is the vorticity tensor. The five basic structure types are: eddy—$\Sigma < -1/3$ and $p < 0$; donor eddies—$\Sigma < -1/3$ and $p > 0$; shear—$-1/3 \leq \Sigma \leq 1/3$; convergence—$\Sigma > 1/3$ and $p > 0$; and streaming—$|u| > u_{\mathrm{RMS}}$ and $|\Omega| < \frac{1}{2}|\Omega|_{\mathrm{RMS}}$, $|S| < \frac{1}{2}|S|_{\mathrm{RMS}}$. (Note that for all regions except streams the level of strain must be above a certain threshold, i.e. $|\Omega| \geq \frac{1}{2}|\Omega|_{\mathrm{RMS}}$, $|S| \geq \frac{1}{2}|S|_{\mathrm{RMS}}$.) Thus eddies are regions of high vorticity with closed streamlines about a low pressure core; donor eddies are regions of high vorticity and positive pressure fluctuation without closed streamlines, shear structures are regions of shear straining; convergence structures are regions of irrotational straining characterised by converging streamlines and positive pressure fluctuation; and stream structures are regions of low deformation and high kinetic energy. This algorithm identifies about 80% of the initial flow. This approach differs from that suggested by Wray & Hunt (1990) in that the conditions are absolute, based on the dimensionless parameter Σ (apart from a lower limit of 1/2 RMS values of S and Ω) which means that similar regions of different flows will be identified as the same structure type; donor regions are recognized; the two cases of small second invariant are distinguished (i.e. shear when the S and Ω are both large, but approximately equal, and unstructured or stream when S and Ω are both small). These differences mean that the identifications correspond well with a qualitative visual classification of the flow, and that about 50% more of the flow is classified than in the Wray & Hunt (1990) algorithm.

Dividing a flow into different structures is a way of characterizing the flow. Each structure has specific properties (eg. eddies trap particles, convergence region increase the efficiency of mixing) and by seeing how the number of these regions changes with the type of applied distortion one can understand how the overall properties of the flow changes.

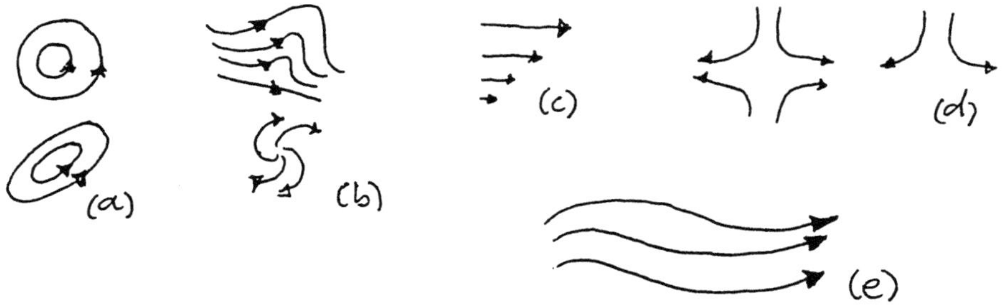

Figure 5: Schematic diagram of the five structure types. (a) Eddy, (b) donor eddy, (c) shear, (d) convergence, (e) streaming.

Jets of turbulent water are sometimes used to control dust, and turbulence is often used to accelerate mixing. If the types of distortions that increase eddy and convergence regions can be found then these distortions could be applied to increase the efficiency of the working fluids in dust control and mixing respectively.

Distortions with large-scale $S + \Omega = 20.0$ and $t = 0.1$ (giving a 'β' of 2.0) were applied to initial KS flow with both the Von Karman spectra,

$$E(k,0) = 1.196k^4/(0.558 + k^2)^{17/6}, \tag{59}$$

and exponential spectra,

$$E(k,0) = k^2 \exp(-0.75k^2). \tag{60}$$

It was found that the vorticity aligned with the intermediate eigenvector of the small-scale rate of strain matrix, but with the largest eigenvector of the large-scale (applied) strain (see figure 6). This alignment increased with the proportion of irrotational straining in the applied distortion. The alignment with the largest eigenvector of the applied strain is clearly expected from an examination of the vorticity equation,

$$\frac{D\omega}{Dt} = \omega \cdot (\mathbf{S}^{S-S} + \mathbf{S}^{L-S}) - (\mathbf{U} \cdot \nabla)\omega, \tag{61}$$

where $\mathbf{S}^{S-S}$ is the small-scale rate of strain tensor and $\mathbf{S}^{L-S}$ is the large-scale rate of strain tensor. The less obvious alignment with the small-scale rate of strain tensor can be shown to increase exponentially with time using an RDT analysis for large time.

This result shows that the alignment of vorticity with the local intermediate eigenvector of the rate of strain observed in DNS is not caused by any particular nonlinear

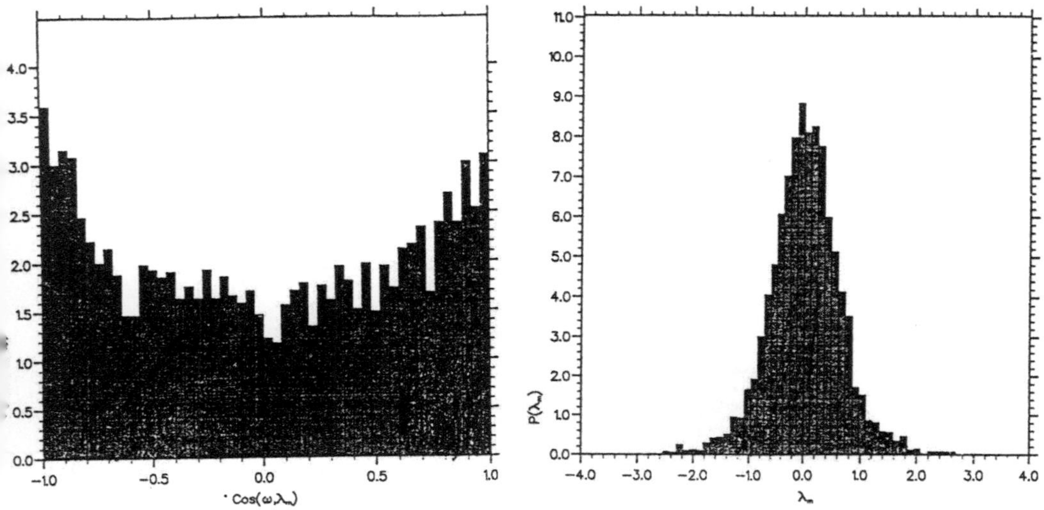

Figure 6: PDF showing the alignment of vorticity with the intermediate eigenvector of the small-scale rate of strain tensor. (a) Before distortion (Gaussian profile), (b) after distortion, $(S^{L-S} - \Omega^{L-S})/(S^{L-S} + \Omega^{L-S}) = 0.8$, $t = 0.10$.

effects, but largely arises simply from irrotational straining. This irrotational straining also tends to produce the long 'worm-like' vortex tubes containing most of the high intensity vorticity also seen in DNS (Vincent & Meneguzzi 1991). Thus, irrotational straining could lead both to the creation of long, thin vortex tubes and associated with this, the alignment between vorticity and the intermediate eigenvector of the rate of strain tensor.

Local vorticity production was measured by calculating $\omega_i \omega_j S_{ij}$ which is the source term in the enstrophy equation,

$$\frac{D(\frac{1}{2}\omega^2)}{Dt} = \omega_i \omega_j S_{ij} - \nu \left(\frac{\partial \omega_i}{\partial x_j} \right)^2 . \tag{62}$$

Initially there is no net vorticity production (Gaussian PDF), and no distortion produced any overall vorticity production. It is found, however, that vorticity production is dramatically increased under high applied irrotational distortion in regions with a positive intermediate eigenvalue (note that because of incompressibility the sum of the eigenvalues is zero) of the small scale rate of strain tensor, and vorticity was destroyed in regions with a negative intermediate eigenvalue. This result was essentially predicted by Betchov (1956).

Figure 7 shows the positively skewed PDF of $\omega_i \omega_j S_{ij}$ (indicating net production of vorticity) under high applied irrotational strain in these 'Betchov' regions. It can be shown using an RDT calculation for large times that this production (or destruction)

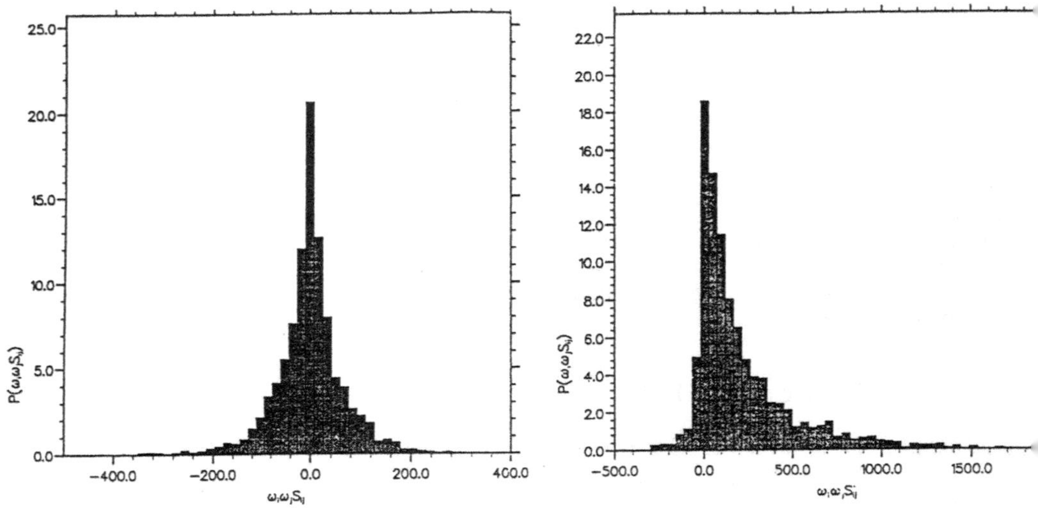

Figure 7: PDF's of vorticity production, $\omega_i\omega_j S_{ij}$. (a) Initially (Gaussian profile), (b) after large-scale irrotational distortion $((S^{L-S} - \Omega^{L-S})/(S^{L-S} + \Omega^{L-S}) = 0.8$, $t = 0.10$.

becomes exponentially large with time under irrotational strain,

$$\langle \omega_i\omega_j S_{ij}\rangle = \langle b\, \omega_{10}^2\rangle \exp(2S^{L-S}t) \quad \text{as } t \to \infty, \tag{63}$$

where S^{L-S} is the large-scale strain, ω_{10} is the initial velocity in the direction of the positive eigenvector of applied strain, and b is the intermediate small-scale eigenvalue. This destruction of vorticity in some strained regions and its production in other strained regions (depending on local values of the eigenvalues of the rate of strain tensor) will eventually lead to highly intermittent vorticity. Thus, in a turbulent flow where the vorticity is initially evenly distributed, linear straining effects will tend to lead to the concentration of vorticity in small regions– -thin vortex tubes. Again, we see linear effects largely responsible for a fundamental aspect of turbulence, the highly intermittent distribution of vorticity.

The alignment between velocity and vorticity retained its Gaussian PDF for all distortions and within all structure types. The number of 'cigar' regions (negative intermediate eigenvalue) and 'pancake' (positive intermediate eigenvalue) regions remained roughly constant at 50% each under all distortions.

The change in the relative proportion of each of the structures with different types of large-scale distortion is shown in figure 8. Figure 8 shows that the percentage of eddies, convergence and shear regions increased with the proportion of rotational distortion, while the percentage of streams and donor eddies increased with proportion of irrotational distortion. The average percentage of eddies over the range of distortions about equals the initial value, while convergence, shear and donor regions are *suppressed* relative to the initial amount for all, but high rotational distortion. Stream

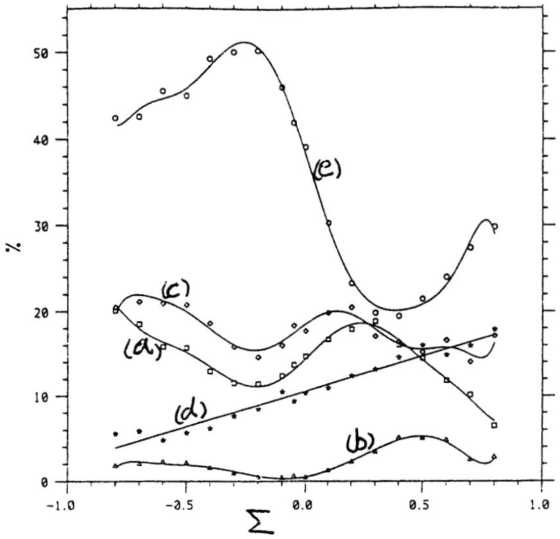

Figure 8: Percentage change in number of structures as a function of type of applied distortion. $\Sigma = (S-\Omega)/(S+\Omega)$. (a) Eddies, (b) donor eddies, (c) convergence regions, (d) shear regions, (e) stream regions.

regions, on the other hand, are *increased* relative to the initial value for all distortions except at high rotational distortion. An interesting point is that for eddies and shear regions a pure shear distortion does not change their number. Changes in percentage of convergence regions mirrors the changes in eddy regions. This is expected since convergence regions are associated with the area between eddies. Shear regions underwent the greatest overall change (30%). These results indicate one should apply rotational distortion to the turbulence used in dust curtains and mixing fluids!

All results in this section except the change in the percentage of the different structure types were independent of the energy spectrum (Von Karman or exponential) used. The changes in the number of structure types with applied distortion was significantly less, but qualitatively the same, for the Von Karman spectrum. There is much more energy in the large wavenumber part of the spectrum in the Von Karman spectrum, indicating most of the structural effects are due to changes in the small-wavenumber (large-scale) part of the spectrum.

5. Conclusions

As explained in the introduction and in §2 Rapid Distortion Theory is formally a method of calculating turbulent flows changing from a defined state under the influence of a linear distortion that has a stronger effect on the inertial term than on the nonlinear interactions of the turbulence. Most of the literature of RDT is concerned with such problems, which may be quite complex in inhomogeneous distortion, boundaries, etc.

However, starting with studies of Townsend (1976), it has long been recognised that this linear theory also gives some insight into the structure of turbulence even when it is slowly changing and approximately independent of initial or boundary conditions. This is the aspect of RDT that has been emphasised in this review. RDT has demonstrated how, only if the mean flow is approximately like a shear flow, do the energy containing eddies become approximately independent of their initial conditions and that the turbulence tends to a constant form of self-similar spectrum. We have not *proved* that the linear theory demonstrates this property, the nonlinear turbulence must also exhibit the same property; but it is likely on physical grounds and it is observed. RDT has also provided a straight forward method for estimating the corrections to isotropic high Reynolds number turbulence for different kinds of distortion. This approach could be developed further with more detailed calculation and different hypotheses about the relaxation. More detailed comparison with other models such as EDQNM could be interesting. The results are sufficiently interesting to stimulate new experimental measurement.

Thirdly, RDT now provides an interesting method for simulating how the eddy structure of turbulence evolves both under the action of mean distortion, (following Lee *et al.* 1990), and even during the mutual random interaction of eddy motion between different scales. The preliminary results presented here show that this kind of interaction explains some of the salient nonlinear processes in fully developed turbulence and indicates how different types of eddy motion evolve. Clearly further analysis in this direction is possible.

Acknowledgements

N. Kevlahan was supported during this research by a British Government ORS award and a British Gas research grant.

References

Adrian, R.J. & Moin, P. 1988. Stochastic estimation of organized turbulent structure: homogeneous shear flow. *J. Fluid Mech.*, **190**, 531-559.

Ashurst, W.T., Kerstein, A.R., Kerr, R.M. & Gibson, C.H. 1987. Alignment of vorticity and scalar gradients with strain rate in simulated Navier-Stokes turbulence. *Phys. Fluids* **30**(8), 2343-2353.

Barcilon, A., Brundley, J., Lessan, M. & Mobbs, F.R. 1979. Marginal instability in Taylor-Couette flows at a very high Taylor number. *J. Fluid Mech.* **94**, 453-463.

Batchelor, G.K. 1953. *Homogeneous turbulence.* Cambridge University Press.

Batchelor, G.K. 1955. The effective pressure exerted by a gas in turbulent motion. In *Vistas in Astronomy* (ed. A. Beer) **1**, 290-295. Pergamon.

Batchelor, G.K. & Proudman, I. 1954. The effect of rapid distortion of a fluid in turbulent motion. *J. Mech. & App. Math.* **7**, pt. 1, 83-103.

Betchov, R. 1956. An inequality concerning the production of vorticity in isotropic turbulence. *J. Fluid Mech.* **1**, 497-504.

Bevilaqua, P.M. & Lykoudis, P.S. 1978. Turbulence memory in self-preserving wakes. *J. Fluid Mech.* **89**, 589-606.

Cambon, C. 1992. Ercoftac Workshop Proceedings. To appear.

Carruthers, D.J., Fung, J.C.H., Hunt, J.C.R., Perkins, R.J. 1991. The emergence of characteristic (coherent?) motion in homogeneous turbulent flows. In *Turbulence and coherent structures* (eds. M. Lesieur & O. Métais), pp. 29-44. Kluwer.

Champagne, F.H., Harris, V.G. & Corrsin, S. 1970. Experiments on nearly homogeneous turbulent shear flow. *J. Fluid Mech.* **41**, 81-139.

Courant, R. & Hilbert, D. 1953. *Methods of mathematical physics*, **1**. New York: Interscience.

Craik, A.D.D. & Criminale, W.O. 1986. Evolution of wavelike disturbances in shear flows: a class of exact solutions of the Navier-Stokes equations. *Proc. R. Soc. Lond.* A. **406**, 13-26.

Derbyshire, S.H. & Hunt, J.C.R. 1992. Structure of turbulence in stably stratified atmospheric boundary layers; comparison of large eddy simulations and theoretical models. In *Proc.* 3rd *I.M.A. Conf. on Stably Stratified Flows, 'Waves and Turbulence in Stably Stratified Flows'* (ed. S.D. Mobbs). Clarendon Press: Oxford.

Durbin, P.A. & Zeman, O. 1991. Rapid distortion theory for homogeneous turbulence with application to modeling. Submitted to *J. Fluid. Mech.*

Falco, R.E. 1991. A coherent structure model of the turbulent boundary layer and its ability to predict Reynolds number dependence. *Phil. Trans. R. Soc. Lond.* A. **336**, 103-129.

Frisch, U. 1991. From global scaling *à la* Kolmogorov, to local, multifractal scaling in fully developed turbulence. *Proc. R. Soc. Lond.* A. **434**, 89-99.

Fung, J.C.H., Hunt, J.C.R., Malik, N.A. & Perkins, R.J. 1991. Kinematic simulation of homogeneous turbulent flows generated by unsteady random Fourier modes. *J. Fluid Mech.* (In the press.)

Gibson, C.H. 1991. Kolmogorov similarity hypotheses for scalar fields: sampling intermittent turbulent mixing in the ocean and galaxy. *Proc. R. Soc. Lond.* A. **434**, 149-164.

Goldstein, M.E. 1978. Unsteady vortical and entropic distortion of potential flow round arbitrary obstacles. *J. Fluid Mech.* **89**, 433-468.

Hunt, J.C.R. 1973. A theory of turbulent flow round two-dimensional bluff bodies. *J. Fluid Mech.* **61**, 625-706.

Hunt, J.C.R. 1992a. Developments in computational modelling of turbulent flows. ERCOFTAC workshop introductory lecture. To appear.

Hunt, J.C.R. 1992b. Rapid Distortion Theory. Lecture notes for the ERCOFTAC Summer School. Cambridge University Press.

Hunt, J.C.R. & Carruthers, D.J. 1990. Rapid distortion theory and the 'problems' of turbulence. *J. Fluid Mech.*, **212**, 4997-532.

Hunt, J.C.R., Stretch, D.D. & Britter, R.E. 1988. Length scales in stably stratified turbulent flows and their use in turbulence models. In *Stably stratified flow and dense gas dispersion* (ed. J.S. Puttock). Oxford: Clarendon Press.

Hunt, J.C.R. & Vassilicos, J.C. 1991. Kolmogorov's contributions to the physical and geometrical understanding of small-scale turbulence and recent developments. *Proc. R. Soc. Lond.* A, **434**, 183-210.

Hussain, A.K.M.F. 1986. Coherent structures and turbulence. *J. Fluid Mech.* **173**, 303-356.

Kaltenbach, D.-J., Gertz, T. & Schumann, U. 1989. Transport of passive scalars in neutrally and stably stratified homogeneous turbulent shear flows. In *Proc. 7th shear flows conference.* Springer.

Kelvin, Lord (W. Thomson) 1887. Stability of fluid motion: rectilinear motion of viscous fluid between two parallel plates. *Phil. Mag.* **24**(5), 188-196.

Kida, S. & Hunt, J.C.R. 1989. Interaction between different scales of turbulence over short times. *J. Fluid Mech.* **201**, 411-445.

Kolmogorov, A.N. 1941. The local structure of turbulence in incompressible viscous fluid for very large Reynolds numbers. *Dokl. Akad. Nauk SSSR*, **30**(4), p. 301.

Komori, S., Ueda, H., Ogino, F. & Mizushina, T. 1983. Turbulence structure in stably stratified open-channel flow. *J. Fluid Mech.* **130**, pp. 13-26.

Lee, M.J., Kim, J. & Moin, P. 1990. The structure of turbulence at high shear rate. *J. Fluid Mech.* **216**, 561-583.

Lumley, J. 1965. The structure of inhomogeneous turbulent flows. *Proc. Intl. Coll. on Radio Wave Propagation* (ed. A.M. Yaglom & V.I. Takasky), *Dokl. Akad. Nauk. SSSR*, 166-178.

Monin, A.S. & A.M. Yaglom. 1971. *Statistical theory of turbulence*, **II**. MIT Press.

Mumford, J.C. 1982. The structure of the large eddies in fully developed turbulent shear flows. Part 1. The plane jet. *J. Fluid Mech.* **118**, 241-268.

Reynolds, W.C. *Symp. Combustion Model Recip. Eng.* New York: Plenum Press, 41-66.

Rogallo, R.S. 1981. Numerical experiments in homogeneous turbulence. *NASA Tech. Memo.*, **81315**.

Shtilman, L., Spector, M. & Tsinober, A. 1992. On some kinematic versus dynamic properties of homogeneous turbulence. (submitted to *J. Fluid Mech.*)

Taylor, G.I. 1935. Turbulence in a contracting stream. *Z. angew. Math. Mech.*, **15**, p. 91.

Townsend, A.A. 1958. Turbulent flow in a stably stratified atmosphere. *J. Fluid Mech.* **3**, 361-372.

Townsend, A.A. 1976. *The structure of turbulent shear flow*. Cambridge University Press.

Tucker, H.J. & Reynolds, A.J. 1968. The distortion of turbulence by irrotational plane strain. *J. Fluid Mech.* **32**, 657-673.

Van Atta, C. 1991. Local isotropy of the smallest scales of turbulent scalar and velocity fields. *Proc. R. Soc. Lond.* A. **434**, 139-148.

Vincent, A. & Meneguzzi, M. 1991. The spatial structure and statistical properties of homogeneous turbulence. *J. Fluid Mech.* **225**, 1-20.

Wray, A.A. & Hunt, J.C.R. 1990. Algorithms for classification of turbulent structures. In *Proc. IUTAM Symp.* 1989, *Topological fluid mechanics* (eds. H.K. Moffatt and A. Tsinober), 95-104. Cambridge University Press.

Wyngaard, J.C. & Coté, O.R. 1972. Modelling buoyancy driven mixed layers. *J. Atmos. Sci.* **33**, 1974-1988.

Addresses of the authors

Prof. J.C.R. Hunt
Meteorological Office
London Road
Bracknell
Berkshire
U.K. RG12 2SZ

N. Kevlahan
University of Cambridge
Dept. of Applied Maths. and Theoretical Physics
Silver St.
Cambridge
U.K. CB3 9EW

Discussion

Betchov: Let me give you a glimpse on the question of relativistic solitons. In 1956 De Broglie wrote in a book (L. De Broglie, 1956, Une Tentative d'Interprétation Causale et Non-linéaire de la Mécanique Ondulatoire, Gauther Villars) that the addition of non-linear terms to the Dirac equations might lead to the appearance of isolated regions where the non-linear terms area all important. He does not call them solitons, but the concept is clearly present in his mind. These regions would tell the location and the velocity of a particle, thus eliminating the Heisenberg uncertainty.

In 1968, in his textbook on elementary particles, Bernstein (J. Bernstein, 1968, Elementary Particles and their Currents, published by Freemand and Co.) states that interaction between quantized fields leads to intricate non-linear problems. After Hasimoto's discovery of a non-linear Schrödinger equation governing the twisting of vortex lines, I was tempted to follow De Broglie's idea and to look for cubic terms in the Dirac equations. They should be norm-preserving, isotropic and relativistic.

To start, let me write the following set of eight equations between eight functions of x, y, z, and t, where Pt means partial dP(x,y,t)/dt, etc.

$P_t = A_x + B_y + C_z$ $\qquad$ $A_t = P_x - H_y + G_z$

$B_t = H_x + P_y - F_z$ $\qquad$ $C_t = -G_x + F_y + P_z$

$F_t = -Q_x + C_y - B_z$ $\qquad$ $G_t = -C_x - P_y + A_z$

$H_t = B_x - A_y - Q_z$ $\qquad$ $Q_t = -F_x - G_y - H_z$

You can think of P and Q as scalars while A,B,C form a vector and F,G,H form another one. If you form the second derivatives, you find that each function simply obeys a wave equation with unit wave velocity.

Note that P is tied to A,B,C as a fluid density to the velocity A,B,C. Ditto, for Q tied to F,G,H, but with -t instead of +t. As for the relations between the set A,B,C and the set

F,G,H, in the absence of P and Q, you simply find the Maxwell equations between electric and magnetic fields.
They combine acoustics and Maxwell type vibrations.
The integral of the sum of these squared functions over a large volume depends only upon "energy" coming in or out. The norm is preserved; the system cannot explode or fade away. It can only radiate in all directions. Who cares about such equations?
Well, once upon a time, there was a gentleman named Dirac who worked with four complex functions. His first function ψ is my P - iH, etc. You can take his four complex equations and splitting real and imaginary terms simply come to the above set of eight equations. They are Dirac's equations.
In the Dirac formulation the z-axis plays a special role. As a set of eight relations, they become clearly isotropic.
If you write -C instead of +C, it is F that plays the role of a density, with velocity components Q, C, B. Thus, there is no reason to think of P and Q as absolute scalars. It is a matter of definitions (gauge transformation). The above linear equations have two relativistic quadratic invariants involving sums of squares or the product P*Q+A*F+B*G+C*H.
Various isotropic non-linear cubic terms can be constructed using one of the available invariants and one of the eight functions. Equations of this type have been proposed by Durr and studied by Heisenberg (H. P. Durr and W. Heisenberg, 1961, Theory of the Strange Particles, Proc. Conf. Aix en Provence on Elementary Particles. W. Heisenberg, 1966, Introduction to the Unified Field Theory of Elementary Particles, Interscience Publisher). The matter of relativistic invariance raises difficulties that I have not yet completely eliminated. The problem can be solved by introducing a second set of eight functions. Thus, the first set could describe an electron and the second a positron. Then non-linear relativistic interaction can be postulated between these two sets and numerical experiments can be performed as in the case of turbulence. My preliminary results indicate a trend towards time stationary islands of non-linearity. I also found that the non-linear terms have skew probability distributions. I think that such equations enriched with cubic terms could lead to interesting results. Can solitons tell the location and speed of particles? This would eliminate the quantum mechanical uncertainty. A physical constant must sit in front of the non-linear term, having the dimension of a cross-section. Is it the size of a neutrino?

Herring: We will now begin the discussion of all the papers.

Antonia: I made the following observations in turbulent boundary layer experiments. The non-uniqueness and the non-universality are particularly evident if you do the measurements at very low Reynolds numbers. Then, of course, you are very sensitive to

initial conditions. It is difficult to reproduce the "normal" Coles wake function at low Reynolds numbers - this is also evident in the direct numerical simulations of Spalart (Spalart, 1988, J. Fluid Mech., **187**, 61-98).
If you do the measurements at high Reynolds numbers, you also have problems. For example, you can achieve a certain Reynolds number based on momentum thickness ϑ by either varying x, keeping the free stream velocity constant or keeping x constant and varying the free stream velocity. The boundary layer characteristics will not necessarily be identical in the two cases.

Moin: There is a question of whether the boundary layers are unique or not. Far enough downstream, one would expect that the boundary layer would select the natural instability modes. Let me just give you an example: All of the direct numerical simulations start with random numbers in contrast to real flow in a wind tunnel. Eventually, in due course, the calculation reaches a statistically steady state. You recover the law of the wall. The profile evolves out of random numbers. Turbulent intensities evolve out of random numbers.

Hunt: You are talking about the channel flow calculations.

Moin: All boundary layers.

Hunt: I'm very sceptical about whether you have described all boundary layers. Although the mean profiles follow the law of the wall, if you look at the low frequency horizontal velocity fluctuations, they always reflect the outer structure of the boundary layer, which differs depending on the initial and outer boundary conditions. That's where of course there is a bit of a struggle with Prof. Yaglom. Recent research on convective boundary layers shows that generally the depth of the large eddies extends right through the boundary layer; this is observed in the atmosphere, even in the neutral boundary layer. So, the large scale structures in the outer part of a boundary layer will always have some memories of the initial conditions.

Moin: You cannot change the rules of the game. The rules are that you fix the boundary conditions. If you then go sufficiently far down-stream or in time; then eventually you have to forget where they came from.

Hunt: In the channel I agree. The essential travel time is much larger than the large memory time of turbulence. But if the memory time can keep getting bigger and bigger as it is in the boundary layer, then I believe that you will always have some residual memory, which will have effect on some components, on some parts of the spectrum even near the surface. But there will be many features that will be little affected. So I think there is no certain answer to this question (S. E. Belcher, W. S. Weng & J. C. R. Hunt, 1991, Structure of turbulent boundary layers perturbed over short length scales. Eighth Symposium on Turbulent Shear Flows, Munich, 12-2-1 - 12-2-5). We've got to decide which these components are. Then we can probably be a little bit more specific.

Aref: I think Dr. Zaslavsky convinced us yesterday that bursting, for example, was something that was fully acceptable in a dynamical systems description, and the model by Lumley and coworkers would tend to bear out that maybe there is some description of bursting within such a framework. This morning we saw a lot of different mechanistic approaches to bursting and there may be others that haven't been presented at this meeting. I would be curious to know from some of the boundary layer experts what they think bursting is and do they think there is a consensus on a model that gives a type of bursting that we see in turbulent boundary layers?

Landahl: I think you do see something like bursting in the final stages of transition. You can see these high frequency fluctuations that Mike Gaster was talking about. You do see them sometimes, but you also do see things that look like concentrated outflows from the boundary layer. Perhaps it's something universal when you have a shear flow - and this can happen - but of course it only happens on the right conditions.

Yaglom: I have a question related to what was said by Prof. Gaster. I don't know any reliable proof for the statement that the Reynolds stress tensor (i.e. the cospectrum $E_{1,3}(k)$ of the horizontal and vertical velocity fluctuations must be proportional to $k^{-5/7}$ in the inertial range. I know the old paper by J. C. Wyngaard and O. R. Cote (Quart. J. Roy. Meteor. Soc., 1972, Vol. 98, 590 - 608) containing a sketch of such a proof, but their arguments don't seem to be convincing to me. I know that most of the experimental data related to the cospectrum E_{13} agree satisfactorily with the "-7/3 power law", but I know no theoretical reasons for this. This law clearly contradicts the Kolmogorov theory since all the off-diagonal terms of the Reynolds-stress tensor $\overline{u_i u_j}$ must vanish in the case of a locally isotropic velocity field and hence $E_{13}(k) = 0$ according to Kolmogorov's theory. Therefore the non-zero values of $E_{13}(k)$ show the limitations of the Kolmogorov theory. It is interesting to understand why the Kolmogorov prediction about the shapes of the spectra in the inertial range agree so well with the data related to spectra of three velocity components but disagree with the experimental results related to cospectrum E_{13} in turbulent wall flows.

Hunt: One of the big problems is why laboratory experiments show extreme sensitivity to the Reynolds stress spectrum with stratification whereas a lot of high Reynolds number experiments do not. There is an argument constructed on these lines.

Moin: I would like to return to Aref's question. I lately looked at the results of Aubry et al. I really don't think bursts occur the way it was described originally Kline. For instance, the observed high intermittent activity of the wall shear stress during the so-called bursting event is nothing more than just a streamwise vortex passing a fixed probe. I think if you look at measurements in the early stage, nobody really knew what a burst was. It was decided from flow visualisation coupled with hot-wire anemometry. You can reproduce the same signature with a streamwise vortex just passing a probe. In simulations you see the same event with a similar signature as in the experiments. I doubt very much that Aubry et al. actually calculated this event. Their calculation didn't have much of a streamwise variation and therefore it wouldn't have these short streamwise intermittent vortices. Burst was a very simple observation and now it has become a monster. It really is not, it's just a vortical structure passage.

Moffatt: Landahl said at finite-time singularity this regime had something to do with bursts. I was going to ask him anyway to expand a little bit on this finite time singularity, which seems to be terribly interesting. What exactly goes singular and how does it go singular with time, and what's the physical interpretation of this?

Landahl: I'm somewhat scared of physical interpretation. The physical interpretation is: fluid elements are coming together in a horizontal plane and since there is a wall, there is no other way, the fluid can only go up. The whole flow structure is controlled by continuity so that whatever pressure gradient in the vertical and in the horizontal is caused or required, it will develop. So, what I'm trying to say is that we have to sort out cause and effect. I think there is a bit of confusion because of the cause and effect. I don't think vortices cause an effect. The vortices are a product of dynamical effects that give rise to local disturbances that are strong, and then they live for a while in the fluid. I think what one should try to understand is how these local vortex pieces are formed and then try to see what happens once you have them. Because once you have them, it's quite clear that they are going to stay there for a while. If they are sitting in a fix position, they will produce something that looks like a burst. But that's not the creative process that is behind the formation of the burst.

Jimenez: I'd like to comment on the question of bursting. Bursting is something that we looked at quite carefully in this minimal channel. The channel that contains basically one structure. And there there was bursting. Vortices that change with time at some point change quite violently. This seems to happen at the moment where one vortex runs into another. Then they begin to interact and they start violent activity.

Moin: Many years ago Klebanov did some experiments. I think it was the first time that he got these very sharp spikes and determined hairpin vortices. Quite an exchange occured between F. Hama and P. Klebanov. Klebanov identified very sharp spikes with his hot-wire signal and said that was a hairpin vortex. F.Hama was using flow visualisation or he put his hot-wire in a different place and saw high frequency oscillations - My argument is that they are the same phenomena. You get a high shear layer which is associated perhaps with a hairpin vortex. If you put your hot-wire there, you won't see the passage of the vortex itself (which you might well see at another part of the flow), you'll see some response from this, like a shear layer instability. So I think a lot depends on the way you look and how you look at what you see.
The other point I'd like to make about finite time singularity: Some years ago Stuart and Stuardson looked at the nonlinear wave packet evolution, there was a finite time singularity. What that really means is that you've got to rescale the problem. If your mathematics is telling you it's singularity, that means that you've done mathematics incorrectly. But in that region something else is going to happen and find a scale. So, I think the presence of singularity is a strong indication of something worth looking at in greater detail mathematically.

Landahl: I think that the basic problem involved in transition and in fully developed boundary turbulence is to understand how the mean shear is transformed into local fluctuations. Mike Gaster showed some examples of this forming. But I think one should look primarily for the inviscid mechanism that could be behind this. Because there are all sorts of more rapid mechanisms rather than the growth of the Tollmien-Schlichting waves, which is a very slow thing and may or may not be interesting at all.

Hunt: I think it would be nice to be able to find a way to use this knowledge of structures, to find some better explanations for this scaling - I think a beautiful scaling argument is the spectrum at very high Reynolds numbers proportional to k -1. It may be explained physically in terms of pairing of eddies.
If you've got turbulent boundary layer here, consider the spectra E_{33}, the spectra of E_{11}. For the wave numbers k1 this is going to be greater than 1/z, where z is the distance to the surface, you have a Kolmogorov -5/3 spectrum. (No blocking at the surface.) Here ε is a function of z. For k1 less than 1/z the blocking of the eddies causes the spectrum E_{33} to be roughly constant. Here you've got eddy scales that are much larger than the distance from the surface. But they are much less than the height of the boundary layer h, and therefore the only possible scaling is that E must depend upon u*, and therefore you have to get a spectrum of k^{-1}. Following the idea of Perry, for the E_{11} spectrum the only possible scaling is $u_*^{\,2}k^{-1}$.

Looking closely at recent atmospheric boundary layer data very close to the ground, I agree with what Perry has been saying. It's rather an interesting spectrum because it suggests that from a statistical point of view you've got structures that are strained by some du/dz that's proportional to 1/z. The time sclae T_L depends on z and therefore is proportional to z/u*. The total strain Tdu/dz is the same throughout the logarithmic layer. I think it is an appealing physical idea. So all these eddies are bigger than the distance from the surface. They are very long streaky eddies. (D. J. Carruthers, J. C. R. Hunt & R. Holroyd, 1989, Airflow and dispersion over complex terrain. In Air Pollution Modelling and its Applications, VII, ed. H. van Dop Plenum, 515-529)

Yaglom: I spoke a little bit about this in my talk. There is a short wave number range where the spectrum $E_{33}(k)$ of the vertical velocity component is also proportional to k^{-1}. This range is much shorter than the range of velocity of the "-1 power - law" for spectra E_{11} and E_{22} of horizontal velocity components, but for $E_{33}(k)$ it is also clearly seen in the atmospheric surface-layer data, which is a flow with much greater value of Re than those which are typical for the laboratory experiments; see, e.g. the paper by B. A. Rader and A. M. Yaglom in "Turbulence and Coherent Structures" (O. Métai and M. Lesieur, eds. Kluwer, 1991, 388-412). However, in the laboratory boundary layers with relatively large values of Re, the -1 - power range of the E_{33} spectrum is also observable. The difference with the spectrum $E_{11}(k)$ of the horizontal velocity fluctuation is that in the case of E_{11} the low wave number limit of the -1 power range is scaled by the boundary layer thickness δ and thus is proportional to δ^{-1} while for E_{33} this lower limit of the -1 power range is determined by some other parameters (probably it is proportional to z^{-1} where z is the distance from the wall).

Landahl: The streaks near the wall tend to contain most of the Reynolds stress and it is the component $k_1 = 0$ that is dominating. So, I feel a little uncomfortable about anything going like k_1^{-1}. But of course there must be a cut-off at some place.

Hunt: These structures are long compared to the distance from the ground, that's the point. They are streak-like structures. I think we are on the same track.

Novikov: I have a question about what you said about k = 0: does that mean that you assume that your stress is in a finite regime?

Landahl: No. It's looking at an individual eddy which has a length growing like t before viscosity sets in. There is a cut-off with viscosity. But before the viscosity cut, the $k_1 = 0$ component is the one growing like t and therefore will dominate in the statistics. But I showed that you have to include the fact that these streaks have a finite life time, then everything decays.

Novikov: So the finite time singularity will disappear?

Landahl: No. The finite time singularity might still be there because I'm just talking about the streaks; the finite time singularity is not directly associated with streaks. A streak is a consequence of this outer break-time instability that makes these structures grow continuously in stream-wise direction, but the finite time singularity is due to fluid elements colliding in a way and having to go up. Everything is based on the assumption that you have infinite horizontal boundaries on the top, so that you can ignore the effect of non-parallelism and finite widths of channel.

Moffatt: This $k_1 = 0$ linear growth rate - the actual result of the Reynolds stress - must depend on a sort of spectral density around $k_1 = 0$.

Landahl: Remember, there is a second wave number component you have to include, k_3. That's the controlling thing.

Herring: There is some very interesting work on the second invariant for separating regions in a turbulent flow. Does this emerge from rapid distortion theory calculations.

Hunt: What a number of people have been doing in 2-D and 3-D turbulence is to characterize the different eddy structures (rather like solutions of differential equations) by regions of either high rotational strain or highly irrotational straining motion, i.e. swirling or squashing motion. This is a kinematic analysis. I might give two points: The first point is that we think that this provides the key to explaining how particles move apart in turbulence, and in fact my ex-student, Dr. Malik, is here. His work in Cambridge showed that it very much depends on how much time a particle spends in either the regions where there is pulling apart or swirling around, and that has a strong effect on the coefficient.
Using rapid distortion theory, what we did is take a random flow field and then strain it and then look to see how these regions of eddying motion or straining motion are affected by shear (J. C. R. Hunt, 1989, Some changes in the doing and the application of applied mathematics, 1964-1989, I.M.A. Bulletin **25**, 97-103). It's interesting to see, of course, that the elongated regions are placed at low angles to the flow. The kinematic analysis is a very useful way of looking at how structure changes when you apply shear or some other kind of information. I think it is a useful guide for examining turbulent flows and also for certain kinds of mixing and scalar processes. It is being extended in other ways by Cantwell and other researchers.

Wygnanski: A question to Mike Gaster: You demonstrate that amplification of disturbances when they come from a point source momentary or to a point source when you put in random disturbances are more or less equal. They are very different from the typical wave trains. Can you say anything about the fact that it is a 3-D point source?

Gaster: 3-D are very important, but even in the 2-D case calculations based on limit theories show that the Reynolds stress is greatly different in a wave packet than in a continuous wave train. But all these observations were 3-D, and I think 3-D are more important. So I think you need both effects; the modulation in time and the modulation in span. They both contribute to a much more powerful breakdown.

Hunt: This is the final thing that was done at Nasa Ames. In fact, this work was started in 2-D with oceanographers, who were taking slices through turbulent flows and looking at regions defined by these invariants (du/dx) · (du/dz). You can write this as $S^2 - 1/2\omega^2$, where S is the straining rate. When this is positive, you have straining motions; we also added some other conditions based on pressure. We had assumed that this differentiation of regions of high swirl and high strain rate were very much a dynamical consequence. The interesting thing is that if you should take a random flow field that satisfies continuity (à la Kraichnan) and you then make the same division of the flow, you find that if the random field has the same spectra as an actual turbulent flow, you have a similar division between these regions of straining and high swirling. For example, the shape of these swirl regions had an aspect ratio of about 3:1 for the kinematic simulation and about 5:1 when you look at the non-linear full computations, which suggested to me that the kinematics does quite a lot of organisation long before any dynamics comes into play.
It is interesting to study these fields using rapid distortion theory; and what we will be doing is to take a two-scale spectrum E (k) and then consider a small-scale spectrum. U is the distorted flow field and u the distorting flow field. You start with such a spectrum; you look at random distortion; this begins to elongate these kinds of structures; a quantitative analysis shows how the spectrum and these structures evolve. The other way of looking at this is using wavelet transforms to analyse more details of these local structures.
The method of kinematic simulation also provides new insights into how pairs of particles move apart. If their distance apart is Δ, then $G_\Delta \varepsilon t^3$ in the inertial subrange. G_Δ is the number I want to refer to. In the literature, (P. Monin & A. Yaglom) this coefficient varies by a factor of 100. There are three contrasting arguments: The argument of Dr. Novikov that if a particle moves off in a turbulent flow field, then the deviation of the particle from its initial trajectory is an amount Y. Then he says that if I have another particle that moves off, then this particle will not be correlated to the other particle, so that $\overline{\Delta^2} = \overline{2Y^2}$ is a good estimation. Now, if we know that $\overline{Y^2}$ also goes like εt^3, there is a coefficient equal to $2\pi C^L$. This number is about 10 using data for the Lagrangian spectrum. This gives us one

estimate of G_Δ, which is 10. Now, in Prof. Yaglom's book there is another estimation given by the experiments of Tatarski, which gives an estimate of 0.1. Then there are three theories, one paper by Lesieur and Larchevêque and another paper by Kraichnan and another by Tomson; they all get an estimation in the range of 2 to 4 for this coefficient. So it's very interesting to look at it using this method of creating random Fourier modes with the right spectrum. You have to have a very long spectrum to obtain the inertial range result. Malik (N. A. Malik, 1991, Ph.D. thesis, Cambridge) found that his experimental coefficient G_Δ was about equal to 0.1. This was a numerical simulation using about 1000 Fourier components corresponding to a -5/3 spectrum over an energy scale of $10^4 : 1$. There is a nice explanation for our result compared to those of Lesieur, Kraichnan and Thomson. If you simply say that $d\Delta/dt$ is going to be equal to the structure function of Δ, then $d\Delta/dt$ is going to depend on the rate at which two particles move apart, which is $\varepsilon^{1/3}\Delta^{1/3}$. Straightforward calculation gives you a value for the coefficient G_Δ; you are implicitly assuming that the rate at which particles move apart just depends on the strain motion. If you actually crank in the value of S, you get an answer that is fairly close to those coefficients. But in reality, the particles are only partly doing this. When the particles are swirling around, they are not moving apart. And if you consider the proportion of time they spend in straining regions, you then find that your estimation of Δ^2 depends on this fraction of time to the third power times ε. Now, if this fraction is 1/2, $(1/2)^3 = 1/8$, you find your coefficient of 10 decreases to order 1. In fact, you find in numerical simulations that particles are spending about 1/3 of the time here - you've got $1/3^3$. Simulations show that particles separate much more slowly than you would expect on the basis of simple straining incidents; in fact, they're in these essential topological regions quite a lot of the time, where they swirl around. (T. M. J. Newley, H. J. Pearson & J. C. R. Hunt, 1991, Stably stratified rotating flow through a group of obstacles, Geophys. Astrophys. Fluid Dynamics, **58**, 147-172).

Herring: But your straining effect was computed on the basis of a random field. This is a kinematic effect of difference in size of these regions determined by continuity.

Hunt: Yes, the point being that in a kinematic simulation you've got the possibility for the particles to swirl together, whereas most statistical arguments make the essentially topological assumption that you are always in a sort of straining motion.

Moffatt: There is also the question of the time dependence of this field. These regions of trapping will not persist as regions of trapping for a very substantial time. So, of course, the question is if persistence of strain plays a part.

Hunt: In this simulation strain and vorticity both change with time.

Herring: But another aspect of this is that perhaps regions of high vorticity will concentrate into worms, which are very thin and very intense, and that may affect the trapping.

Hunt: But there is a circulation region around them.

Novikov: I have a short comment on this. My original calculation is actually a very old scheme based on the assumption that increments of velocity are independent, which is incorrect. Years ago I showed that this contradicts the Navier-Stokes equation. Because of this, I have introduced a local relaxation, which gives from the Lagrangian description exact Kolmogorov third order moments with all coefficients. So I think these old schemes have to be strongly corrected.

Hunt: What's your estimation of this number G_Δ?

Novikov: I have no estimation right now. It's quite a difficult problem.

Lesieur: The work with Michèle Larchevêque used EDQNM type theory of turbulence (2-point closure) based on the ideas of Kraichnan, Orszag, Herring. For the EDQNM, just looking at the velocity equation, you have the problem of one adjustable constant, which, in fact, can be related to the Kolmogorov constant. This is very similar to what Yakhot and Orszag do in the so-called RNG theories. The second point: for the passive scalar you have two further adjustable constants - and we spent a lot of years with Herring, Larchevêque and some other people, trying to fit these constants. Now, I would say that I don't believe at these values of G_Δ. I wouldn't bet anything on that, but neither on the value 0.1 with the kinematic simulation. I think your question is totally open. It is a universal parameter of turbulence. You can try to determine it. You can do this either by new experiments; you need to have high Reynolds numbers, and you can also try to obtain them by direct numerical simulations except, as was said, you don't have inertial range up to now. So it's a real open question and it's a very interesting problem.

Herring: There is another problem with Kraichnan's Lagrangian History theory in this case. Up to this point it cannot predict the scalar spectrum very well at high Prandtl numbers. In fact, in a way that's analogous to missing some dynamics for part of the track. It is of course the same thing as Hunt brought. People know this and there are new theories, strain-based theories, which perform better.

Adrian: It seems to me that the thing to do is to use the strain rate appropriate to the separation.

Hunt: Yes, that's right.

Session Six

Chairman J. Jiménez

A New Approach for Large Eddy Simulation of Turbulence and Scalar Transport

Parviz Moin

1. Introduction

The basic philosophy of large eddy simulation (LES) is to explicitly compute only those large-scale motions that are directly affected by the boundary conditions and to model the small scales. LES has always been described as attractive for engineering computations because it is significantly less computer-intensive than direct numerical simulation (DNS), yet promises to be more accurate and robust than single point closures.

The methodology of large eddy simulation is to filter or locally volume average the Navier Stokes equations and model the resulting subgrid scale stresses. The Smagorinsky eddy viscosity model and its variants have been used extensively to parameterize the effects of subgrid scale stresses on the large scale structures. Unfortunately, this model has some of the same difficulties as those prevalent in single point closures. These include variation of "optimal" model coefficient in different flows, inadequate asymptotic behavior near solid boundaries, and the non-vanishing of the model in laminar flows which leads to inaccurate predictions of transition from laminar to turbulent flow.

Some attempts have been made for improving subgrid scale models in analogy with single point closure models by using moment equations for SGS stresses (Deardorff, 1973; Schumann, 1975). These improvements do not address the aforementioned deficiencies. Other attempts for more rigorous parameterization of small scales have appealed to the statistical theories of turbulence (Lesieur, 1990). Although, physically and mathematically more attractive, these models are not applicable to inhomogeneous flows where models formulated in terms of Fourier transformed variables can not be used. Recognizing this important limitation, Metais & Lesieur (1990) proposed an eddy viscosity model, formulated in the physical space, which was consistent with the spectral eddy viscosity model of Chollet & Lesieur (1981) and Kraichnan (1976). Although this represents an improvement over Smagorinsky's model, it does not have the correct asymptotic behavior near walls (Comte, Lee & Cabot, 1990).

The key element that has been missing in most SGS modeling efforts has been the effective utilization of the large scale field which is computed directly. In large eddy simulation the *entire* spectrum of the resolved field is available and it would be unwise not to use this information in parameterization of the small scales being affected by the large scales. The model of Bardina *et al.*(1980) is an example where the difference between the velocity field at two scales was used to represent the subgrid scale velocity fluctuations. The dynamic modeling approach, which is a much more recent development (Germano, Piomelli, Moin & Cabot, 1991), uses a similar extrapolation procedure from the large scale computed field to the subgrid scales. In contrast to Bardina's approach, the extrapolation is used for the turbulent stresses rather than the velocity field. Furthermore, a precious mathematical identity (Germano, 1990) relating the stresses at different scales provides the means for *calculation* of the model coefficient as part of the simulation. It turns out that the model coefficient is a function of space and time, that is, it varies, for example, with different initial conditions and in different flow regimes.

The dynamic model has overcome many of the deficiencies of previous subgrid scale models, and has led to accurate predictions of several transitional and turbulent flows without ad hoc adjustments. This paper outlines the steps in the development of the dynamic model, points out key assumptions in its formulation, its successes, and areas in need of further study and improvement.

2. Dynamic Model

The subgrid scale stresses in the incompressible governing equations for the large scale field are

$$\tau_{ij} = \overline{u_i u_j} - \overline{u}_i \overline{u}_j, \tag{1}$$

where overbar indicates a filtered or large scale flow variable and u_i is the velocity in the $i-$ direction. Similarly the subgrid scalar flux is

$$q_k = \overline{\theta u_k} - \overline{\theta}\overline{u}_k, \tag{2}$$

where θ is a scalar. Both τ_{ij} and q_k must be modeled in terms of the large scale variables to close the LES equations. In the initial formulation of the dynamic model the Smagorinsky eddy viscosity model was used to parameterize the small scale stresses

$$\tau_{ij} - \frac{1}{3}\delta_{ij}\tau_{kk} = -2\nu_t \overline{S}_{ij}, \qquad \nu_t = C\Delta^2|\overline{S}|, \tag{3}$$

where $\overline{S}_{ij}$ is the large scale strain rate tensor,

$$\overline{S}_{ij} = \frac{1}{2}\left(\frac{\partial \overline{u}_i}{\partial x_j} + \frac{\partial \overline{u}_j}{\partial x_i}\right),$$

and $|\overline{S}| \equiv |2\overline{S}_{ij}\overline{S}_{ij}|^{1/2}$, Δ is the filter width, which is generally related to the mesh size of the resolved field in the LES calculation. An analogous eddy diffusivity model can be defined as (Edison, 1985)

$$q_k = -\alpha_t \frac{\partial \overline{\theta}}{\partial x_k}, \qquad \alpha_t = C_\theta \Delta^2 |\overline{S}|. \tag{4}$$

The residual stresses and the scalar flux corresponding to a lower wavenumber cut-off or larger scale filter denoted by $\widehat{(\cdot)}$ are

$$T_{ij} = \widehat{\overline{u_i u_j}} - \widehat{\overline{u}}_i \widehat{\overline{u}}_j \,, \tag{5}$$

$$Q_k = \widehat{\overline{\theta u_k}} - \widehat{\overline{\theta}}\, \widehat{\overline{u}}_k \,. \tag{6}$$

These are modeled in the same manner as the residual stresses in (3) and (4):

$$T_{ij} - \frac{1}{3}\delta_{ij} T_{kk} = -2\widehat{\nu}_t \widehat{\overline{S}}_{ij} \,, \quad \widehat{\nu}_t = C\widehat{\Delta}^2 |\widehat{\overline{S}}| \,, \tag{7}$$

$$Q_k = -\widehat{\alpha}_t \frac{\partial \widehat{\overline{\theta}}}{\partial x_k} \,, \quad \widehat{\alpha}_t = C_\theta \widehat{\Delta}^2 |\widehat{\overline{S}}| \,, \tag{8}$$

where $|\widehat{\overline{S}}| \equiv |2\widehat{\overline{S}}_{ij}\widehat{\overline{S}}_{ij}|^{1/2}$ and $\widehat{\Delta}$ is the test filter width. We have assumed that the model coefficients, C and C_θ are the same for both stresses at the grid level and the test filter level. We will show in the next section that this assumption is not necessary. The interesting observation first made by Germano (1990) is that the stresses at the two levels are related through the simple identity

$$\mathcal{L}_{ij} = T_{ij} - \widehat{\tau}_{ij} = \widehat{\overline{u}_i \overline{u}_j} - \widehat{\overline{u}}_i \widehat{\overline{u}}_j \,. \tag{9}$$

Similarly for the scalar fluxes, we have

$$\mathcal{F}_k = Q_k - \widehat{q}_k = \widehat{\overline{\theta}\, \overline{u}_k} - \widehat{\overline{\theta}}\, \widehat{\overline{u}}_k \,. \tag{10}$$

The utility of these identities stems from the fact that their right hand sides can be computed directly; it is this feature that is used in the dynamic model for computing the model coefficients. For example, the model expressions for τ_{ij} and T_{ij} given by (3), (7) are substituted in (9) to obtain:

$$\begin{aligned} L_{ij} - \frac{1}{3}\delta_{ij} L_{kk} &= -2\widehat{\nu}_t \widehat{\overline{S}}_{ij} + \widehat{2\nu_t \overline{S}_{ij}} \\ &= -2C\widehat{\Delta}^2 |\widehat{\overline{S}}| \widehat{\overline{S}}_{ij} + \widehat{2C\Delta^2 |\overline{S}| \overline{S}_{ij}} \end{aligned} \tag{11}$$

This expression is then used at every point to compute the model coefficient C. However, Eq. (11) is an equation relating second rank tensors, that is, at each point there are six equations for the single unknown, C. Furthermore (11) is strictly an integral equation for C, since C is now a function of space and time, and the test filter is also applied to C in the last term on the right hand side of (11). All investigators to date have assumed that C is a very slowly varying function of space, and thus have removed it from the integral representing the test filter to obtain algebraic equations for C. A priori tests using DNS data have shown that C

is in fact a very rapidly varying function of space, and removing it from the filtering integral represents an inconsistency. If, on the other hand, it is prescribed in (3) and (7) that the model coefficient is only a function of the coordinates where the flow is not filtered, then the analysis does not involve an inconsistency. For example, in large eddy simulations of decaying isotropic turbulence, Moin *et al.* (1991) assumed that C was only a function of time, and in simulations of channel flow Germano *et al.* (1991) assumed that C is a function of time and the direction normal to the walls. In the isotropic decay problem, the flow was filtered in all spatial directions, and in the channel flow simulations the flow was filtered only in the streamwise and spanwise directions.

Germano *et al.* (1991) contracted (11) with $\overline{S}_{ij}$ to obtain a single algebraic equation for C. Lilly (1992) used the least square technique to find the coefficient that minimizes, *at each point*, the sum of the square errors in the six equations. This procedure leads to the following expressions for the model coefficients

$$C\Delta^2 = -\mathcal{L}_{mn}\mathcal{M}_{mn}/\mathcal{M}_{k\ell}\mathcal{M}_{k\ell}\,, \tag{12}$$

$$C_\theta\Delta^2 = -\mathcal{F}_i\mathcal{H}_i/\mathcal{H}_j\mathcal{H}_j\,. \tag{13}$$

where,

$$\mathcal{M}_{ij} = 2\left[(\widehat{\Delta}/\Delta)^2|\widehat{\overline{S}}|\widehat{\overline{S}}_{ij} - \widehat{|\overline{S}|\overline{S}_{ij}}\right]\,, \qquad \mathcal{H}_k = (\widehat{\Delta}/\Delta)^2|\widehat{\overline{S}}|\frac{\partial\widehat{\overline{\theta}}}{\partial x_k} - \widehat{|\overline{S}|\frac{\partial\overline{\theta}}{\partial x_k}}$$

The coefficients in (12) and (13) are functions of space and time and thus, adjust to the flow conditions. The key desirable features of the model, that is, its proper asymptotic behavior near walls, and vanishing of the model in laminar flow, are derived from the $\mathcal{L}_{mn}$ term in the numerator of (12). However, Cabot & Moin (1991) found that, when used in actual simulations, the local coefficient leads to numerical instabilities in LES of channel flow. (The same observation has been made by Lund (1991) in large eddy simulations of isotropic turbulence.) As in Germano *et al.* (1991), they then assumed that the model coefficients are only functions of time and the direction normal to the walls, and used spatial averaging in (11) to obtain global expressions for the coefficients which are the same as (12) and (13) with the exception that the numerators and denominators are spatially averaged over the homogeneous spatial directions. The resulting models have been successfully applied to the channel flow including heat transport in a case of isothermal walls, one heated and one cooled. The profiles of mean velocity, temperature, and root-mean square temperature fluctuations are shown in Figures 1-3. The agreement with the DNS data is good. Very good agreements were also obtained (e.g., Figure 4) with the experimental data and direct numerical simulation data for the decay of isotropic turbulence (Moin *et al.* 1991, Squires, 1991). It is remarkable that such

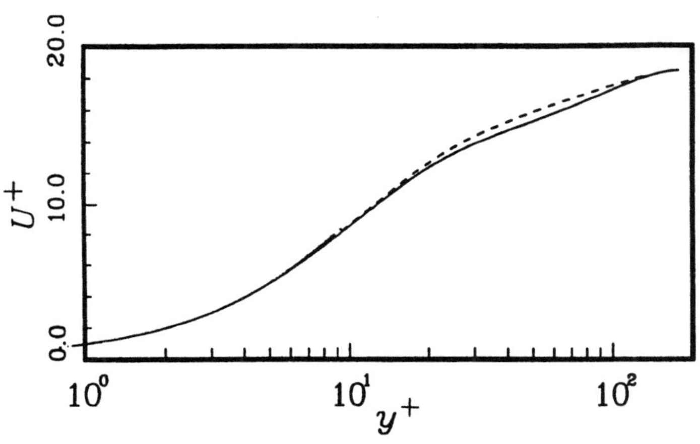

FIGURE 1. Mean velocity profile in turbulent channel flow. (——) from DNS; (----) from LES.

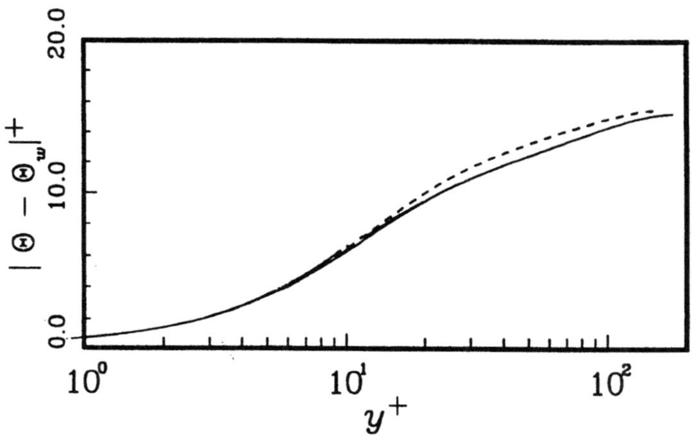

FIGURE 2. Mean temperature profile in turbulent channel flow. (——) from DNS; (----) from LES.

good agreements were obtained for different quantities in different flows without any adjustments to the model parameters. The dynamic modeling procedure has been extended to compressible flows with similar success (Figure 2).

3. Limitations and Further Developments

Several alternative formulations of the dynamic model have been explored. As pointed out in §2 it is not necessary to assume that the model coefficients are the

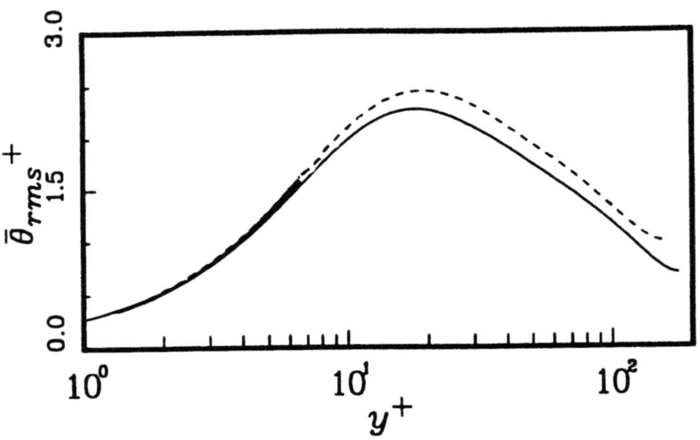

FIGURE 3. Root-mean square temperature fluctuations for a channel flow. (——) from DNS; (----) from LES.

same in (3) and (7). With this restriction removed, the least square technique can now be used to solve for both constants. We will now have six equations for two unknowns instead of six equations for one unknown as in the original formulation. The resulting coefficients are

$$C_1 = \mathcal{L}_{ij}(\alpha_{ij}\beta_{kl}\beta_{kl} - \beta_{ij}\alpha_{kl}\beta_{kl})/(2\Delta^2 D)$$

$$C_2 = \mathcal{L}_{ij}(\alpha_{ij}\alpha_{kl}\beta_{kl} - \beta_{ij}\alpha_{kl}\alpha_{kl})/2\hat{\Delta}^2 D,$$

where C_1 is the model coefficient in (3), and C_2 is the test field model coefficient in (7), and

$$\alpha_{ij} = \widehat{|\overline{S}|\overline{S}_{ij}} \qquad \beta_{ij} = |\hat{\overline{S}}|\hat{\overline{S}}_{ij}$$

and

$$D = \alpha_{ij}\alpha_{ij}\beta_{kl}\beta_{kl} - (\alpha_{kl}\beta_{kl})^2$$

One of the drawbacks of eddy viscosity models is the well known misalignment of the residual stress and the large scale strain rate tensors. This misalignment may have also contributed to the poor numerical performance of the local formulations in (12) and (13). An alternative formulation of the dynamic model that may improve the alignment problem is to use a tensor model coefficient

$$\tau_{ij} = -(C_{ik}\overline{S}_{kj} + C_{jk}\overline{S}_{ki})\Delta^2|\overline{S}| \tag{14}$$

where, C_{ij} is the model coefficient. Using this model instead of Smagorinsky's model in (9) leads to the following six equations for the coefficients

$$\mathcal{L}_{ij} = -(C_{ik}\mathcal{M}_{kj} + C_{jk}\mathcal{M}_{ki})\Delta^2$$

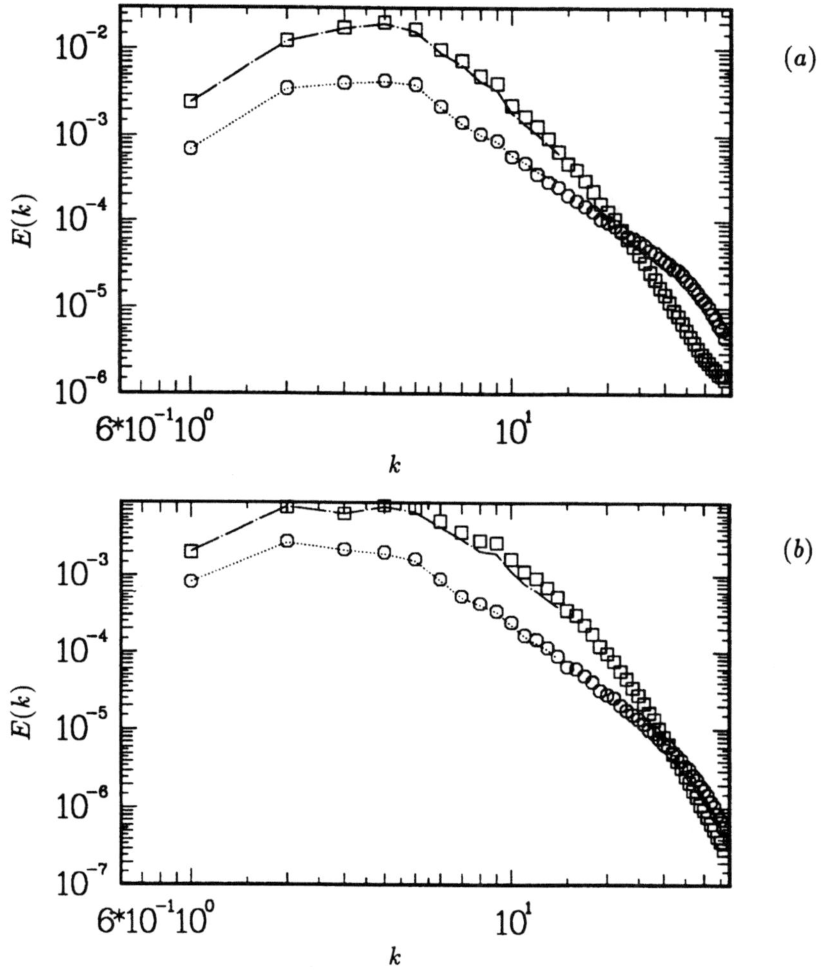

FIGURE 4. Radial energy spectrum from compressible turbulence. (a) $t/\tau_e = 0.72$. (b) $t/\tau_e = 1.98$. □ , solenoidal velocity (DNS, 96^3); ○ , dilatational velocity (DNS, 96^3); —·— , solenoidal velocity (LES, 32^3); ········ , dilatational velocity (LES, 32^3). τ_e is the eddy turn-over time.

All the coefficients can in principle be obtained uniquely from this equation. The fact that several coefficients can be obtained as part of the computations rather than prescribed a priori is clearly one of the main advantages of the dynamic model procedure.

Another interesting possibility is the inclusion of quadratic strain rate terms in the residual stress model:

$$\tau_{ij} - \tau_{kk}\delta_{ij}/3 = -2C_1\Delta^2|\overline{S}|\overline{S}_{ij} - 2C_2\Delta^2(\overline{S}_{ik}\overline{S}_{kj} - \overline{S}_{kl}\overline{S}_{lk}\delta_{ij}/3)$$

Again, both model coefficients can be deduced from the dynamic model procedure.

The main issue concerning the dynamic model at present is the numerical instabilities that result when local formulation of the model is used in large eddy simulations. The numerical instability is in part a result of persistent negative eddy viscosity in some flow regions, and in part due to rapid spatial varions of ν_t including large spikes. The use of finite-differences rather than spectral methods for computation of derivatives in the model significantly enhances the numerical stability of the solution. Another, possibly related, concern is the inconsistency in the treatment of the spatially dependent coefficients of the model as independent of the filtering process (see §2). The consistent treatment requires a solution of an integral equation. Ghosal, Lund & Moin (1992, unpublished) have obtained analytical solutions of the resulting integral equation for homogeneous flows. These alternative formulations are, currently under investigation at the Center for Turbulence Research.

4. Conclusions

The dynamic model methodology is an altogether new approach to subgrid scale modelling which is most promising. This review provided an outline of the development of the dynamic subgrid scale model for large eddy simulation of turbulent flows. The application of the model to date have been impressive. However, its utility in applications of engineering interest will depend on the development of a numerically stable local formulation of the model. At present, this is an active area of research.

Acknowledgements

I am grateful to Bill Cabot, Tom Lund, and Sandip Ghosal for many stimulating and productive discussions. The studies reviewed here have been done in collaboration with them. This effort is sponsored by the U.S. Office of Naval Research, and the Air Force Office of Scientific Research.

REFERENCES

Bardina, J., Ferziger, J. H. & Reynolds, W. C. 1980 *AIAA Paper* No. 80-1357.

Cabot, W. & Moin, P. 1991 *CTR Manuscript 128*, Center for Turbulence Research, Stanford University, Stanford, CA 94305.

CHOLLET, J. P. & LESIEUR, M. 1981 *J. Atmos. Sci.* **38**, 2747-2757.

COMTE, P., LEE, S. & CABOT, W. H. 1990 In *Proceedings of the 1990 Summer Program of the Center for Turbulence Research*, Stanford University and NASA-Ames Research Center, 31-45.

DEARDORFF, J. W. 1973 *J. Fluids Eng.* **95**, 429.

EDISON, T. M. 1985 *J. Fluid Mech.* **158**, 245.

GERMANO, M. 1990 *CTR Manuscript 128*, Center for Turbulence Research, Stanford University, Stanford, CA 94305.

GERMANO, M., PIOMELLI, U., MOIN, P. & CABOT, W. H. 1991 *Phys. Fluids A.* **3**, 1760-1765.

GHOSAL, S., LUND, T., S. & MOIN, P. 1992 *To be published.*

KRAICHNAN, R. H. 1976 *J. Atmos. Sci.* **33**, 1521-1536.

LESIEUR, M. 1990 Turbulence in Fluids, *Kluwer Academic Publishers*, The Netherlands.

LILLY, D. 1992 *Phys. Fluids A.* **4**, 633-635.

LUND, T., S. 1991 In *Annual Research Briefs*, Center for Turbulence Research, Stanford University, Stanford, CA 94305, 177-190.

MÉTAIS, O. & LESIEUR, M. 1990 *Submitted to J. Fluid Mech.*

MOIN, P., SQUIRES, K., CABOT, W. & LEE, S. 1991 *Phys. Fluids A.* **3**, 2746-2757.

SCHUMANN, U. 1975 *J. Comput. Phys.* **18**, 376.

SQUIRES, K. 1991 In *Annual Research Briefs*, Center for Turbulence Research, Stanford University, Stanford, CA 94305, 207-223.

Address

Center for Turbulence Research
Stanford University
Bldg. 500
Stanford CA 94305-3030

On the Contribution of Multi-Point Experimental Techniques to the Study of Turbulence

R. J. Adrian

Abstract. Experimental methods such as planar laser induced fluorescence and particle image velocimetry, offer the ability to measure flow properties, scalar or vector, at many points simultaneously. The manner in which such experimental methods are likely to advance the frontiers of turbulence research, and the challenges that must be surmounted are discussed.

1. Introduction

During the last ten years much of the progress made in our understanding of the structure of turbulent motion was accomplished by means of the direct numerical simulation of turbulent flow in very simple geometries at low Reynolds numbers. These simulations, notably by S. Orszag and co-workers at Princeton, and J. Kim, P. Moin, R. Moser, P. Spalart and co-workers at NASA Ames Research Center, Stanford, have been a source of reliable information whose richness could not be rivaled by experimental techniques. Vector velocity, pressure, vector vorticity, and indeed the entire deformation tensor, and scalars are available as virtually continuous fields in three-dimensional space and time, with resolution approaching the Kolmogorov microscale. Studies of the fields in, for example, the turbulent channel flow numerical simulation of Kim, Moin and Moser (1987) have led to a more unified picture of the wall layer that consists of many of the elements such as sweeps, bursts, hairpin eddies, and wall streaks that have been observed in isolation previously, plus modifications of these elements and understanding of their interactions.

Given the success of numerical simulation, it is natural to inquire into the role of physical experimental turbulence research vis-as-vis structural studies, and especially the role of new techniques for the simultaneous measurement of turbulence quantities at multiple points. These techniques include planar laser induced fluorescence (Hassa et al., 1987; Dahm et al., 1991), particle image velocimetry (Adrian, 1991; Papantoniou and Dracos, 1989), magnetic resonance imaging (Lee et al., 1987) and tomographic interferometry

(Hesselink, 1988). Each of these methods offers a capability to measure a nearly continuous, instantaneous field of data, usually in a two-dimensional plane, but sometimes in a three-dimensional volume. Developments of the optical versions of these methods have been enabled in the broadest sense by the availability of two-dimensional data acquisition, in the form CCD arrays or CID arrays, and paced by data transfer rates and access to large, high speed computers for unique analysis. It would not be unfair to say that the lag between numerical simulation and experimental work is a consequence of the limited computer power that has been available to experimentalists. Such limits are being removed, and the purpose of this paper is to discuss the role that experiments may play in the next decade of turbulence research.

2. Experimental Capabilities

The simplest measurement is that of a scalar field, and for this reason it provides a useful baseline for the characterization of current measurement capabilities. Dahm et al. (1991) have performed elegant measurements of scalar concentration derived from laser induced fluorescence by using a video camera that images a 256 × 256 x-y array of pixels and a swept laser light sheet that illuminates five successive planes stacked in the z-direction. The spatial resolution of this experiment is $\Delta x = \Delta y = 116$ microns, and $\Delta z = 220 \mu m$ and $45 ms$ to scan a set of five planes. The observed volume was $30mm \times 30mm \times 1mm$ thick. Since the experiment is simply a matter of measuring light intensity, the time resolution or alternatively the volume of fluid to be scanned is limited purely by the rate of data transfer to computer, in this case 9.3 M bytes per second. Increased rates up to 100 M bytes per second are now available, extending the volume to, for example, 30 × 30 × $10mm$ or increasing the sample rate to about one 256^2 plane per millisecond.

As another example, Liu et al. (1991) describe a particle image velocimeter experiment in $Re = 2872$ channel flow in which two-dimensional velocity measurements were made on a planar 115 × 100 grid of 1 × 0.30 × $0.8mm$ measurement volumes on a $0.5mm$ spacing in a single plane. Unlike the foregoing example from Dahm et al. (1991), this work used photographic recording because the increased resolution needed to resolve small particles in PIV could not be accommodated by video transfer rates. The penalty is, of course, a loss of temporal resolution. By using cinematic recording, the projected temporal resolution for PIV is 50 frames per second at 140 × 100 independent vectors per frame. This rate is comparable to the scalar measurement rate, but the level of effort required for image analysis is much higher, and considerable time is spent in off-line processing.

The PIV technique can be extended to three-dimensional vector measurements on a plane by means of stereo photography. Fully three-dimensional measurements are possible by means of two or three camera photogrammetry (Maas, 1991) or holography. The requirements for unambiguously matching images from the separate cameras to the same particle limit the number of particles in the volume of measurement to about one-thousand, and this is relatively sparse data. Consequently, three-dimensional photogrammetry uses particle tracking methods to find the velocities of individual particles and interpolates between the measurement points which are located at random points in space. The wavenumber resolution of such interpolation is of the order of the mean spacing between particles, and it corresponds to about one-tenth of the linear dimension of the total volume of measurement when the number of particles is of order one-thousand. Holographic PIV has not yet been fully explored, but resolution is expected to be about one millimeter over fluid volumes of order $100mm^3$.

The spatial resolution of PIV can be improved by using optical magnification to zoom in on a smaller region of flow. Taking one millimeter resolution on a $100 \times 100mm$ grid as the nominal current standard at unity magnification, it is possible to achieve 1000/M micron resolution on a $100/M \times 100/M$ *mm* grid using a M-fold magnification.

To illustrate the fine resolution that can be achieved, Figure 1 shows the vector field measured in the logarithmic region of the radial-axial plane of turbulent pipe flow at Reynolds number equal to 50,000. The fluid was air, the pipe diameter was 127*mm* and the scattering particles were 2 micron oil droplets illuminated by a double pulsed Yag laser. Images were recorded with magnification M=6, and the photographs were interrogated using auto-correlation analysis of the groups of particle images contained in one-millimeter spots, corresponding to 167 micron diameter measurement volumes in the flow. This is small even by the standards of thermal anomometry or laser Doppler velocimetry. The critical factor in achieving such resolution is the concentration of light scatterers which must be enough to produce ten to twenty particles in each measurement volume for the case of auto-correlation analysis.

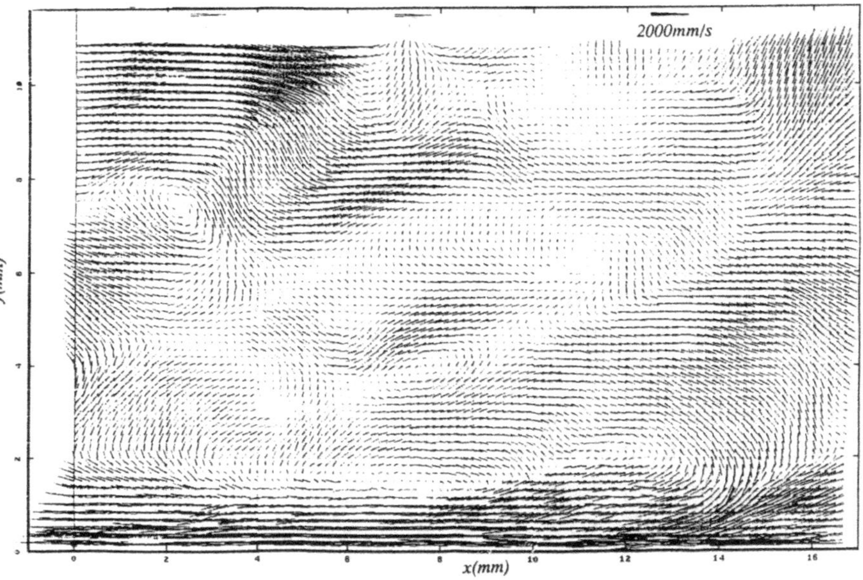

Figure 1 ***Turbulent Log Layer at Reynolds Number 50,000*** The instantaneous velocity field in the axial-radial plane of turbulent pipe flow recorded with a very high resolution particle image velocimeter. Each vector is measured with a spatial resolution of 0.167*mm* on a grid of the same spacing. A constant mean flow velocity of 4960*mm/s* has been subtracted to reveal the turbulent fluctuations. The fluid is air seeded with one-micron diameter oil droplets at one part per million volume concentration. *Courtesy of T. Urushihara.*

Further enhancement of the spatial resolution by a factor of two to four appears possible without increasing concentration by improved image processing. One technique, proposed by Guezennec and Kiritsis (1990), follows conventional auto-correlation of the interrogation spot with particle tracking to identify individual particle motions within each spot. With particle tracking the spatial resolution is defined by the maximum particle image displacement, and this can be as small as fifty microns at five-fold magnification using two micron particles. The available field of view for such an experiment, 20 × 20*mm* would yield a grid of 400 × 400 independent points, or 800 × 800 Nyquist sampled points. Another technique under development involves cross-correlation of a small spot with a larger region, and it promises similar improvements to in-plane resolution. With such improvements to the in-plane resolution, the thickness of the laser sheet must be reduced to provide corresponding resolution in the out-of-plane directory.

The accuracy of the scalar methods is determined by the camera resolution, and this is typically 8 bits with some more accurate (but slower) 12 bit units available. The accuracy of PIV is of the order of one-percent of full scale in turbulent flow. These accuracies permit useful evaluation of the spatial derivations, albeit not with the precision afforded by good numerical techniques.

3. Significance of Multi-Point Methods

Relative to single-point experimental methods, multi-point methods extend our experimental investigative powers in a number of important ways. Foremost, of course, is the ability to observe, quantitatively, the *instantaneous* field of a scalar or vector. This removes the uncertainties of inferray the structure of underlying physical processes from measurements of mean statistics or conditionally averaged fields. It also makes possible measurements with respect to moving boundaries, and identification of flow elements in terms of spatially extended structures. Many of these functions were performed in the past by flow visualization, and in this sense, one might think that the newer methods have only made flow visualization more quantitative. This would not, however, accurately reflect the real significance of the newer methods, because they not only quantify some types of measurements, they also provide *types of spatial information that were heretofore inaccessible by flow visualization.* An obvious case is the instantaneous vector flow field, as opposed to the Lagrangian motion of flow visualization marker fields which is essentially an integral of the velocity field. Perhaps, even more important, is the new information contained in the *gradients* of the measured fields which permits evaluation of the vorticity field, the rate-of-strain field, the scalar flux under Fickian diffusion and associated quantities such as the dissipation of concentration and the viscous dissipation of turbulent kinetic energy.

4. Contribution of Experiments

From the considerations of Section 2, it is clear that experimental techniques find difficulty in providing the rich, three-dimensional, time-resolved fields that one obtains from direct numerical simulation. The most complete, and clearly specified data will continue to be derived from calculations. However, the experimental techniques do have the capability to measure three-dimensional scalar fields with almost the same accuracy and spatial resolution as DNS calculations, and PIV methods can provide resolutions better than DNS on two-dimensional planes, with time resolved measurements or volume measurements available at somewhat reduced resolution.

The advantages of experimental results derive principally from the fact that they are "real." Although spatial resolution acts as a low-pass filter and attenuates small scale motions, it does not affect convergence or accuracy of the measured velocity field as in DNS. Computations must have high accuracy, high spatial resolution, and high temporal resolution (small time-step advancement) in order to arrive at correct solutions. They must also be fully three-dimensional, whereas the experiments can select sub-regions of a three-dimensional flow for investigation, thereby greatly reducing the data that is involved. Two important numerical issues, in-flow/out-flow conditions and periodic side wall boundary conditions are also absent in physical experiments. (They are often replaced by imperfect boundary conditions that can be troublesome, but they are at least natural.)

There is an interesting dualism between the new methods for field measurement and large eddy simulation. When the experiments are setup to observe motion in large scale regions, the instruments average over spatial resolution cells in much the same way and with much the same scale as large eddy simulations. Experiments of this type can be used to verify the resolved scale motions computed from an LES. If the experiment is setup to focus on a smaller region, the instruments are able to resolve the motions corresponding to the non-resolved scales of LES. Experiments of this type are obviously useful to study sub-grid scale closure models and small scale turbulence physics. Experimental studies are able to treat complicated conditions that are beyond reach of current DNS. These include rough surfaces, non-zero pressure gradient, complicated geometries, and complex fluids such as dilute polymer solutions, bubbly liquids, and other two-phase flows.

Finally, the current generation of measurement techniques is able to make measurements in high Reynolds number flows that are presently difficult to simulate numerically. The data in Figure 1 at Reynolds number 50,000, corresponds to Re_θ of approximately 2,500, and higher Reynolds number experiments appear possible. At these Reynolds numbers, the logarithmic layer becomes significant, and turbulent structures not seen in the low Reynolds number simulation begin to appear (Wark et al., 1990). Other interesting phenomena also become observable. For example, it appears that there is a large scale motion whose integral length scale is of the order of several pipe radii. Since this is much larger than the streamwise extents of the computational domains used in most DNS computations, this large scale motion does not appear to have been observed.

5. Summary

Experimental techniques have not yet reached the high level of development seen in the field of direct numerical simulation, but even upon attaining their ultimate potential it is unlikely that they will be able to offer the same completeness of information. On the other hand, they will offer the assurance that is inherent to empirical evidence, and they will allow investigation of complex flows. There is a healthy complementarity between multidimensional experimental technique and direct numerical simulations and large eddy simulations that should be exploited.

6. References

R. J. Adrian (1991) "Particle Imaging Techniques for Experimental Fluid Mechanics," *Ann. Rev. Fluid Mech.*, **23**, 261–304.

W. J. A. Dahm, K. B. Southerland and K. A. Buch (1990) "Four-Dimensional Laser Induced Fluoescence Measurements of Conserved Scalar Mixing in Turbulent Flows," to appear in *Applications of Laser Techniques to Fluid Mech.*, ed. R. J. Adrian et al. (Springer-Verlag: Berlin).

Y. G. Guezennec and N. Kiritsis (1990) "Statistical Investigation of Errors in Particle Image Velocimetry," *Exp. in Fluids*, **10**, 138–146.

C. Hassa, P. H. Paul and R. K. Hanson (1987) "Laser Induced Fluorescence Modulation Techniques for Velocity Measurements in Gas Flows," *Exp. in Fluids*, **5**, 240–246.

L. Hesselink (1988) "Digital Image Processing in Flow Visualization," *Ann. Rev. Fluid Mech.*, **20**, 421–485.

J. Kim, P. Moin and R. Moser (1987) "Turbulence Statistics in Fully Developed Channel Flow at Low Reynolds Number," *J. Fluid Mech.*, **177**, 133-166.

S. J. Lee, M. K. Chung, C. W. Mun and Z. H. Cho (1987) "Experimental Study of Thermally Stratified Unsteady Flow by NMR-CT," *Exp. in Fluids*, **5**, 273–281.

Z. C. Liu, C. C. Landreth, R. J. Adrian and T. J. Hanratty (1991) "High Resolution Measurements of Turbulent Structure in a Channel with Particle Image Velocimetry," *Exp. in Fluids*, **10**, 301–312.

G. Maas (1991) *Proc. IUTAM 279*, Tech. Univ. Delft, June 1991.

D. Papantoniou and Th. Dracos (1989) "Analyzing 3-D Turbulent Motions in Open Channel Flow by Use of Steroscopy and Particle Tracking," *Adv. in Turbulence*, **2** (Springer-Verlag: Berlin), 278–285.

C. Wark, P. W. Offutt and R. J. Adrian (1990) "Structure of Turbulence Using PIV in Wall Bounded Shear Flow," *Bull. Amer. Phys. Soc.*, **35**, 2284 (Abstract only).

Acknowledgements

This work is supported by ONR Grant N00014-90-J-1415 and AFOSR Grant 90-0169.

Address of the author

Professor Ronald J. Adrian
University of Illinois
Department of Theoretical and Applied Mechanics
216 Talbot Laboratory
104 South Wright Street
Urbana, IL 61801-2935
Fax No. 217/244-5707
rja@del.tam.uiuc.edu

December 1991

Direct Numerical Simulations and Hot Wire Experiments : A Possible Way Ahead?

Professor Robert A. Antonia

Abstract. Direct numerical simulations have contributed significantly to the study of the physics of turbulence and to turbulence modelling, albeit at low Reynolds numbers. Hot wire experiments should continue to play a useful part in providing data at higher Reynolds numbers but it is important that their limitations are realised. To this end, hot wire measurements in a fully developed turbulent duct flow have been compared with other types of measurements as well as direct numerical simulations for the same flow and comparable Reynolds numbers. Preliminary hot wire measurements of spatial velocity derivatives in a turbulent channel flow show encouraging agreement with the simulations.

1 Introduction

There is little doubt that direct numerical simulations (DNS), which resolve all the essential scales of the turbulent motion, have uncovered many new kinematical features of turbulent flows and enhanced our ability to improve turbulence models, irrespective of their complexity, especially in the near-wall region. Examples of contributions DNS data bases can make to various facets of our understanding of turbulent flows are given in Moin and Spalart (1989), and Robinson et al. (1990). Use of the data bases for fine tuning and improving turbulence models has been reviewed by Mansour (1991). The extension to non-isothermal flows (Kim and Moin, 1989; Lyons et al., 1991; Kasagi et al., 1991) has yielded useful insight into the heat transfer characteristics of the near-wall region.

Many of the available simulations, with and without heat transfer, have been obtained in the same flow (a fully developed turbulent channel flow) and at nominally the same Reynolds number (Kim et al., 1987; Kuroda et al., 1989). The value of this Reynolds number ($h^+ \equiv hU_\tau/\nu \simeq 180$, where h is the half-width of the channel, U_τ is the friction velocity and ν is the kinematic viscosity of the fluid; the superscript + will denote normalisation by wall variables) is insufficient to achieve a clear separation between the largest and smallest length scales of the turbulent motion so that an inertial subrange is non-existent. Simulations have also been made at a slightly larger Reynolds number in this flow (Kim, 1989) and for a range of Reynolds numbers in a zero pressure gradient turbulent boundary layer (Spalart, 1988). The range of all existing simulations falls nevertheless within that for which low Reynolds number effects are known to be important. Extrapolation to large Reynolds numbers is risky, especially since the cause for low Reynolds number effects is not entirely clear. In this context, the DNS data bases have, together with experimental data, provided some insight into low Reynolds number effects (Antonia et al., 1991,1992). Antonia et al. (1992), observed generally good agreement between the hot wire and the DNS data, except for $\overline{v^2}$ (v is the velocity fluctuation normal to the wall), especially in the wall region.

Despite this, as yet unexplained difference, it is expected that hot wire measurements will continue to play an important role in terms of providing data at realistic Reynolds numbers and over as wide a range of length scales as possible. In this regard, the usefulness of fine scale turbulence measurements can be enhanced by using the DNS data as a yardstick for hot wire measurements of velocity derivatives. Following a brief review of low Reynolds number effects and of the comparison between hot wire and DNS data, results from preliminary measurements of velocity derivatives in a fully developed channel flow are described.

2 Comparison Between Data Obtained by Different Methods

Recent advances in turbulence measurement techniques, including the advent of new techniques, allow a more critical evaluation of data bases than has been possible hitherto.

A comparison between hot wire measurements and results obtained from direct numerical simulations and by LDV (laser Doppler velocimetry) and PIV (particle image velocimetry) is shown in Figure 1[1]. All data sets were obtained in a fully developed channel flow and comparable values of h^+.

Figure 1a indicates quite reasonable agreement for u'^+ (the prime denotes the rms value). While the LDV, PIV and DNS data for v'^+ show good agreement (Figure 1b), the measured hot wire values are significantly larger, everywhere across the flow but especially close to the wall. The discrepancy is in the same direction as that previously noted by Kim et al. (1987) when comparing their DNS values with the hot film data of Eckelmann (1974) and Kreplin and Eckelmann (1979). It was suggested that signals from an X-probe also sense the velocity component normal to the plane of the probe, such cross contamination resulting in larger values of v' and w'. The error is expected to be largest for v' in view of the relatively small magnitude of v.

There are several possible sources of error in connection with X-wire measurements.

[1]Other DNS, LDV and PIV data apart from those used in Figure 1 are also available. The LDV data of Niederschulte et al. (1990) and the PIV data of Liu et al.(1991), also obtained for a fully developed turbulent channel flow, are in good agreement with the corresponding data in Figure 1. Similarly, the DNS data of Kuroda et al. (1989) are in good agreement with the DNS data of Figure 1.

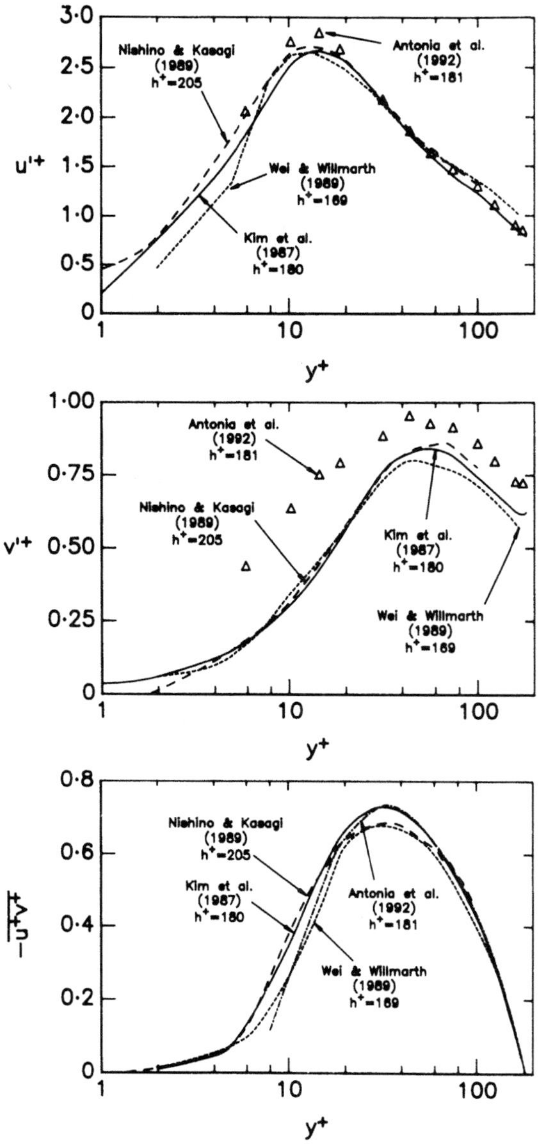

Figure 1: Comparison between hot wire measurements of u'^{+}, v'^{+} and $-\overline{u^{+}v^{+}}$ in a turbulent channel flow and results obtained by three other methods. (a) u'^{+}; (b) v'^{+}; (c) $-\overline{u^{+}v^{+}}$. Δ, (a), (b) or — - —(c), hot wire (Antonia et al., 1992); ——, DNS (Kim et al., 1987); ----, LDV (Wei and Willmarth, 1989); — —, PIV (Nishino and Kasagi, 1989).

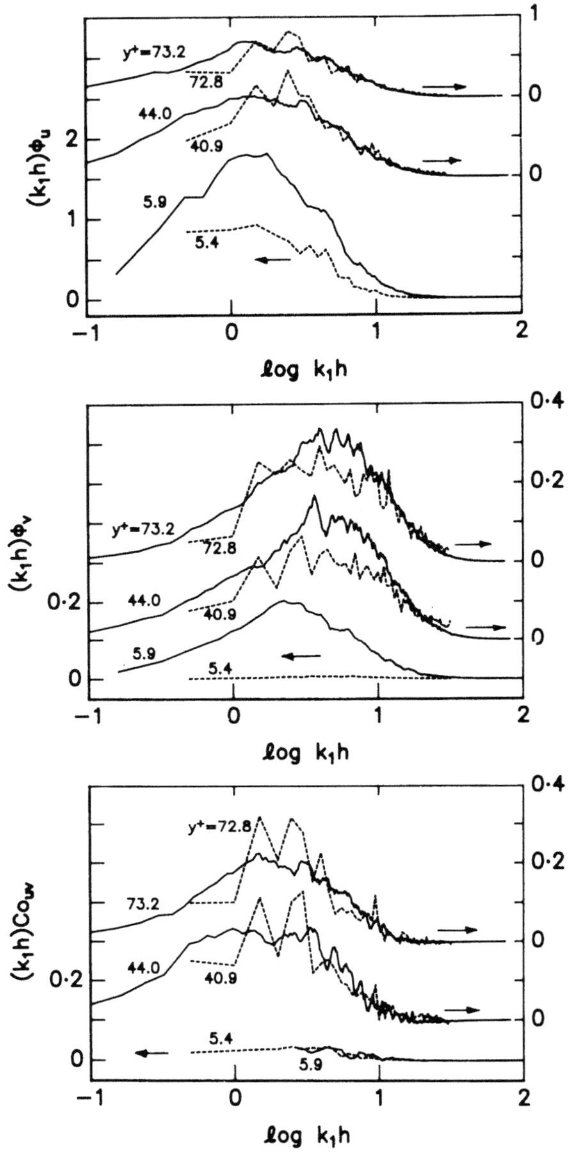

Figure 2: Comparison between hot wire and DNS spectra and cospectra of u and v in a turbulent channel flow. (k_1 is the streamwise wavenumber). (a) ϕ_u; (b) ϕ_v; (c) Co_{uv}. ——, hot wire; - - - -, DNS.

Some of these were considered by Antonia et al. (1992), who eliminated such possibilities as the inadequacy of a 90° included angle (a 90° X-probe was used for the data in Figure 1), inadequate wire length ($\simeq 0.3$ mm, the wire diameter was 1.2 μm). No light was shed however on the possible contamination by w. The use of DNS data bases to evaluate the response of an X-probe has not helped to clarify this problem. Moin and Spalart (1989) suggested that the error caused by neglecting w was important while Suzuki and Kasagi (1990) found it to be negligible. These latter authors found that the lateral spacing between the wires causes v'^+ to be overestimated (but only in the near-wall region).

Clearly more work is needed before the X-probe v data can be trusted. Some encouragement can however be drawn from the comparison in Figure 2 between the X-probe and the DNS spectra and cospectra at three locations in the flow. Allowing for the scatter in the DNS and experimental data, quite good agreement is obtained for the uv cospectrum, even at the edge of the sublayer. There, the v spectrum shows the largest discrepancy (distributed over all wavenumbers) but as y^+ increases, the departure is confined to a smaller wavenumber range centered near the peak of the spectrum.

3 Low Reynolds Number Effects

Spalart's (1988) direct numerical simulations of a zero pressure gradient boundary layer (for several values of the momentum thickness Reynolds number R_θ, the highest being 1410) indicated that, in the wall region, several quantities (for example, $\overline{u^2}$, $\overline{w^2}$, the pressure variance $\overline{p^2}$ and mean square vorticity $\overline{\omega^2}$) did not scale on wall variables. This result was confirmed by the measurements of Wei and Willmarth (1989) and the direct numerical simulations (see Antonia et al., 1991,1992) in a fully developed turbulent channel flow. The experimental uncertainty becomes large near the wall but the DNS data give a clear indication of the significant influence of the Reynolds number in this region. Particularly large is the dependence on Reynolds number of the wall values of $\overline{p^2}$, $\overline{\omega_x^2}$ and $\overline{\omega_z^2}$. The systematic increase of $\overline{\omega_x^2}$ and $\overline{\omega_z^2}$ with h^+ in the wall region of the channel is illustrated in Figure 3 and is consistent with Wei and Willmarth's (1989) suggestion of increased vortex stretching with h^+. Interestingly, practically no influence of h^+ can be discerned on $\overline{\omega_y^2}$.

Before considering possible causes for the low Reynolds number effects, attention should be drawn to the fact that these effects have been observed in both a boundary layer and a channel flow, thus discounting the possibility that the turbulent/non-turbulent interface is primarily responsible for these effects. Several possible causes for low Reynolds number effects have been advanced. Spalart (1988) attributed the effect to the inactive motion (Townsend, 1961; Bradshaw, 1967). Wei and Willmarth (1989) suggested a "geometry" effect (direct interaction between inner regions on opposite channel walls). If this latter effect were correct, one would need to find a separate explanation for the boundary layer — a rather unattractive challenge. The geometry effect in the channel was investigated in some detail — by introducing a small amount of heat at one of the walls in experiments and simulations — by Antonia et al. (1992). It was observed that thermal excursions can indeed reach the opposite wall, the extent of penetration

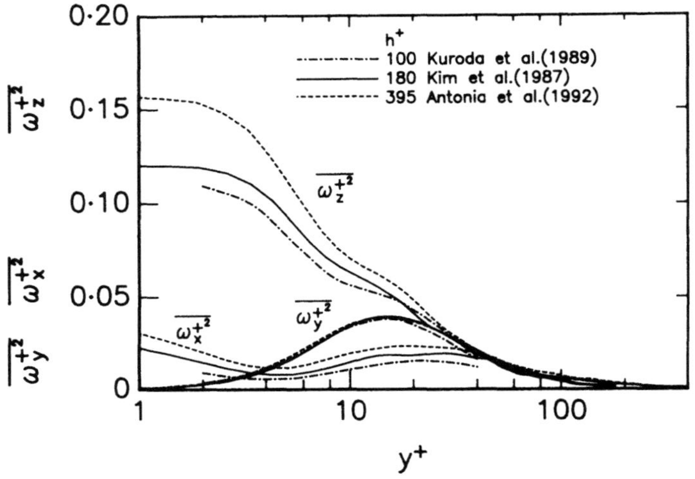

Figure 3: Reynolds number dependence of the three components of the mean square vorticity in a turbulent channel flow, obtained from direct numerical simulations. — - —: $h^+ = 100$, Kuroda et al. (1989); ——: $h^+ = 180$, Kim et al. (1987); - - -: $h^+ = 395$, Antonia et al. (1991).

becoming more significant at low Reynolds numbers. There was no evidence however to support Wei and Willmarth's speculation that opposite near-wall regions constantly exchange opposite-signed vorticity and shear stress.

Antonia et al. (1992) also found no significant evidence for relating low Reynolds effects to the inactive motion. It was argued however that the significant evolution (in the inner region) of $\overline{u^+v^+}$ and $\overline{v^{+2}}$ with h^+ was more consistent with the notion that the active motion (which contributes practically all of $\overline{u^+v^+}$ and $\overline{v^{+2}}$) depends on the Reynolds number. The notion would be readily transportable to other flows.

4 Scalar Transport

The extension of direct numerical simulations to a passive scalar field has provided more reliable near-wall information on the scalar transport than has been possible experimentally. It has also confirmed and extended our knowledge of kinematical features associated with the organised motion.

The use of temperature as a passive marker of various features of the organised motion has been effectively exploited in experiments in different turbulent shear flows. The DNS scalar data of Kim and Moin (1989) have confirmed several characteristics of quasi-coherent structures which had previously been identified primarily on the basis of velocity signatures. An important asset of the DNS is the ability to view the scalar field in conjunction with the three-dimensional velocity field. Views of projections of the

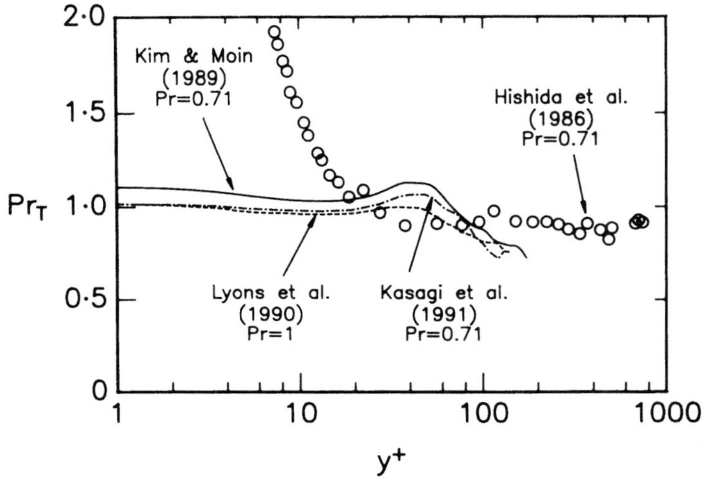

Figure 4: Comparison between measured and DNS values of the turbulent Prandtl number. Measurement : **o**, Hishida et al. (1986), Pr = 0.71. DNS : ——, Kim and Moin (1989), Pr = 0.71; ----, Lyons et al. (1991), Pr = 1; — - —, Kasagi et al. (1991), Pr = 0.71.

instantaneous velocity vectors in the (y, z) plane are especially instructive in terms of identifying vortical structures. The data have, inter alia, underlined the importance of quasi-streamwise vortices in terms of their likely contribution to momentum and heat transport in the near-wall region. Important scalar concentrations can be identified in association with streamwise vortices (though not necessarily in counter-rotating formation).

A quantitative measure of the relative importance of momentum and heat transport is given by the turbulent Prandtl number Pr_T. There has been considerable uncertainty in the measurment of Pr_T in the wall region. The wall behaviour of Pr_T has been clarified by the DNS data. In particular, these data indicate, to a good first approximation, that Pr_T approaches a constant value ($\simeq$ 1.1) at the wall (nearly independently of Pr, the molecular Prandtl number). This behaviour is consistent with the nearly perfect correlation that exists between u and the temperature fluctuation θ.

The three different simulations of Figure 4 yield essentially the same distribution of Pr_T (and a correlation coefficient $\rho_{u\theta} = \overline{u\theta}/u'\theta'$ that is close to 1). The fully developed pipe flow measurements of Hishida et al. (1986) — obtained with a single hot wire, a single cold wire and a V-shaped hot wire — yielded smaller values of $\rho_{u\theta}$ near the wall ($\rho_{u\theta} \simeq 0.75$ at $y^+ = 10$) and an anomalous increase in Pr_T in this region (Figure 4).

This discrepancy probably reflects the difficulty of measuring v reliably in the wall region. The clarification, by the DNS data, of the behaviour of Pr_T in the wall region should provide a sufficient incentive to improve near-wall measurements of $\overline{uv}$ and $\overline{v\theta}$.

In the context of wall measurements, it should be noted that reliable data of $\rho_{u\theta}$ have been obtained (Antonia et al., 1988) using a pair of single wires (one hot, one cold). Also,

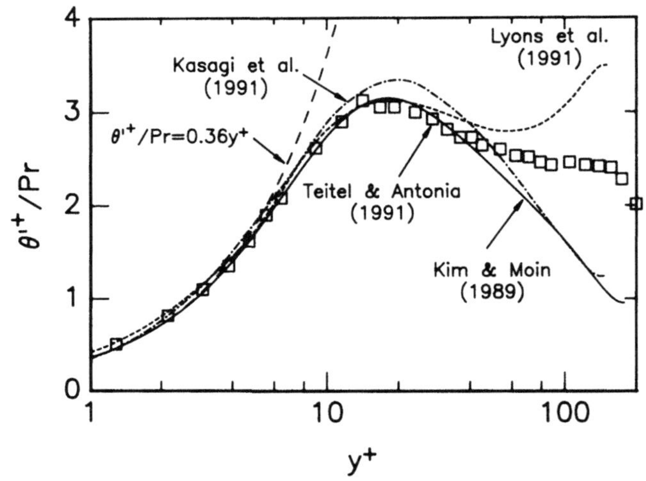

Figure 5: Comparison between cold wire and DNS values of the rms temperature in a turbulent channel flow. □, Teitel and Antonia (1991), $Pr = 0.72$; ----, Lyons et al. (1991), $Pr = 1$; ——, Kim and Moin (1989), $Pr = 0.71$; — - —, Kasagi et al. (1991), $Pr = 0.71$.

a single cold wire has proved to be quite effective in determining the mean temperature and the various statistics of θ. Figure 5 shows that measured values of $\theta^{+\prime}$ are in quite good agreement with the DNS data over most of the sublayer and buffer region. The differences at larger y^+ are due to different thermal boundary conditions used in the experiment and in each of the three simulations. The consequences of these different conditions have been addressed by Teitel and Antonia (1991).

5 Measurement of Velocity Derivatives

Velocity derivative data bases generated by direct numerical simulations are particularly suitable for studying the fine scale structure of turbulence since all the velocity derivatives which contribute to the turbulent energy dissipation and vorticity fluctuations are available. For the DNS data bases of Kim et al. (1987) for $h^+ = 180$ and Kim (1989) for $h^+ = 395$, the spatial resolution is such that velocity derivatives in all three spatial directions can be determined with sufficient accuracy. The measurement of velocity derivatives in turbulent shear flows is, in general, not straightforward. A list of the major difficulties associated with the use of hot wires for measuring velocity derivatives was presented and discussed in some detail in Antonia et al. (1986). Apart from the possibility of errors due to the use of Taylor's hypothesis and the finite length of the hot wire, a more significant difficulty is the judicious selection of the spatial separation between hot wires or, alternately, the determination of the appropriate correction that one needs to apply to spatial velocity derivatives obtained from the use of parallel single hot wires or parallel X-wires. Even when measurements are carried out for a range of separations, the extrapolation to zero separation requires some care (spectral corrections are possible, but they involve making certain assumptions, e.g. local isotropy and a form of the three-dimensional energy spectrum, e.g. Antonia et al., 1987). The availability of the DNS small sale data base thus provides a very useful opportunity for checking (fine-tuning) the experimental procedure used for obtaining the components of the average turbulent energy dissipation $\overline{\epsilon}$ and $\overline{\omega^2}$.

Measurements of $\partial u/\partial x$ and $\partial u/\partial y$ are currently being carried out at $h^+ = 180$, i.e. the Reynolds number of the Kim et al. (1987) data base. The low Reynolds number provides favourable experimental conditions in terms of spatial resolution requirements for the measurement of velocity derivatives. The magnitude of the Kolmogorov length scale η — as inferred from the DNS dissipation values — is constant ($\simeq$ 0.19 mm) in the wall region up to $y^+ \simeq 5$ and increases to about 0.39 mm at the channel centreline. For the measurement of $\partial u/\partial y$, the separation Δy between the wires (Pt-Rh wires of 1.27 μm diameter) is 0.7 mm while the wire length is 0.28 mm. These latter dimensions were selected so that corrections (e.g. Wyngaard, 1969; Antonia et al., 1987) to the high wavenumber part of the $\phi_{\partial u/\partial x}$ and $\phi_{\partial u/\partial y}$ could be neglected, at least as a first approximation. For greater accuracy, spectral corrections would be necessary, especially in the outer region of the flow. A possible correction procedure has been outlined in Antonia and Mi (1991). When Δy is less than about 3η (this value was inferred from separate experiments where the separation between the wires was varied), there is a systematic (and spurious) increase in $\overline{(\Delta u/\Delta y)^2}$ due to errors associated with the determination of Δy and wire calibrations (e.g. Mestayer and Chambaud, 1979). The present separation ($\simeq$ 0.7 mm) was chosen so that the ratio $\Delta y/\eta$ ($\simeq$ 3.7) was marginally greater than 3 in the near-wall region and errors in $\overline{(\partial u/\partial y)^2}$ would be small. Near the channel centreline, this separation ($\Delta y/\eta \simeq 1.8$) would result in $\overline{(\partial u/\partial y)^2}$ being overestimated by about 30%. Obviously, the use of a fixed separation makes it impossible to optimise the ratio $\Delta y/\eta$ everywhere in the flow; this difficulty applies to any derivative data obtained with fixed geometry multiple wire probes (e.g. Dracos et al., 1990; Balint et al., 1991).

Measured values of $\overline{(\partial u/\partial x)^2}$ (inferred from $\overline{(\partial u/\partial t)^2}$ using Taylor's hypothesis) and $\overline{(\partial u/\partial y)^2}$ (inferred from the approximation $\Delta u/\Delta y$), are compared with the DNS distributions in Figures 6 and 7 respectively. The difference in $\overline{(\partial u^+/\partial x^+)^2}$ at small y^+ may be due partly to the use of Taylor's hypothesis for the hot wire measurements and partly to the relatively poorer spatial resolution in the x direction for the DNS data. The agreement in $\overline{(\partial u^+/\partial y^+)^2}$, especially at small y^+ (where $\Delta y/\eta$ is probably optimum for the hot wire experiment), is quite encouraging.

Antonia et al. (1991) presented distributions, across the channel, of all the components of the homogeneous turbulent energy dissipation

$$\overline{\epsilon^+} \equiv \overline{\omega^{+2}} = \overline{u^+_{i,j}u^+_{i,j}} \quad , \tag{1}$$

where standard tensor notation is used. The assumption of isotropy allows considerable

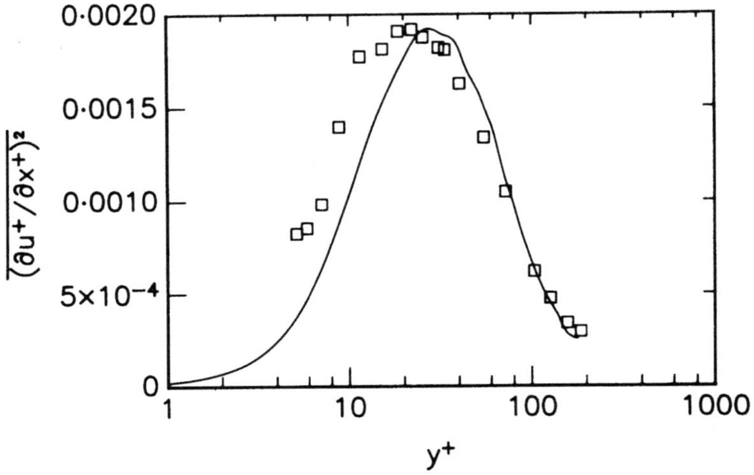

Figure 6: Comparison between measured and DNS distributions of $\overline{(\partial u^+/\partial x^+)^2}$ in a turbulent channel flow ($h^+ = 180$). □, measurement; ——, DNS (Antonia et al., 1991).

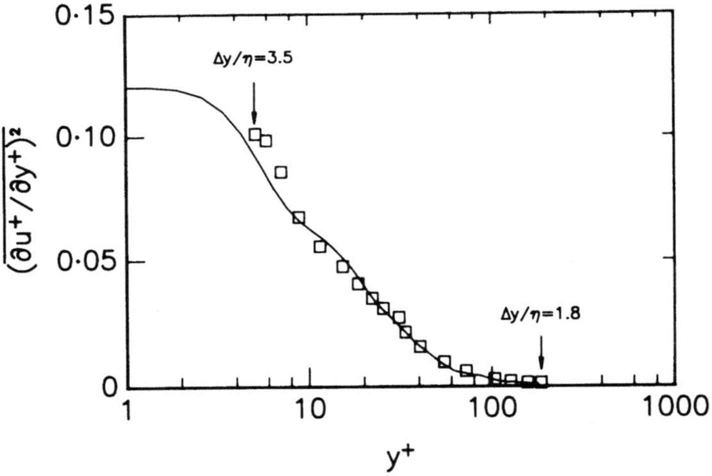

Figure 7: Comparison between measured and DNS distributions of $\overline{(\partial u^+/\partial y^+)^2}$ in a turbulent channel flow ($h^+ = 180$). □, measurement; ——, DNS (Antonia et al., 1991).

simplification to Eq. (1), viz.

$$\overline{\epsilon^+} = 15\overline{\left(\frac{\partial u^+}{\partial x^+}\right)^2} \quad . \tag{2}$$

This assumption is however only approximately valid near the channel centreline. Anisotropy invariant maps of the dissipation and vorticity tensors indicated that, outside the wall region, axisymmetry provided a more plausible approximation to the DNS data than isotropy. Axisymmetric forms of $\overline{\epsilon}^+$ were recently given by George and Hussein (1991) for invariance of properties about the x axis. One of these, viz.

$$\overline{\epsilon}^+ = 2\overline{\left(\frac{\partial u^+}{\partial y^+}\right)^2} + 2\overline{\left(\frac{\partial v^+}{\partial x^+}\right)^2} + 8\overline{\left(\frac{\partial v^+}{\partial y^+}\right)^2} - \overline{\left(\frac{\partial u^+}{\partial x^+}\right)^2} \tag{3}$$

was found to be a good approximation to Eq. (1) outside the buffer region ($y^+ \gtrsim 40$). An interesting feature of Eq. (3) is that $\overline{(\partial v/\partial x)^2}$, $\overline{(\partial v/\partial y)^2}$ and $\overline{(\partial u/\partial x)^2}$ have approximately similar distributions across the channel, differing only in magnitude. It was found that the sum of the second, third and fourth terms on the right side of Eq. (3) could be replaced by $11\overline{(\partial u/\partial x)^2}$ so that the resulting approximation to the dissipation, viz.

$$\overline{\epsilon}^+ \simeq 2\overline{\left(\frac{\partial u^+}{\partial y^+}\right)^2} + 11\overline{\left(\frac{\partial u^+}{\partial x^+}\right)^2} \quad , \tag{4}$$

could be measured using a pair of parallel wires. It was also found that the spanwise mean square vorticity

$$\overline{\omega_z^{+2}} = \overline{\left(\frac{\partial u^+}{\partial y^+}\right)^2} + \overline{\left(\frac{\partial v^+}{\partial x^+}\right)^2} - 2\overline{\left(\frac{\partial u^+}{\partial y^+}\right)\left(\frac{\partial v^+}{\partial x^+}\right)} \tag{5}$$

could be approximated by

$$\overline{\omega_z^{+2}} \simeq \overline{\left(\frac{\partial u^+}{\partial y^+}\right)^2} + 2\overline{\left(\frac{\partial u^+}{\partial x^+}\right)^2} \tag{6}$$

over a significant part of the channel. The measured values of $\overline{(\partial u^+/\partial x^+)^2}$ (Figure 6) and $\overline{(\partial u^+/\partial y^+)^2}$ (Figure 7) were used in Eqs. (4) and (6). The resulting values of $\overline{\epsilon}^+$ (Figure 8) and $\overline{\omega_z^{+2}}$ (Figure 9) are indeed in quite close agreement with the DNS homogeneous dissipation and spanwise vorticity distributions in the region $y^+ \gtrsim 30$. For $y^+ < 30$, Eq. (4) overestimates $\overline{\epsilon}$ but Eq. (6) remains a reasonable approximation to $\overline{\omega_z^{+2}}$ (the slightly lower measured values of $\overline{\omega_z^{+2}}$ reflect the slightly lower values of $\overline{(\partial u^+/\partial y^+)^2}$ in Figure 7).

Also included in Figures 8 and 9 are the measured isotropic values of $\overline{\epsilon}^+$ [Eq. (2)] and $\overline{\omega_z^{+2}}$ ($= 5\overline{(\partial u^+/\partial x^+)^2}$). They grossly underestimate $\overline{\epsilon}^+$ and $\overline{\omega_z^{+2}}$ except near the channel centreline. It has previously been pointed out (Antonia et al., 1984,1986) that mean square values of velocity derivatives do not represent a sensitive test for local isotropy (or isotropy of the small scale motion) since they are likely to be influenced by

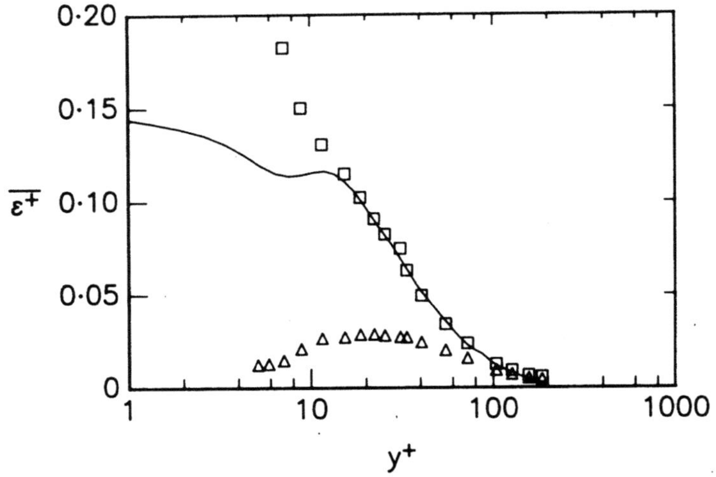

Figure 8: Comparison between measured estimates of $\bar{\epsilon}$ and DNS (homogeneous) values of $\bar{\epsilon}$. Measurements : □, Eq. (4); Δ, Eq. (2), isotropy. DNS (Antonia et al., 1991) : ——, Eq. (1).

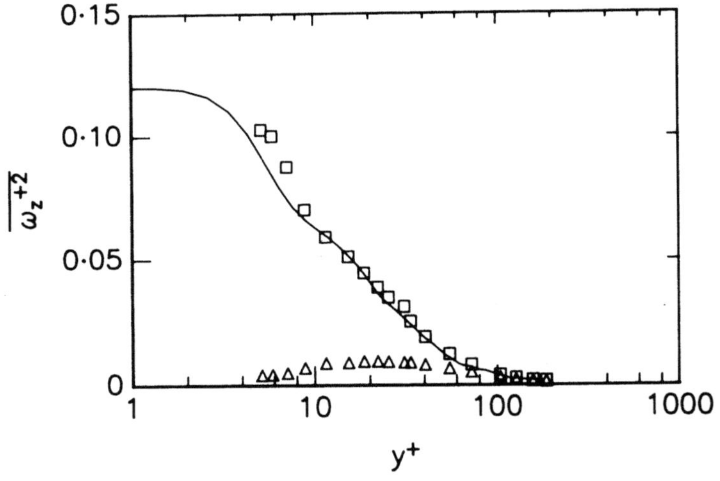

Figure 9: Comparison between measured approximations to $\overline{\omega_z^2}$ and DNS (homogeneous) values of $\overline{\omega_z^2}$. Measurements : □, Eq. (6); Δ, $\overline{\omega_z^{+2}} = 5\overline{(\partial u^+/\partial x^+)^2}$, isotropy. DNS (Antonia et al., 1991) : ——, Eq. (5).

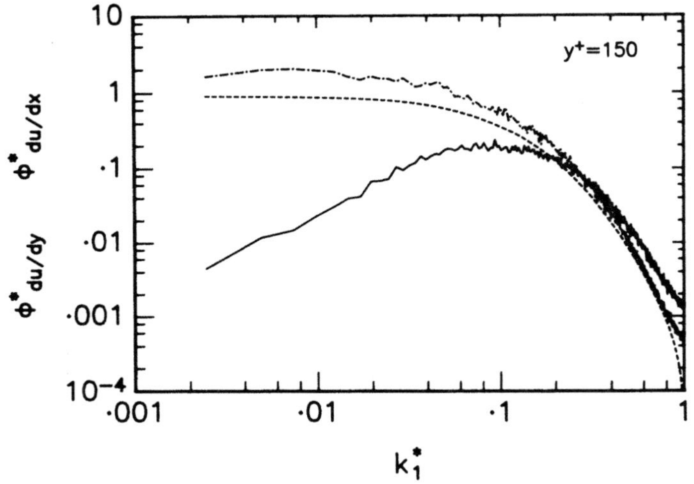

Figure 10: Measured spectrum of $\partial u/\partial y$ and comparison with isotropic calculation based on the measured spectrum of $\partial u/\partial x$ (also shown). The asterisk denotes normalisation by Kolmogorov scales. $\phi_{\partial u/\partial y}$: — - —, measured; ----, calculated using Eq. (7) and measured $\phi_{\partial u/\partial x}$ (——).

the anisotropy of the large scale motion. All the available evidence supports the idea that the high wavenumber part of velocity derivative spectra is indeed consistent with isotropy. This consistency does not seem to be dependent on the Reynolds number. A recent investigation carried out in collaboration with Dr J. Kim (NASA-Ames) has shown that the high wavenumber part of the DNS vorticity (or dissipation) spectrum satisfies isotropy. Figure 10 shows that the high wavenumber part of $\phi_{\partial u/\partial y}$, is also consistent with isotropy. The broken curve in Figure 10 represents the calculation of $\phi^*_{u_{1,2}}$ (the asterisk denotes normalisation by Kolmogorov scales) using the isotropic result

$$\phi_{\partial u/\partial y}(k_1) = 2\int_{k_1}^{\infty} k^{-1}\,\phi_{\partial u/\partial x}\,dk \tag{7}$$

and the measured distributions of $\phi^*_{\partial u/\partial x}$ (also shown in Figure 10). Eq. (7) follows from the relation given in Antonia et al. (1984). The measured and calculated distributions of $\phi_{\partial u/\partial y}$ are in close agreement in the range $0.2 \lesssim k_1\eta \lesssim 0.8$. For $k_1\eta < 0.2$, the difference reflects the low wavenumber isotropy while for $k_1\eta > 0.8$, the discrepancy is likely to be due to the noise in the measurement of $\phi_{\partial u/\partial y}$ (the calculation may also be affected by the noise in the $\phi_{\partial u/\partial x}$ measurement).

$\phi_{\partial u/\partial y}$ has not been corrected for the possible high wavenumber attenuation due to the finite separation between the wires since $\Delta y \simeq \eta$ at this flow location. Corrections may however be necessary at other locations or at higher Reynolds numbers. The observed consistency of the spectra with local isotropy is important since it permits corrections to be applied.

6 Conclusions

Comparison with the DNS data or with LDV or PIV measurements indicates that the hot wire measurement of $\overline{v^2}$ is seriously overestimated, especially near the wall.

While the available direct numerical simulations have been carried out under conditions for which low Reynolds number effects are important, they have also made a significant contribution in terms of documenting and improving our understanding of these effects.

Apart from providing reliable near-wall data, for isothermal and non-isothermal conditions, direct numerical simulation data bases have provided useful approximations to quantities such as the energy dissipation and mean square vorticity. These approximations are amenable to measurement and encouraging results using hot wires have been obtained in a turbulent channel flow at a similar Reynolds number. Hot wire measurements of comparable quality should be possible at higher Reynolds numbers.

Acknowledgements

I am grateful to Dr J. Kim for making the DNS data available to me. I would also like to thank Dr M. Teitel and Mr. Y. Zhu for their contributions to the channel flow experiments. The financial support of the Australian Research Council is acknowledged.

References

Antonia, R. A., Anselmet, F. and Chambers, A. J. (1986), Assessment of Local Isotropy Using Measurements in a Turbulent Plane Jet, *J. Fluid Mech.*, **163**, 365-391.

Antonia, R. A., Browne, L. W. B. and Chambers, A. J. (1984), On the Spectrum of the Transverse Derivative of the Streamwise Velocity in a Turbulent Flow, *Phys. Fluids*, **27**, 2628-2631.

Antonia, R. A., Browne, L. W. B., and Kim, J. (1991), Some Characteristics of Small Scale Turbulence in a Turbulent Duct Flow, *J. Fluid Mech.*, **233**, 369-388.

Antonia, R. A., Browne, L. W. B. and Shah, D. A. (1988), Characteristics of Vorticity Fluctuations in a Turbulent Wake, *J. Fluid Mech.*, **189**, 349-365.

Antonia, R. A. and Mi, J. (1991), Corrections for Velocity and Temperature Derivatives in Turbulent Flows, *Expts. in Fluids* [submitted]

Antonia, R. A., Shah, D. A. and Browne, L. W. B. (1987), Spectra of Velocity Derivatives in a Turbulent Wake, *Phys. Fluids*, **30**, 3455-3462.

Antonia, , R. A., Teitel, M., Kim, J. and Browne, L. W. B. (1992), Low Reynolds Number Effects in a Fully Developed Turbulent Duct Flow, *J. Fluid Mech.*, [to appear].

Balint, J-L., Wallace, J. M. and Vukoslavcevic, P. (1991), The Velocity and Vorticity Vector Fields of a Turbulent Boundary Layer. Part 2. Statistical Properties, *J. Fluid Mech.*, **228**, 53-86.

Bradshaw, P. (1967), "Inactive" Motion and Pressure Fluctuations in Turbulent Boundary Layers, *J. Fluid Mech.*, **30**, 241-258.

Dracos, T., Kholmyansky, M., Kit, E. and Tsinober, A. (1990), Some Experimental Results on Velocity-Velocity Gradient Measurements in Turbulent Grid Flows, in K. H. Moffatt and A. Tsinober eds. *Topological Fluid Mechanics*, Cambridge, C.U.P. [to appear]

Eckelmann, H. (1974), The Structure of the Viscous Sublayer and the Adjacent Wall Region in a Turbulent Channel Flow, *J. Fluid Mech.*, **65**, 439.

George, W. K. and Hussein, H. J. (1991), Locally Axisymmetric Turbulence, *J. Fluid Mech.* [to appear]

Hishida, M., Nagano, Y. and Tagawa, M. (1986), Transport Processes of Heat and Momentum in the Wall Region of Turbulent Pipe Flow, *Proc. Eighth International Heat Transfer Conference*, San Francisco, Vol. 3, 925-930.

Kasagi, N., Tomita, Y. and Kuroda, A. (1991), Direct Numerical Simulation of the Passive Scalar Field in a Two-Dimensional Channel Flow, *Proc. Third ASME-JSME Thermal Engineering Joint Conference*, Reno.

Kim, J. (1989), On the Structure of Pressure Fluctuations in Simulated Turbulent Channel Flow, *J. Fluid Mech.*, **205**, 421-451.

Kim, J. and Moin, P. (1989), Transport of Passive Scalars in a Turbulent Channel Flow, in *Turbulent Shear Flows 6*, Berlin, Springer, 85-96.

Kim, J., Moin, P. and Moser, R. (1987), Turbulence Statistics in Fully Developed Channel Flow at Low Reynolds Number, *J. Fluid Mech.*, **177**, 133-166.

Kreplin, H. and Eckelmann, H. (1979), Behavior of the Three Fluctuating Velocity Components in the Wall Region of a Turbulent Channel Flow, *Phys. Fluids*, **22**, 1233.

Kuroda, A., Kasago, N. and Hirata, M. (1989), A Direct Numerical Simulation of the Fully Developed Turbulent Channel Flow at a Very Low Reynolds Number, *Proc. ISCFD*, Nagoya, 1174-1179.

Liu, Z-C., Landreth, C. C., Adrian, R. J. and Hanratty, T. J. (1991), High Resolution Measurements of Turbulent Structure in a Channel with Particle Image Velocimetry, *Expts. in Fluids*, **10**, 301-312.

Lyons, S. L., Hanratty, T. J. and McLaughlin, J. B. (1989), Turbulence-Producing Eddies in the Viscous Wall Region, *AIChE Jnl.*, **35**, 1962-1974.

Lyons, S. L., Hanratty, T. J. and McLaughlin, J. B. (1991), Direct Numerical Simulation of Passive Heat Transfer in a Turbulent Channel Flow, *Int. J. Heat Mass Transfer*, **34**, 1149-1162.

Mansour, N. N. (1991), The Use of Direct Numerical Simulation in Turbulence Modelling, *Paper AIAA-91-0221*, presented at 29th Aerospace Sciences Meeting, Nevada.

Mestayer, P. and Chambaud, P. (1979), Some Limitation to Measurements of Turbulence Micro-Structure with Hot- and Cold-Wires, *Boundary-Layer Meteorol.*, **16**, 311-329.

Moin, P. and Spalart, P. R. (1989), Contributions of Numerical Simulation Data Bases to the Physics, Modeling and Measurement of Turbulence, in W. K. George and R. Arndt (eds.) *Advances in Turbulence*, New York, Hemisphere, 11-38.

Niederschulte, M. A., Adrian, R. J. and Hanratty, T. J. (1990), Measurements of Turbulent Flow in a Channel at Low Reynolds Numbers, *Expts. in Fluids*, **9**, 222-230.

Nishino, K. and Kasagi, N. (1989), Turbulence Statistics Measurement in a Two-Dimensional Channel Flow Using a Three-Dimensional Particle Tracking Velocimeter, *Proc. Seventh Symposium on Turbulent Shear Flows*, Stanford, CA, 22.1.1-22.1.6.

Robinson, S. K., Kline, S. J. and Spalart, P. R. (1990), Quasi-Coherent Structures in the Turbulent Boundary Layer. Part II : Verification and New Information from a Numerically Simulated Flat-Plate Layer, in S. J. Kline and N. H. Afgan (eds.) *Near Wall Turbulence*, New York, Hemisphere, 218-247.

Spalart, P. R. (1988), Direct Simulation of a Turbulent Boundary Layer up to $R_\theta = 1410$, *J. Fluid Mech.*, **187**, 61-98.

Suzuki, Y. and Kasago, N. (1990), Evaluation of Hot-Wire Measurements in Turbulent Wall Shear Flows Using a Direct Numerical Simulation Data Base, in W. Rodi and E. N. Ganic (eds.) *Engineering Turbulence Modelling and Experiments*, New York, Elsevier, 361-370.

Teitel, M. and Antonia, R. A. (1991), Heat Transfer in Fully Developed Turbulent Channel Flow : Comparison Between Experiment and Direct Numerical Simulations, *Int. J. Heat Mass Transfer*, (submitted)

Townsend, A. A. (1961), Equilibrium Layers and Wall Turbulence, *J. Fluid Mech.*, **11**, 97-120.

Wei, T. and Willmarth, W. W. (1989), Reynolds-Number Effects on the Structure of a Turbulent Channel Flow, *J. Fluid Mech.*, **204**, 57-95.

Wyngaard, J. C. (1969), Spatial Resolution of the Vorticity Meter and Other Hot-Wire Arrays, *J. Sci. Instrum*, **2**, 983-987.

Address of the Author

Professor R. A. Antonia,
Department of Mechanical Engineering,
University of Newcastle,
NEW SOUTH WALES 2308 AUSTRALIA
meraa@cc.newcastle.edu.au

Small-Scale Structures in Turbulence: Their Implications for Turbulence Closures

Jackson R. Herring and Robert M. Kerr

Abstract. We compare high-resolution direct numerical simulations (DNS) to the predictions of moment closure for two- and three-dimensional turbulence. Our goal is to assess the impact of vortical structures on two-point covariances, through a comparison of DNS to simple closures, which make no explicit reference to structures at all. In two dimensions, we note that there is a profound effect, and this is associated with the development of isolated vortex structures. We interpret their presence in terms of a shutting down of the enstrophy transfer process, and give phenomenological estimates in terms of a modified closure. In three dimensions, the DNS spectra at $R_\lambda \sim 100$ is in good agreement with the Direct Interaction Approximation (DIA). The agreement is poorer for Markovian theories, such as the Eddy Damped Quasinormal Markovian (EDQNM) approximation. We also comment on the possible development of finite time singularities in the vorticity for inviscid flows.

1. Introduction

Recent direct numerical simulations (DNS) have demonstrated that incompressible Navier Stokes turbulence develops intense and compact small-scale vortex structures (vortex tubes, etc.). Such structures, if sufficiently intense and spare, are inhospitable to moment closure. A measure of the intensity of vorticity structures is the vorticity kurtosis ($\mathcal{K}(t) \equiv \langle(\nabla\times\mathbf{u})^4\rangle/\langle(\nabla\times\mathbf{u})^2\rangle^2$). Thus, the two-dimensional calculations of McWilliams (1984) show late time $\mathcal{K}(t) \sim 60$, and are in substantial disagreement with closures (Herring and McWilliams, 1984). The three-dimensional computations by Kerr (Herring and Kerr, 1991) show a peak $\mathcal{K}(t) \sim 10$. It is as yet unclear whether these structures are the byproduct of large-scale dynamics, or are themselves the dynamics driving the large-scale features.

In the present paper, we examine high-resolution DNS for both two- and three-dimensional flows, in an effort to delineate the role of structures in simple flows and to determine the extent to which they are irreconcilable with traditional closure ideas. Such closure models, in the forms of the Test Field Model (TFM) (Kraichnan, 1971) or the Direct Interaction Approximation (DIA) (Kraichnan, 1959), may be regarded as structureless models, embodying the same dynamical constraints as Navier-Stokes (such as inviscid conservation laws, and equipartitioning among available degrees of freedom). Comparisons of DNS to them may then be useful as a measure of the extent to which structures are essential in describing the dynamics at the two-point moment level.

For two-dimensional turbulence, we are able to partially reconcile the large disparity between DNS and TFM by the *ansatz* that the scaling parameter for the Green's function is related to $\mathcal{K}(t)$. This result is interpreted in terms of the effects of vortex tubes in inhibiting energy and enstrophy transfer.

The comparisons of DNS to closure calculations is frequently obscured by the fact that the latter usually makes a Markovian assumption (for strictly practical reasons), and this inhibits the theory's ability to follow rapidly evolving developments of the flow. Two examples illustrating this point are presented: (1) the study of scalar variance transfer at low Prandtl number, and (2) the development of three-dimensional isotropic inviscid flow from smooth initial data. In both these cases, we show that the non-Markovian (DIA) theory gives better results than the Markovianized theory (either the TFM or the eddy-damped quasi-normal Markovian, or EDQNM, approximation). We also discuss the nature of the three-dimensional singularity (whether $sup \mid \nabla \times \mathbf{u}(\mathbf{x},t) \mid$ is bounded as $\nu \to 0$) and the issue of whether the DIA develops a finite-time enstrophy singularity.

2. Two-Dimensional Turbulence

Perhaps the most spectacular failure of moment closures occurs for two-dimensional turbulence, where the development of intense, isolated structures signals a shutting down of the traditional form of enstrophy transfer to small scales. The failure of the two-point closure for this problem was convincingly demonstrated in the comparisons of DNS to closure by Herring and McWilliams (1985). We recall how dramatic is the enstrophy cascade diminished by isolated vortices in Fig. 1, which compares the energy spectrum of DNS with that of the closure (TFM or $\sim$ EDQN) at 20 large-eddy circulation times into the decay. Note that the TFM energy spectrum, $E_{TFM}(k)$, has a substantial k^{-3} range, whereas $E_{DNS}(k)$ is more like $k^{-4+\cdots}$. Also, a large fraction of $E_{TFM}(k)$ resides at the largest available scale, $k = 1$. Fig. 2 shows the vorticity field at this time. Yet Benzi *et al.* (1987) have shown that if the isolated vortices are subtracted out by a suitable procedure, the remaining flow field has an energy spectrum consistent with $\sim k^{-3}$. One interpretation is that isolated vortices are the "black holes" of turbulence, contributing to the energy spectrum but disdaining to engage in the transfer process. We may effect this idea in the EDQNM by modifying the energy transfer equation to

$$\{\partial_t + 2\nu k^2\}E(k,t) = \lambda^{-1}(t)T\{E(k),k\} \ , \tag{2.1}$$

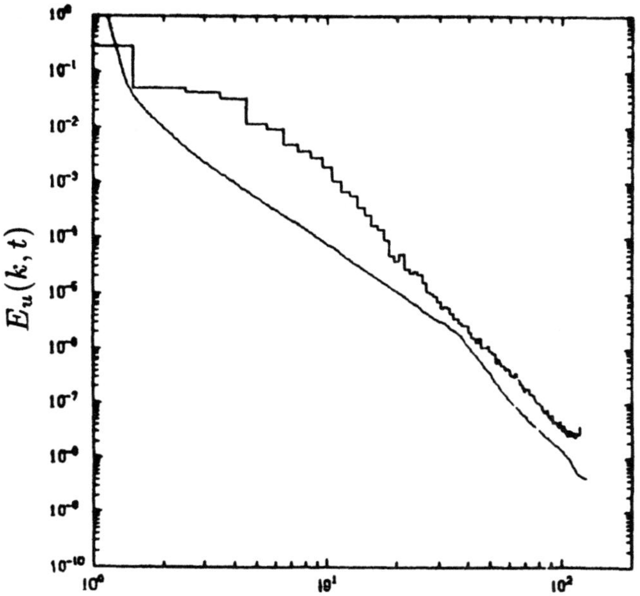

Fig. 1. Comparison of $E(k,t)$ of McWilliams DNS to TFM for $t = 20$. Here, $E(k,0) = 1/(6^4 + k^4)$, $\nu = 1. \times 10^{-4}$, and DNS resolution is 256^2. For TFM, (see (2.1)–(2.2)) $\lambda(t) = .65$. Histogram is the DNS, and smooth curve, TFM. From Herring and McWilliams (1985).

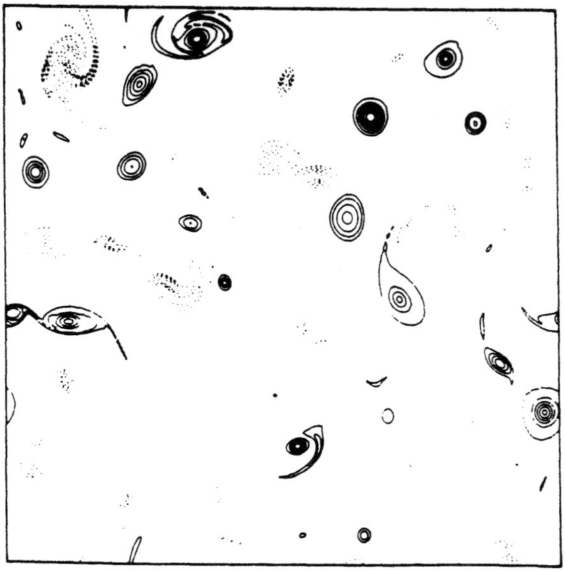

Fig. 2. Vorticity contour map for initial value problem described in Fig. 1 caption. From McWilliams (1984).

where $\lambda(0)T\{E(k),k\}$ ($\lambda(0) = .65$) is the standard EDQNM transfer function, and $T(k)$ is given by

$$\mathcal{T}(k) \equiv (T/(\pi k)) = \int_{\Delta} dpdqU(q)\{U(p) - U(k)\}\Theta(k,p,q) \ , \tag{2.2}$$

where—as usual—(Herring *et al.*, 1974)

$$U \equiv (E(k)/(\pi k)) \ ,$$

and

$$B(k,p,q) = (p^2 - q^2)(k^2 - q^2)(1 - x^2)/k^2 \ . \tag{2.3}$$

If $\lambda(t)$ is

$$\lambda(t) = 0.65(1 + .0497\ \Omega_0\ t) \ , \tag{2.4}$$

then the comparison of theory to DNS (at $t = 20$) improves to that given in Fig. 3, with progressively better agreement as $t \to 0$. The corresponding enstrophy ($\mathcal{E}(t) \equiv \int_0^\infty E(k,t)k^2\ dk$) decay is given in Fig. 4, with the modified theory shown by the solid line, the unmodified by the dashed line, and the circles by the McWilliams experiment. The modification is—so far—completely *ad hoc*, and is frankly designed to make theory and simulation agree, insofar as possible, with a single parameter, $\lambda(t)$. Note, however, that the modified equation has still a k^{-3} inertial range at large Reynolds numbers; the effective steepening of the slope to $\sim k^{-4.5}$ is due simply to a progressive diminution of the nonlinear interaction, $T(k,t)$, with time, in accordance with (2.2)–(2.4). Asymptotics of total enstrophy, $\mathcal{E}(t)$ (in Fig. 3), for the modified theory (2.4) may be deduced by recalling that the TFM with $\lambda \sim \mathcal{O}(1)$ agrees with Batchelor's theory (1967) (Lesieur and Herring, 1985); and the modified theory is but a rescaling of that theory, with a new time variable, $d\,\tau = dt/(1 + \Omega_0 t)$. Thus,

$$\mathcal{E}(t) \sim \{ln(1 + .0497\ \Omega_0 t)\}^{-2} \ . \tag{2.5}$$

However, it would be a mistake to imagine—even if our conjectures were right—that any such law could be seen in the DNS, since, as noted by Lesieur and Herring (1985), Batchelor's decay law assumes sufficiently large-scale resolution to support a $E(k) \sim k^3$ range as $k \to 0$. But the DNS shown in Fig. 1 has already substantial amplitude at $k = 1$. We would be equally in error to think that by matching the DNS spectrum for $t \leq 20$, we could on this basis deduce the $t \to \infty$ asymptotics of $\mathcal{E}(t)$.

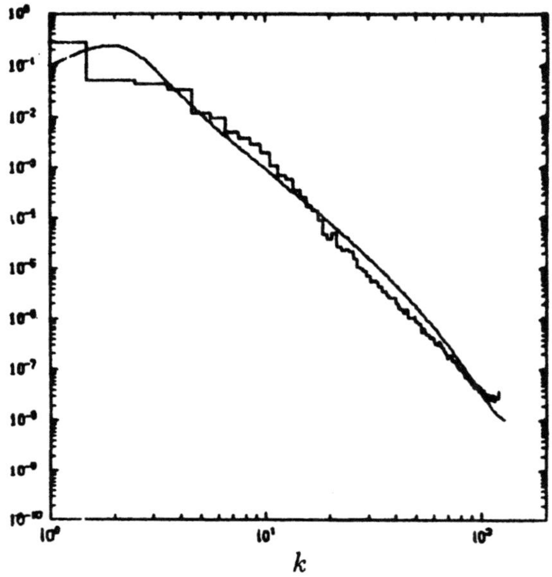

Fig. 3. Same comparison as in Fig. 1, except that $\lambda(t) = .65/(1 + .0497\ \Omega_0\ t)$, with Ω_0 the initial (rms) enstrophy.

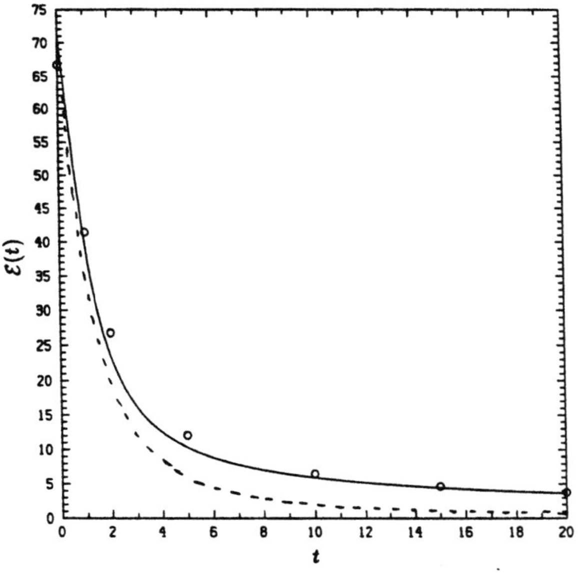

Fig. 4. Enstrophy, $(\mathcal{E}(t) \equiv \int_0^\infty E(k,t)k^2\ dk)$. Modified theory, solid line; unmodified, dashed line; and circles, the McWilliams experiment.

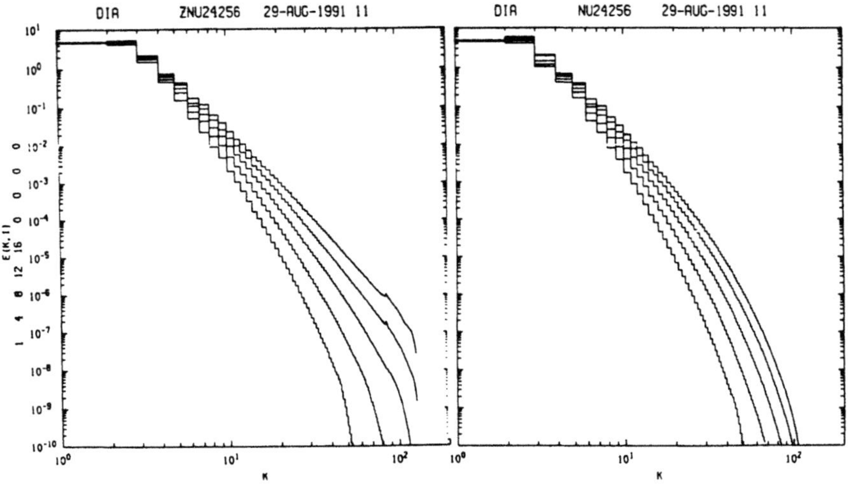

Fig. 5. Evolution of DNS $E(k,t)$ spectrum, whose initial shape is (3.2). Times shown are $t = (.150, .175, .200, .225, .250)$. (a) for $\nu = 0$, (b) $\nu = .0014$.

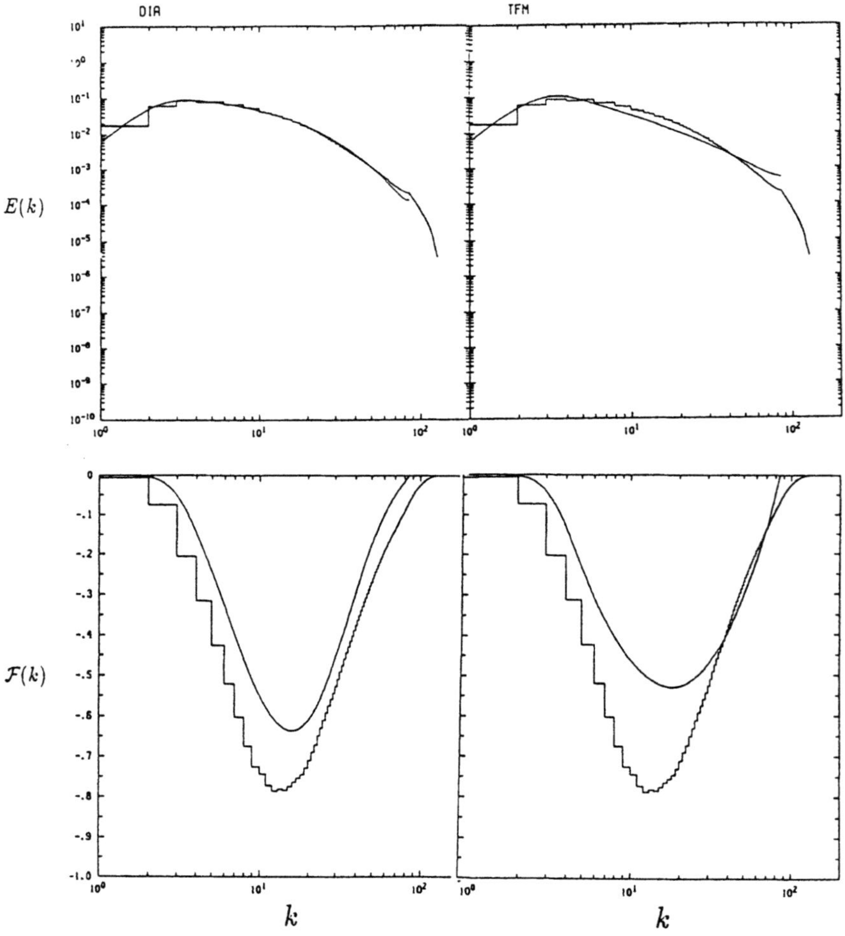

Fig. 6. Comparisons of DNS (histograms), DIA (left panel), and TFM (right panel) for an initially Gaussian initial value problem for which $E(k,0)$ is (3.3). Shown are energy spectra, $E(k,t)$, (top panel), and energy flux, $\mathcal{F}(k)$ (defined by (3.4)) (bottom). Time of evolution is $t = 1.0$

1989) does not eliminate the secondary vortices entirely. This suggests that if an initial random spectrum covers all wavenumbers, there is more room for secondary vortices that might suppress the singularity to appear than with the spectrum (3.2), which might explain why attempts to do inviscid calculations when the initial spectrum covers all wavenumbers have not yielded any sign of singular behavior while the spectrum used here (3.2) does.

Figure 6 presents comparisons of DNS (points), DIA (solid line), and TFM (dashed line) for an initially Gaussian initial value problem for which the energy spectrum $E(k,0)$ is

$$E(k,0) = Ck^4 exp(-2(k/k_0)^2) \, , \tag{3.3}$$

where $k_0 = 4.7568$, $C \ni E(0) = 1.5$. The lower panel of this figure shows the energy flux, $\mathcal{F}(k,t)$, defined as

$$\mathcal{F}(k,t) \equiv \int_0^k dpT(p,t), \; T(k,t) = \{\partial/\partial t + 2\nu k^2\} E(k,t) \, . \tag{3.4}$$

For this run, $R_\lambda(0) = 300$, while at $t = 1.0$, $R_\lambda \simeq 70$. At earlier times than shown in Fig. 6, we may note an approximate power-law behavior, as in Fig. 5(a). Both DIA and TFM show early-time power-law behavior.[1] At $t \geq .5$, the TFM develops a significant $k^{-5/3}$ range, and its Kolmogorov coefficient, $\mathcal{C}$, ($\ni E(k) \simeq \mathcal{C}\epsilon^{2/3}k^{-5/3}$) is near its asymptotic value $\mathcal{C} = 1.77$. The agreement of DIA and DNS for $E(k,t)$ is remarkable. Such agreement between a "structureless" theory (DIA) and "worm-filled" DNS suggests that these tubes may have little impact on second moments for $R_\lambda \sim 100$—this provided the flow remains substantially three-dimensional. The agreement with respect to $\mathcal{F}$ seems less satisfactory, but the ordinate scaling here is linear as opposed to the logarithmic scaling of the energy spectra. Some imagination is required to read into any of the $\mathcal{F}(k)$ shown the flat segment characteristic of an inertial range (although less imagination, perhaps, for TFM).

4. Concluding Comments

The free decay two-dimensional turbulence poses special problems for two-point closures in that the flow develops—at late time—an extreme form of spatial intermittency. If we wish simple closure (such as the TFM or EDQNM) to follow the actual spectra evolution, we must progressively suppress the overall nonlinear transfer, roughly as (2.4). In this case, Batchelor's decay law is modified from $\mathcal{E}(t) \sim t^{-2}$

[1] Notice that the closures are computed on a wavenumber span $(2/3) \times 256$, whereas the span for DNS is 256, with the 2/3 rule for de-aliasing. This is appropriate, since the closures utilize a de-aliased wavenumber integration scheme.

to something much slower, such as (2.5). We should note that a recent scaling theory for two-dimensional turbulence of Carnavale *et al.* (1991), which deals explicitly with the intense vortex tubes comprising the late stage flow field, and models their decrease in number by vortex merger as time proceeds, predicts $\mathcal{E}(t) \sim t^{-\xi/2}, \xi \sim 0.75$, a somewhat different form from (2.5). But our form should not be taken too seriously, especially since it is empirically based on early time behavior. Perhaps a suitable approach would be to fix the parameter λ by the scaling theory so that $\mathcal{E}(t)$ is properly modeled.

We have said nothing about a possible theoretical justification of $\lambda(t)$ as in (2.4). We recall in this connection that the physics underlying the TFM is to estimate dephasing of triple moments entering (2.1) by pressure scrambling effects. That is to say, it estimates memory times of triple moments by the rate at which the solenoidal field is converted into a longitudinal component were the pressure absent. It is reasonable to suppose that if the flow is bound up in intense vortices, then the strain field, which is primarily the means of such dephasing, has little chance to act.

It is clear that vortex tubes survive in three dimensions, but it is not clear that their involvement in the dynamics of decay is as critical as in two dimensions. We know that for DNS at $R_\lambda \sim 100$, $\mathcal{K}(t) \sim 10$, and that the vortex tubes have cross-sectional diameters of the order of the Kolmogorov scale. Perhaps it is useful in this connection to note that the decay law for $\mathcal{E}(t)$ in three dimensions is well predicted by closure, in contradiction to the two-dimensional case.

Our three-dimensional comparison of closure to DNS suggests that non-Markovian effects may well be an important ingredient in an accurate closure. This may be the reason that the DIA, despite its lack of invariance to random Galilean transformation, is somewhat more accurate than TFM at $R_\lambda \sim 100$. It would be important to examine the accuracy of a properly invariant theory, such as that of Kaneda (1983). Preliminary results indicate, however, that this theory is quite similar to TFM. Finally, we note that attempts to estimate the Kolmogorov constant, $\mathcal{C}$, from $R_\lambda \sim 100$ DNS seem always to yield too high a value, ~ 2.5 instead of $\in(1.4–1.8)$. In the present case, it is ~ 3.5. One interpretation of the present study is that this R_λ is too small for inertial effects to become manifest, and the emergence of an inertial range in the TFM is attributable to Markovianization.

Acknowledgments

The National Center for Atmospheric Research is sponsored by the National Science Foundation. This work was supported through NSF Interagency Agreement No. AFOSR-ISSA-90-0016.

References

Benzi, R., S. Paternello, and P. Santangelo, 1987. *Europhys. Lett.*, **3**, 379.

Batchelor, G. K., 1969: Computation of the energy spectrum in homogeneous two-dimensional turbulence. *Phys. Fluids Suppl.*, **12** II, 233–239.

Brissaud, A., U. Frisch, J. Léorat, M. Lesieur, A. Mazure, A. Pouquet, R. Sadourny, and P. L. Sulem, 1973: Catastrophe énergétique et nature de la turbulence. *Annales de Géophysique (Paris)*, **29**, 539–546.

Carnavale, G. F., J. C. McWilliams, Y. Pomeau, J. B. Weiss, and W. R. Young, 1991: Evolution of vortex statistics in two-dimensional turbulence. Preprint.

Herring, J. R. 1969: Statistical theory of thermal convection at large Prandtl number. *Phys. Fluids*, **12**, 39–52.

Herring, J. R., and R. M. Kerr, 1991: Contribution of two-point closure to large eddy simulations. *Proceedings of the International Workshop on Large Eddy Simulations*, University of South Florida, Tallahassee, B. Galperin, Ed., Cambridge University Press, MA, in press.

Herring, J. R., and J. C. McWilliams, 1985: Comparison of direct numerical simulation of two-dimensional turbulence with two-point closure: The effects of intermittency. *J. Fluid Mech.*, **158**, 229–242.

Herring, J. R., and J. C. McWilliams (Eds.), 1990: *Lecture Notes on Turbulence*, World Scientific, Singapore/New Jersey/London/Hong Kong, 371 pp.

Herring, J. R., S. A. Orszag, R. H. Kraichnan, and D. G. Fox, 1974: Decay of two-dimensional homogeneous turbulence. *J. Fluid Mech.*, **66**, 417–444.

Kerr, R. M. 1991: A scenario for the initiation of turbulence. Preprint.

Kerr, R. M. 1992: Evidence for a singularity of the incompressible, three-dimensional Euler equations. Preprint.

Kraichnan, R. H., 1959: The structure of isotropic turbulence at very high Reynolds numbers. *J. Fluid Mech.*, **5**, 497–543.

Kraichnan, R. H., 1971: An almost-Markovian Galilean-invariant turbulence model. *J. Fluid Mech.*, **47**, 513–524.

Lesieur, M., 1987: *Turbulence in Fluids*, Martinus and Nijhoff Publishers, Dordrecht, The Netherlands. 286 pp.

Lesieur, M. and J. R. Herring, 1985: Diffusion of a passive scalar in two-dimensional turbulence. *J. Fluid Mech.*, **161**, 77–95.

McWilliams, J. C, 1984: The emergence of isolated vortices in turbulent flows. *J. Fluid Mech.*, **146**, 21–43.

Melander, M. and F. Hussain, 1989: Cross-linking of two antiparallel vortex tubes. *Phys. Fluids A*, **1**, 633–634.

Proudman, I., and W. H. Reid, 1954: On the decay of a normally distributed and homogeneous turbulent velocity field. *Phil. Trans. Roy. Soc., Vol. A*, **247**, 163–189.

Pumir, A., and E. Siggia, 1990: Collapsing solutions to the 3-D Euler equations. *Phys. Fluids A*, **2**, 220–241.

Tavoularis, S., J. C. Bennett, and S. Corrsin, 1978: Velocity-derivative skewness in nearly isotropic turbulence. *J. Fluid Mech.*, **88**, 63–69.

Wray, A. A., and Hunt, J. C. R., 1990: Algorithms for classification of turbulent structures. In *Topological Fluid Mechanics*, H. K. Moffatt and A. Tsinober, eds., Cambridge University Press, 95–104.

Address of the first author

Dr. Jackson R. Herring
National Center for Atmospheric Research
P. O. Box 3000
Boulder, Colorado, 80307, USA
herring@ncar.ucar.edu

A Note on Local Isotropy in High-Reynolds-Number Turbulence

Srinivas V. Veeravalli, Seyed G. Saddoughi,
Alexander A. Praskovsky & Peter Bradshaw
Center for Turbulence Research, Stanford University

Communicated to the Monte Verita meeting
By W. C. Reynolds

Abstract. Detailed measurements of the velocity field were made in the wall boundary layer of the 80' by 120' facility at NASA Ames. The Reynolds Number R_λ, based on the Taylor microscale λ at the measurement location, was approximately 1450, one the largest attained in laboratory flows. The data indicate that to within measurement accuracy, the w-spectrum follows, but the v-spectrum deviates from, the isotropic relation in the inertial subrange. No definite statement can be made regarding local isotropy for the dissipating scales because the spectral measurements were contaminated by high-frequency electrical noise, but it appears that the inertial-subrange anisotropy persist in the dissipation region.

1. Introduction

The hypothesis of local isotropy at high Reynolds number proposed by Kolmogorov (1941, 1962) states that the small-scale structures of turbulent motions are independent of large-scale structures and mean deformations. This hypothesis is an integral element of most approaches to understanding turbulence be they theoretical, modeling, or even computational methods like large-eddy simulation.

The total turbulent energy dissipation, ε, is given by $15\nu\,\overline{(\partial u/\partial x)^2}$ in locally isotropic turbulence. In the high-wavenumber region of the spectrum, Kolmogorov's universal equilibrium hypothesis states that $E_{11}(k_1)/(\varepsilon\nu^5)^{1/4}$ is a universal function of $(k_1\eta)$, where

$$\int_0^\infty E_{11}(k_1)\,dk_1 = \overline{u^2}\,, \qquad (1)$$

k_1 is the longitudinal wavenumber and $\eta = (\nu^3/\varepsilon)^{1/4}$ is the Kolmogorov length scale. If the motion is isotropic, the transverse spectra $E_{22}(k_1)$ (for the velocity component normal to the wall) and $E_{33}(k_1)$ (for the spanwise component) are uniquely determined from the longitudinal spectrum (Batchelor 1953):

$$E_{22}(k_1) = E_{33}(k_1) = \frac{1}{2}\left(1 - k_1 \frac{\partial}{\partial k_1}\right)E_{11}(k_1) \tag{2}$$

In the inertial subrange, the 3D spectrum takes the form (Kolmogorov 1941)

$$E(k) = C_1\, \varepsilon^{2/3}\, k^{-5/3}, \tag{3}$$

where k is the wavenumber magnitude, and, assuming isotropy, the one-dimensional longitudinal and transverse spectra are

$$E_{11}(k_1) = C_2\, \varepsilon^{2/3}\, k_1^{-5/3}, \tag{4}$$

and

$$E_{22}(k_1) = E_{33}(k_1) = C'_2\, \varepsilon^{2/3}\, k_1^{-5/3}, \tag{5}$$

respectively. The constant C_1 is equal to $55C_2/18$, and equation (2) evaluated in the inertial subrange gives $C'_2/C_2 = 4/3$.

There have been many experiments, conducted in variety of flows, in which the concept of local isotropy has been investigated (see e.g. Townsend 1948; Townsend 1954; Mestayer 1982; Browne, et al 1987; Karyakin, et al 1991; Van Atta 1991 etc. Some additional experiments are listed in figure 1). There have recently been a number of theoretical and computational studies stressing the importance of *non-local* (in Fourier space) interactions in the energy cascade process, and this brings into question Kolmogorov's concept of a self-similar cascade (see e.g. Domaradzki & Rogallo 1988; Yeung & Brasseur 1991; etc). Also, Durbin & Speziale (1991) show that local isotropy is inconsistent with the presence of mean strain. Veeravalli & Saddoughi (1991) reviewed the above literature and concluded that there was no consensus regarding the local-isotropy hypothesis. Therefore, a fresh experimental investigation into this question was undertaken.

2. Apparatus and measurement techniques

The experiments described here, were conducted in the boundary layer on the test-section ceiling of the full scale aerodynamics facility at NASA Ames. The test section is 80' high, 120' wide, and approximately 155' long. All four walls of the test section are lined with acoustic paneling, yielding a rough-wall boundary layer. The measurement station was located towards the end of the test section on the centerline of the tunnel. These experiments were conducted while NASA engineers were investigating the flow around an F-18 fighter aircraft in the central region of the working section.

The full details of measurement techniques are given by Veeravalli & Saddoughi (1991); however very briefly, the hot-wire instrumentation used consisted of Dantec model P51 cross-wire probes, modified to support 2.5 μm Platinum plated Tungsten wires with an etched length of approximately 0.6 mm, TSI model 1050 hot-wire bridges, and model 1052 signal conditioners. The hot-wire output voltages were digitized on a micro computer equipped with an Adtek AD830 12-bit analog-to-digital converter. The data were obtained in three spectral bands using a band-pass filtering (Rockland model 452 and 852 units) technique with approximately one decade of overlap between neighboring bands. The spectral densities of the hot-wire signals were computed by a fast-Fourier-transform algorithm. To convert frequencies to wavenumbers, Taylor's hypothesis was used.

3. Results and Discussion

The overall features of the boundary layer were measured in detail and it was verified that the outer part of the boundary layer followed the standard behavior and the interference from the F-18 aircraft was not significant (see Veeravalli & Saddoughi 1991). For the present we will focus on the longitudinal and transverse spectra obtained at one position in the boundary layer. The relevant flow parameters are summarized in table 1. In figure 1, using Kolmogorov scaling, a longitudinal spectrum of the present study is compared with a compilation of some experimental work taken from Chapman (1979) with later additions. The agreement is very good. The present spectrum spans five decades in wavenumber and eight decades in amplitude and has a -5/3 slope over a region spanning approximately two decades in wavenumber. This is one of the longest -5/3 ranges seen in laboratory flows. It must be noted that although the measurements extend up to $k_1\eta \approx 1$, the data are reliable only up to $k_1\eta \approx 0.3$ because the combined error in the spectrum, due to lack of sufficient spatial resolution (Wyngaard 1968) and electronic noise, exceeds 20% for $k_1\eta > 0.3$. Note that the behavior of the return-channel data of Karyakin et al. (1991) at low-wavenumbers perhaps indicates the presence of large-scale unsteadiness in their flow.

To investigate the isotropy of the inertial subrange, we use equations (4) and (5) and present in figure 2 the compensated spectra $k_1^{-5/3}E_{ii}(k_1)$, where i = 1, 2 or 3 corresponds to u, v or w respectively. In the inertial subrange, these compensated spectra should be independent of wavenumber, and the v- and w-spectra should be equal to each other and larger than the u-spectrum by a factor 4/3. Also shown in figure 2(a) are the seventh-order least-square polynomial fits to the u-spectra obtained from the uw and uv X-wires. These u-spectra are in good agreement, giving some indication that the differences between v- and w-spectra (figure 2b and 2c) in the inertial-subrange region may be real. Taking the dissipation as

$$\varepsilon = 15\nu\int k_1^2E_{11}(k_1)\, dk_1, \tag{6}$$

Townsend, A. A. (1954), Q. J. Mech. Appl. Math., 7, 104-127.

Van Atta, C. (1991), Proc. Roy. Soc. Lond. A 434, 139-147.

Veeravalli, S. V. & Saddoughi, S. G. (1991), Center for Turbulence Research, Stanford University/NASA Ames Annual Res. Briefs, 3-20.

Wyngaard, J. C. (1968), J. Sci. Instrum. 1, 1105-1108.

Yeung, P. K. & Brasseur, J. G. (1991), Phys. of Fluids A3, 884-897.

The freestream velocity, U_e (m/s)	≈ 40
The boundary layer thickness, δ (m)	≈ 1.0
The momentum thickness Reynolds number, R_θ	$\approx 300{,}000$
Measurement location, y/δ	≈ 0.4
Local mean velocity, U (m/s)	$= 35.1$
Local turbulence intensity, $\sqrt{\overline{u^2}}/U$	$= 0.08$
Microscale Reynolds number, R_λ	≈ 1450
Kolmogorov length scale, η (mm)	≈ 0.1
Ratio of X-wire span, $l/\sqrt{2}$ to η	≈ 4.3

TABLE 1: Flow Parameters.

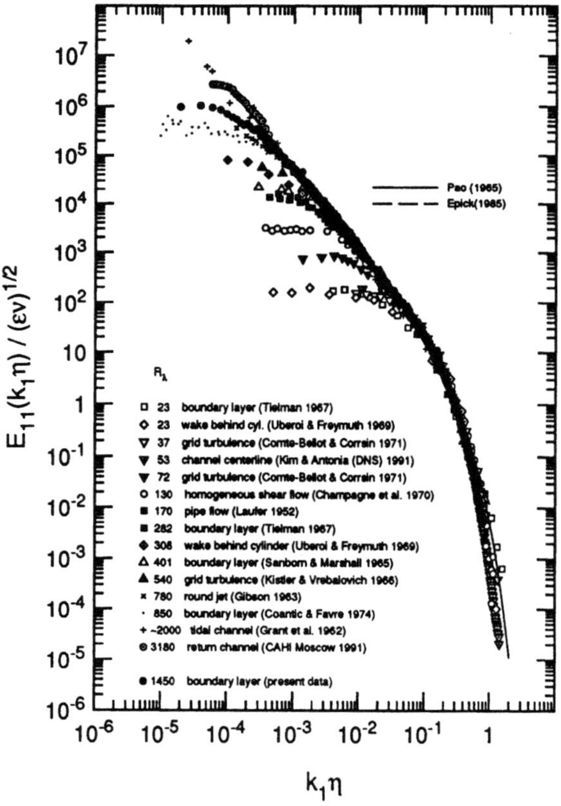

Figure 1. Normalized longitudinal spectrum compared with data from other experiments. This compilation is from Chapman (1979) with later additions.

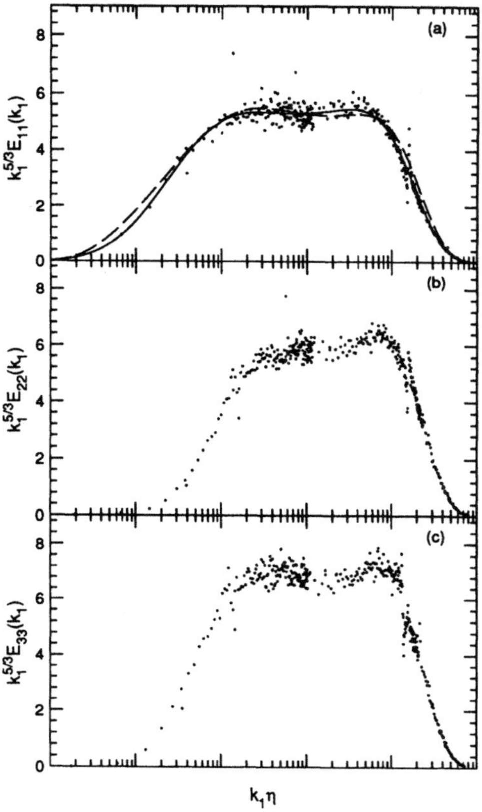

Figure 2. Compensated longitudinal and transverse spectra (symbols show the unsmoothen data). (a) u-spectra (•, unsmoothen data from uv X-wire; —— polynominal fit to data from uv X-wire; ------, polynominal fit to data from uw X-wire); (b) v-specrum; (c) w-spectrum.

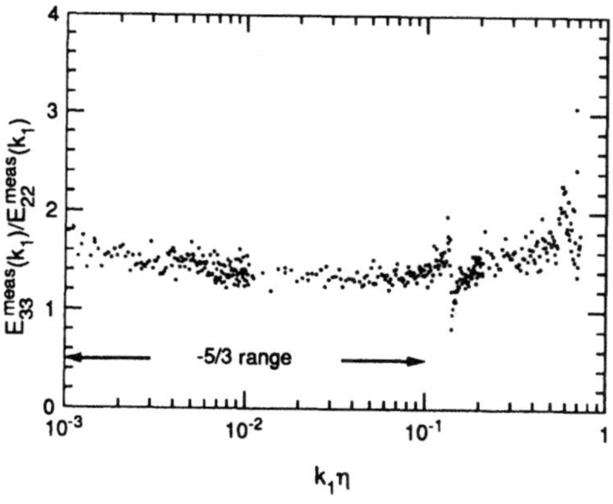

Figure 3. Spectral distribution of the ratio of the measured w-spectrum to v-spectrum.

$C_2 \approx 0.58$ and $C'_2 \approx 0.75$ were obtained from u- and w-spectra respectively. These both yield $C_1 \approx 1.75$. Therefore, it appears that w-spectrum follows, but the v-spectrum deviates from, the isotropic relation in the inertial subrange.

The spectral distribution of the ratio of the measured w-spectrum to v-spectrum in the inertial and the dissipation ranges is shown in figure 3. For isotropic turbulence, this ratio should be equal to 1. In the -5/3 region, $10^{-3} < k_1\eta < 10^{-1}$, the average value of this ratio is about 1.43. Also, it appears that the inertial-subrange anisotropy persists in the dissipation range; however, as mentioned above, the data are not reliable for $k_1\eta > 0.3$. To resolve the viscous sub-range more accurately, further experiments are going to be conducted in the same facility.

Acknowledgements

We wish to thank the Full-Scale Aerodynamics Research Division at NASA Ames for permitting us to use their facility, and to thank the F-18 test team for their cooperation during the experiments. We also wish to thank Prof. P. Moin, Prof. W. C. Reynolds, Dr. J. Kim and Dr. R. S. Rogallo for their encouragement and for valuable discussions during the course of this project.

References

Batchelor, G. K. (1953), *The Theory of Homogeneous Turbulence*. Cambridge University Press.

Browne, R. A., Antonia, A. A. & Shah, D. A. (1987), J. Fluid Mech., 179, 307-326.

Chapman, D. (1979), AIAA J., 17, 1293.

Domaradzki, A. J., & Rogallo, R. S. (1988), Proc. Summer Program, Center for Turbulence Research, Stanford University/NASA Ames. CTR-S88, 169-177.

Durbin, P. A. & Speziale, C. G. (1991), Recent Advances in Mechanics of Structured Continua, 117, 29-32.

Karyakin, M. Y., Kuznetsov, V. R. & Praskovsky, A. A. (1991), Izv. AN SSSR, Mech. Zhidk. i Gaza, 5.

Kim, J. & Antonia, R. A. (1991), Submitted to J. Fluid Mech.

Kolmogorov, A. N. (1941), C. R. Acad. Sci U.R.S.S., 30, 301.

Kolmogorov, A. N. (1962), J. Fluid Mech.,13, 82-85.

Mestayer, P. (1982), J. Fluid Mech., 125, 475-503.

Townsend, A. A. (1948), Austral. J. Sci. Res., 2, 161-174.

Addresses of the authors

Dr. Srinivas V. Veeravalli,
Dept. of App. Mech.,
Indian Inst. of Tech.,
New Delhi, India, 110016.

Dr. Seyed G. Saddoughi,
Center for Turbulence Research,
Bldg. 500,
Stanford Univ. CA 94305, USA.
SGS@leland.stanford.edu

Dr. Alexander A. Praskovsky,
Central Aero-Hydrodynamic Inst.,
Moscow, USSR, 140160.

Prof. Peter Bradshaw,
Dept. of Mech. Eng.,
Bldg. 500,
Stanford Univ. CA 94305, USA.

Discussion

Zaslavsky: My comment concerns mainly Moin's talk, but it can refer to several others. It concerns the errors that arise in computer simulations with some discretization scheme. Let us consider the simple equation of a pendulum $x+\omega^2 \sin x = 0$. Its solution is well-known. Let us introduce the second order difference instead of the second derivative and let Δt be a time step. Then we have the equation

$$x_{n+1} - 2x_n + x_{n-1} + \omega^2 \Delta t^2 \sin x_n = 0$$

which can be rewritten in the form

$$P_{n+1} = P_n - k \sin x_n$$

$$x_{n+1} = x_n + P_{n+1}$$

$$k = \omega^2 \Delta r^2 << 1$$

This is the well-known standard mapping, which generates stochastic layers in the phase space. We obtain absolutely differnt properties of the discretized system. Furthermore, it is possible to introduce an explicit form of the potential V(x,t), which is responsible for the discretization and which corresponds to a high frequenzy perturbation. Some general statements are: 1) discretization is equivalent to a high frequency perturbation. 2) The discretization produces stochastic layers and stochastic webs, which can be infinite and which tile the whole phase space of a model. For simple models this can be exactly proved. 3) This statement concerns the cut-off and round-off operations which can generate stochastic webs in the real space of a system. 4) The regular grids which are used for the simulation can produce a process like Lévy flights. Due to Lévy flights there is anomalous transport and this anomalous transport along the stochastic webs is responsible for an instability of the computational schemes. At last a proposal: One can use other kinds of grids which have no square symmetry or another periodical symmetry. This can diminish the effect of the scheme instability due to stochastic webs and to Lévy flights.

Moin: Your comment apparently applies to any computation of turbulence.

For each difference scheme it is possible to write down the Hamiltonian, which is particular for this difference equation, and to derive an explicit form for this potential perturbation to discretisation.

Yaglom: Could that not be done by reducing the step?

Zaslavsky: It's not a good idea. It seems necessary to change strongly the symmetry of the grid in order to diminish the flight instability. Flights are just responsible for the instability of finite grid schemes. It doesn't depend on the size of the step it depends on symmetry. If you have square symmetry then you get Levy-flights. It is very close to the phenomenon of anomalous transport, which was observed in experiments for passive particles by Jerry Gollub, which was published recently. They form a square pattern on the oscillating surface of a liquid under the condition of thermal convection.

Aref: This is a very interesting problem and there are lots of people who have been working on this, on the idea that chaos comes into the condition for numerical instability. However, you find that the time step you would take, which is embodied in your k, is so large that on other grounds you would never use it. Furthermore that particular numerical method is one of the time-stepping methods that you would not use because it has what are called parasitic solutions. And if you do a second order Runge-Kutta, which is the same order, all of this goes away. So the particular example is not so good, but the general idea is certainly interesting and people like York in Maryland and Edward Lorenz and others have been looking at this particular thing. Whether its an issue in turbulence or not I think is very, very difficult to say. I think there may be actually a converse issue, which is at least as interesting, and that is when you do a simulation on a finite computer, every solution is ultimately periodic. If you look, for example, at attempts at making stochastic processes on a computer - like random number generators - they are very difficult to construct.

Zaslavsky: This example is basic because it is an exact one. That is why it is good. But we observed on a given scheme that if we introduce more diffusivity, there will be stronger effects.

Novikov: I have a suggestion in connection with the comparison of direct numerical simulations with closure type models, which I've been trying to promote for more than ten years without much success. The suggestion is very simple. Not only solve equations produced by closure and compare the results with direct numerical simulations, but check directly the closure. Each closure is a connection between say third order moment and second order moment. Second and third order moments are known from direct numerical simulations. We can compare them with closure and just look at what happens. First it is simpler and secondly we can estimate if the closure is good or not. I am not denying simulations based on closure, but this gives us some additional information and very simple tests.

Herring: I think we do that when we look at transfer functions, which are a form of third order cumulants. Another way to do this is to look at Bispectra, and compare closure to simulations (as in Herring and Métais, 1992, J. Fluid Mech., **235**, 103-121). This is partially done if you compare closure predictions to the numerical simulation of Navier-Stokes. It's also partly done with the equivalent comparison of bispectra, which are more directly assessable from the experimentalist point of view than the energy transfer function.

Yaglom: The question about the possibilities to reproduce the experimental results by means of numerical simulation is of very great importance for the future development of the turbulence theory. A very interesting remark by Zaslasvky shows that the discretization, which is an inevitable part of numerical simulation, can sometimes lead to some strange parasitic effects. This remark is related to direct numerical simulation, but it is clear that in the case of large eddy simulation (LES), where the effect of the small-scale fluctuations on the large-scale motions is described by some empirical model, there are even more chances to discover some parasitic effects. Nevertheless, I think that at present LES is the most promising method to study high Reynolds number turbulent flows. It has been impossible so far to directly simulate turbulent flows with large Reynolds numbers and, as I understand, there is no hope that such simulations will become realizable in the near future. Simultaneously, for the time-being there aren't too many possibilities to reach really high values of the Reynolds number in laboratory installations while atmospheric flows are affected by a great number of various factors, that cannot be studied in isolation. Some atmospheric results differ misteriously from the similar results found in the laboratory conditions and it is quite difficult to understand the reason for this. For example, I mentioned in my talk at the beginning of this symposium that statistical characteristics of the temperature fluctuations differ greatly in an atmospheric surface layer at practically neutral thermal stratification and in laboratory boundary layers over slightly heated walls; the dimensionless (normalized by the appropriate temperature scale) value of the mean square temperature fluctuation exceeds in the atmosphere the same value for laboratory

flows by 3 to 4 times. It was assumed by some authors (see, e.g., the paper by B. A. Kader and A. M. Yaglom, 1990, J. Fluid Mech., Vol. 212, 637-662) that this difference is mostly due to the thermal inhomogeneities of the earth surface, but some other factors also exist that can affect the temperature fluctuations in the atmosphere. Therefore, maybe it is worth applying the LES method to study turbulent flows with heat transfer for different thermal boundary conditions at the wall. There are also some discrepancies in the results of the measurements of turbulence characteristics in an unstable stratified atmosphere surface layer performed by various research groups; here also it is very interesting to find the possibility to compare the experimental results with those from numerical simulation.
The last small remark which I want to make is related to what Bill Reynolds talked about. It is unclear to me what reasons can lead to anisotropy of the smallest turbulent fluctuations in the dissipation subrange without any distortion of isotropy for fluctuations in the inertial range of scales. If this is so, then it is necessary to give a physical explanation of the small-scale anisotropy; without such an explanation it is natural to think that the observed anisotropy is due to some experimental errors.

Frisch: How can one physically reconcile this idea of having an isotropic inertial range and an anisotropic dissipation range? Suppose you begin by having large eddies like you have in simple-minded numerical simulations at time equal zero and then certainly as long as you don't have fine structure, the viscosity plays no role; the first thing you will observe (lots of people have observed) is that things are flattening and pancakes are forming. Now, the orientations of these pancakes, reflect the structure of the large eddies that you prepared initially, and will be as anisotropic as these. These sheet-like structures have widths of the order of some power of the viscosity and a spatial extension which is comparable to the integral scale. However, the sheets don't want to stay like that; they will undergo all kinds of instabilities, many of which may still not be fully understood; they will start rolling up, reconnecting,etc. Eventaully, the resulting structures are going to fill out the intermediate scales in a bottom-up fashion. By "bottom-up" I mean that the process starts at small dissipation scales and leads to larger (inertial) scales. The latter are likely to be isotropic. As to the former, it is not clear how they are able to forget the basically anisotropic mechanism referred to above.

Yaglom: I cannot understand the fact that the small scales are affected much stronger than the intermediate scale.

Novikov: The shape of the spectrum can be affected, but anisotropy is very questionable because the time of isotropisation is very small for small-scale motion.

Frisch: If you envisage the smallest scales as being the result of many, many, almost independent steps of energy cascading down, then you have a problem, but with the other scenario (first mentioned in Brachet et al., 1983, J. Fluid. Mech., **130**, 411) you don't have a problem. Because if you have a turbulence which is maintained there are constantly maintained anisotropies and these are constantly regenerating pancakes and other structures which may reflect those anisotropies.

Novikov: And then they mix, they rotate very quickly. Experimental results are averaged over long time. What is your averaging time in comparison with the small scale time?

Frisch: You mustn't forget one thing - the flattening process is a very quick process, its time-scale is the large eddy turn-over time, but its an exponential process it is very fast. After they have formed, of course they would be eaten by viscosity very quickly, but then the smaller scales do not have the time to undergo the process, it's the intermediate scales that will be isotropised by this process, not the very small scales; the very small scales will die quickly, and my suggestion is that these very small scales are not formed by a cascade, but are formed directly by relatively simple deterministic processes which emanate from the large scales.

Reynolds: At the edges of the large scales?

Frisch: That's a possibility.

Herring: It seems to me that large eddy simulations are an attempt to map a large Reynolds number problem down to a modest Reynolds number. As such, you introduce different things like eddy viscosity, which may differ somewhat from the form of the bare viscosity. The exact form of the eddy viscosity may not be too important. In any case, it would seem to me that the new system would have different fluctuation characteristics, it would be more sluggish than real life, and lack the intermittency of the latter. You may, in fact, entirely damp out some instabilities, and be unable to maintain a steady state, unless the resolution exceeds some threshold value. It may be that another strategy - which should be pursued simultaneously with large eddy simulations - is to do a series of direct numerical simulations, with progressively increasing Rayleigh or Reynolds numbers, and try to see if transport profiles become asymptotic as these numbers increase.

Novikov: Partially I agree with what Jack was saying , but I think that the idea which was presented by Parvis is very good because it includes additional information for correcting the scheme. But the distance between these filters is not very big, it's fixed, so you can correct, but not entirely correct. And of course the more systematic and more difficult way to make a real large eddy simulation is to include not just this modelling scheme, but much more knowledge about the internal structure of turbulent flow, of real statistic, of small-scale motion. It is not easy, but I think this is a real prospect for the future for large-eddy simulations. Smagorinsky's scheme is good, it works to a certain extent, but it does not contain much physics in it. I think that the real prospective for simulation of turbulent flow at high Reynolds number is large eddy simulation, otherwise how could we reduce trillions of degrees of freedom which exist. Probably a lot of research on real local properties of the statistics on small-scale motion have to be done, both theoretical and experimental.

Moin: I think it is very important to reiterate why one uses large eddy simulations. The objective is not to understand the physics of small scales, the objective is to do practical engineering calculations, to get reasonable answers for the velocity profile and the heat-transfer coefficients, etc. In that context you would be happy to correctly obtain the average effects of the small scales on the large scales. That is what large eddy simulation is. It is trying to find a more robust way than, say, the standard k-ε model to get quantitative practical results quickly.

Dracos: I have a comment on what Adrian did observe in his PIV measurements in a channel flow. We have done some 3-D measurements by particle tracking velocimetry. We observed the flow field in a Lagrangina frame. The observation volume spans from z = 5 mm to z = 20 mm in a channel flow of total depth 55 mm. In the top view the observation area covers approximatly 120 x 120 mm. You see those large eddies there. Their size is of the order or the depth and they spann over the entive observation depth. The existance of this eddies can explain the acceleration and deceleration of the flow observed by Adrian in vertical cuts in a very similar flow.

Frisch: One of the ways to achieve a higher Reynolds number without making larger facilities is to resort to a suitable combination of the three following factors: low temperature, high pressure and somewhat exotic fluids. There are all kinds of projects in this direction, but when you achieve high Reynolds numbers in that way it means of course that the smallest scales are going to be very small. So, instead of having, for example, a Kolmogorov scale somewhere between 1mm and 1/3 mm, which is a very common thing to have, we may have a Kolmogorov scale which is between 10 and 100 microns. Would you comment on the feasibility of visualisation techniques 2- or 3-D when you want to have resolutions which do not completely spoil those very small scales.

Adrian: I think that you can get down to resolutions in the neighbourhood of several tens of microns if you are willing to give up field view. As far as the physics go, you can certainly get down to looking at small volumes. But the problem is that you want to have a ten thousand to one dynamic range in the smaller scales.

Frisch: I have a braoader question to which perhaps we can come back on Friday, but still, I'd like to formulate it: Given that there are these novel exotic techniques that look promising for multidimentional analysis possessing a resolution which is well below what most hot wire probes can do. If we want to learn more about high Reynolds number flow in a well controlled environment, should the priority be on well controlled, relatively large facilities or on the more exotic ones?
The wind tunnel data presented by Reynolds, Castaing, etc. come from rather monstrous wind tunnels. We are speaking of tens of meters. These are all industrial facilities which are not well controlled. Scientific experiments are then almost exclusively parasitic, and I'm really afraid that there are some very bad limitations on what one can do in such circumstances. At some point in the future I can see exotic facilities overtaking the "monsters" as far as the experimental physics is concerned. I don't insist, of course, that they should fit into my hand, but perhaps on a table.

Group Work

Group Work

Group I **Prospects of physical and numerical experiments for the fundamental understanding of turbulence**

Participants R. Adrian (Chairman), R. Antonia, T. Dracos, U. Frisch, M. Landahl, M. Lesieur, P. Moin, W. Reynolds, I. Wygnanski, A. Yaglom

The potential of numerical and physical experiments to address important questions in turbulence research has increased dramatically over the last decade. Both are now able to provide multidimensional data on scalar and vector fields, although the capabilities of physical experiments are somewhat different from those of numerical experiments. These two approaches will be discussed, and compared, followed by a discussion of the experiments that should be done with them, and the important issues that might be answered by these means.

Numerical Simulation. The principle tools of numerical simulation, by which we mean calculations that are fully resolved down to the small scales, and large eddy simulation. Both methods provide full three-dimensional fields of scalars, including pressure vectors such as velocity and vorticity and higher order tensors such as rate-of-strain. They also provide time series of these fields, but the costs of numerical integration linit the duration of the times series relative to that attainable with physical experiments. The virtue of DNS is the accuracy with which the small scales are computed and the consequent reliability of the large scales. In one sense, DNS can be viewed as a source of highly reliable solutions of the Navier-Stokes equations, while in a more limited sense it can be viewed as a means of simulating the small scales of turbulence. The distinction depends on the domain of the simulation and the accuracy with which it and the boundary conditions provide a realistic environment for the mechanisms that generate large-scale motions and energy-containing motions.

The maximum turbulence Reynolds number that can be simulated is limited to relatively low values in current DNS work. Some increase can be achieved for studies of small-scale mechanisms by trading realistic simulation of the large scales for better small-scale resolution. Ultimately, DNS computations are desired at Reynolds numbers that are high enough to create an inertial subrange and or a logarithmic layer of significant extent, e.g. at least one decade wide. DNS simulations have been applied mainly to simple geometries, but much of the work in the future is expected to address more complicated geometries. For example, flows over riblet surfaces can be computed now. A capability that is difficult, if not impossible, to achieve in physical experiments is the simulation of artificial flows that explore the effects of flow modification. The minimal flow module computation of Jimenez and Moin is one such example, and one should expect a whole new class of "numerical gedanken experiments" like this.

Large eddy simulations may not produce the same accuracy as DNS, but they are capable of revealing the general character of various physical effects. For example, low speed wall streaks are predicted by LES, albeit with some inaccuracy compared to observations. LES can be used to explore and model flow fields which are made complex by the presence of additional forces such as rotation or variables such as humidity and radiation in cases that are not accessible to DNS because of large Reynolds numbers. Ultimately, some form of LES is needed for practical turbulence calculations sicne DNS is not likely to be able to treat flows at practical Reynolds numbers. Improved LES theory is needed to establish the actual Reynolds number of an LES computation.

Both DNS and LES provide huge volumes of simulation information, but these are often used very inefficiently in current studies. Better methods of visualizing and analyzing large three-dimensional vector and tensor data bases are needed. In part, this calls for better graphical tools, but even more importantly, it calls for better conceptual tools, either in the form of analytical descriptors such as fractal geometry and wavelet analysis, or better posed dynamical questions. Physical experiments have similar requirements, but their need is not immediately pressing because they do not provide the same volume of data as numerical simulations.

Physical Experiments. Physical experiments offer capabilities that are, at present, quite complimentary to those of numerical experiments. Physical experiments can be performed for long durations, for many parameter values, and for Reynolds numbers having values that are well beyond those accessible to DNS. They can be used for flows with complex geometry such as flows that are inhomogeneous in three-dimensions, boundary layers with non-zero pressure gradient, or flows over rough surfaces, and there is no question about the "reality" of the results. Current diagnostic capabilities, on the other hand, are limited in comparison to those of numerical simulation. Single-point probes for measurement of velocity such as hot-wires and LDV are very good, but probes for even a simple scalar-like temperature are less well developed, and probes for in-flow pressure continue to want improvement. Recent work on hot-wire vorticity probes suggests that the three-dimensional vorticity vector can now be measured reliably under favorable experimental conditions.

Optical techniques for the measurement of instantaneous two-dimensional and three-dimensional fields of velocity, vorticity, concentration and temperature have evolved considerably, and they now constitute a diagnostic capability that lies somewhere between single-point probe methods and the complete data set determination offered by DNS. Particle image velocimetry and particle tracking velocimetry provide good velocity fields on two-dimensional domains with accuracy sufficient to permit differentiation to determine fields of vorticity and rate-of-strain. The dynamic spatial range (ratio of the field of view to the smallest resolvable scale) offered by PIV permits measurements at Reynolds numbers that exceed those of DNS computations. The measurement may be focused on small-scale motions without affecting the reality of the large-scale motions. Good quality measurements of the three-dimensional velocity vector can be made on two-dimensional light sheet domains using stereo photography. Finally, three-dimensional vector fields can be captured using photogrammetric methods or holographic methods. Similar diagnostic capabilities are available for scalar fields using, for example, planar laser-induced fluorescence or tomographic interferometry. While all these methods are developing rapidly, and show great promise, they cannot currently offer the same richness of diagnostic information that is available from DNS, nor are they likely to ever fully compete with DNS in this regard without major developmet efforts to create experimental facilities on a scale comparable to large computer centers. Thus, experiments and numerical simulations will overlap increasingly, but each will continue to offer capabilities that are complimentary to the other.

Experiments. The state of turbulence theory seems not to have progressed to the point of defining critical experiments which are meeded to decide between competing theories or models. However, a number of experiments are needed to address important issues. These are principally experiments at high Reynolds number, where "high" means large enough to investigate convincingly the phenomenon in question. While a single value cannot be given for obvious reasons, a Reynolds number based on the rms velocity and the Taylor microscale of the order of 10^3 would permit one to address most issues. However, even Reynolds number five or ten times those of current DNS results would be valuable, and would probably yield good logarithmic ranges and inertial subranges. Some of the issues that should be addressed by high Reynolds number experiments are the structure of eddies in the logarithmic layer; the structure in the intertial subrange, and in the dissipation range, including measurements of anisotropy of the small-scales; tests of Reynolds number similarity, and measurements of interactions between large-scale and small-scale structures. Ecxperiments at very high Reynolds number can be performed in special facilities that make use of largeness of scale, or low temperature, low viscosity fluids, or very high density fluids. The atmosphere is a good flow facility in which to explore small-scale structure at very high Reynolds number, but it may not provide the constancy of conditions needed for very high quality, reliable experiments.

Comparisons of physical experimetns done in close conjunction with numerical simulations should be done if there are reasons to expect that the numerical simulations do not perfectly model the flow, e.g. LES computations or DNS with marginal resolution or domain size. It is not, however, particularly rewarding to compare a high quality, fully resolved DNS computation to an extensive companion experiment, although some experimental confirmation is always prudent, and one should always pay careful attention to the discrepancies in such comparisons.

Lastly, it should be noted that numerical simulations are able to create ideal flow conditions in perfect, mathematically precise environments. This capacity is a strength that should be exploited to address questions that require clean definition. Physical experiments, on the other hand, create the real flow fields that exist in imperfect environments, e.g. flow that is acoustically contaminated by wide spectrum noise, or flow in non-infinite domains. This characteristic should be exploited to address practical questions such as the behaviour of a flow in an engineering application, or more subtle questions such as the selection of modes under conditions of non-uniqueness.

In summary, the modern tools of numerical simulation and experimental flow diagnostics give us the ability to expand our empirical knowledge base more in a few years than in the entire previous history of turbulence. The future looks very bright, indeed. The primary challenge is to assimilate what this data is telling us, and in this regard, new and extended theoretical methods and concepts are likely to be essential.

Group II **Future implications of chaos theory, dynamical systems and prospects of low-dimensional descriptions of turbulent flows**

Participants H. Aref (Chairman), J. Jimenez, N. Malik, F. Rys, G. M. Zaslavsky

We decided that there were essentially five areas in which chaotic behaviour in dynamical systems might contribute to turbulence research.

The first area is the success story. The ability of ideas from chaos and dynamical systems to elucidate bifurcations that lead to transition to turbulence. It was clear to us that there will be more work along these lines and that hopefully there will be both a broadening in terms of the possible scenarios of transition to turbulence and investigations of a more extended range of flows. I think that a lot of work has been done on things like the Taylor-Cuette problem and the convection within a relatively small box and that we are beginning to see different investigators branch-out a little, looking at slugs in pipe flow, vortices in wakes, and maybe even spots in boundary layers. I think that we will see more and more bold attempts to extend this kind of modeling to additional flow situations in turbulence.

Then the discussion shifted to what probably is one of the basic blocks to the advocation of chaos. We seem to understand how to find chaos, temporal chaos, by sticking a probe in at one point in space and recording what happens. And we can also understand how rather complex patterns have evolved from flows that are very simple in the Eulerian sense. But putting those two ideas together and getting into the range of spatio-temporal chaos is still something for the future. We see an extension of this description to systems of many loosely coupled degrees of freedom as viable. Also, instead of just investigating pattern formation in relatively weakly nonlinear situations it was felt that one should investigate pattern formation in strongly interacting systems that do not have clear separations of scale because such systems are more relevant to the type of situation that one encounters in turbulence. A lot of the pattern formation studies have very well separated scales and in that sense are easier to deal with. One would like to understand turbulence, which is very obviously both spacial and a temporal type of problem and things like the dynamics of sheer layers rather than just the dynamics of a few dislocations that exist in transition processes.

Then we discussed the notion that was put before the meeting yesterday by Zaslavsky, which was the idea of "computational chaos". What does chaos have to say about large numerical simulations of turbulent flows? And now we are not talking about what kind of chaos might be elucidated by these computations, but what might be the effect of the chaotic nature of the equations of the numerical schemes in assessing the accuracy of such large scale computations. One thing that we discussed, which is kind of interesting, was that maybe there is a criterion not yet identified, that is somewhat like the Courant-Friedrichs-Levy criterion in the integration of hyperbolic systems, that dictates what one must really do. One must match the Lyapunov exponents in the real flow to the Lyapunov exponents in the computation. Somehow the computation does not capture all the Lyapunov exponents that are present in the real flow and one must match these up to have an honest simulation. This is just a dynamical systems' way of saying things that are really quite well-known in other cases.

The other thing is that we noticed that there are a number of very large computations that have led to dimension estimates for big simulations that are in the range of a few hundreds. And this suggests that (maybe) there is a route to follow for some kind of reduced description. It was interesting that maybe when you do a 256^3 simulation, you get the effective dimension of the system in the hundreds. Whether you trust this or not is another issue. When you get such a result, that does suggest to you that there are ways of taking advantage of it and maybe there are ways of compressing this spatio-temporal simulation down to something involving at most a few hundred degrees of freedom. If that should be possible, that might be one of the ways of getting around the limitations of present computer technology. Such limitations will persist for the next decade or two so that reaching really high Reynolds numbers in a direct numerical simulation sense is absolutely impossible. This is another statement of those made in the context of large-eddy simulation. Closely related to these is the idea of trying to identify the effective degrees of freedom taking some kind of renormalisation type approach so that one can compress the description. Is it really true? When you look at the pictures of what comes out of the 256^3-simulations, such as sheets that form strong vortex tubes, do you still need 256^3 degrees of freedom to describe that?

The next thing we discussed was the notion of intermittency. It was felt that there were no very good explanations forthcoming from conventional turbulence theories to describe how intermittency arises, what are the mechanisms in intermittency. "Chaos theory" might be a place to search for such mechanisms. Loosely speaking, intermittency is associated with structures of and patterns in various phases. A couple of specific ideas arose in the discussion. One is that intermittency might be addressed in a few-degree-freedom subsystem residing within the larger system that describes the turbulence, and that the analysis of such a few-degree-freedom subsystem is something that is very well suited for dynamical systems theory. Another idea is that there are localised spatial structures, and that a kind of kinetics of these spatial structures is a description of intermittency, and that the way these structures arise and their kinetics again are something that are best described by dynamical systems' ideas. Then there was a suggestion that maybe the cascade between different scales of space and time is again something that can be described using ideas from dynamical systems.

These are all, of course, rather open questions. All that was suggested was that these issues be looked at from the dynamical systems' point of view.

Finally, we talked about the idea of Lagrangian chaos, that is, that if you see much of these previous topics as being Eulerian chaos, then Lagrangian chaos is sort of a counterpart in terms of spatial structure. There was a feeling that the high-Reynolds-number limit of turbulent flows should in some sense be more Lagrangian. Therefore, it makes sense to look at that spatial structure, and it is easy to understand that when we get back to the interaction of vortices, the idea of Lagrangian chaos has a role to play. We were also reminded that there is a lot of interest in the connection between fractal structure and the isosurfaces that are found within turbulent flow.

We also discussed the notion that there may be a similarity between turbulent mixing and what you see as chaotic mixing in laminar flows. Just to remind you what that is all about, I will show some pictures from experiments showing fine-scale mixing in a turbulent flow, and some numerical results on the fine-scale mixing in a simulated shear layer. This is essentially the same type of thing that you see in the chaotic mixing of laminar flows. And that was about where we stopped the discussion.

Group III **Reduction of non-linearity in turbulent flows - further lines of research and their use in the development of the theory of turbulence**

Participants R. Betchov, H.-W. Bewersdorff, U. Frisch, M. Gaster, A. Gyr, C. B. Hustard, T. Kambe, E. Kit, H. K. Moffatt (Chairman), L. Shtilman

This is a subject that was not actually covered explicitly in the discussions of the last three days. I think I should explain first what we mean by reduction of non-linearity because there are different interpretations of this term. If we adopt a *dynamical systems* viewpoint, then a turbulent flow can be thought of in terms of a trajectory in the function space of all solenoidal vector fields satisfying certain weak constraints (e.g. boundedness of $|\mathbf{u}|$). The *fixed points* in this function space are then fields $\mathbf{u}^E(\mathbf{x})$ that are *steady* solutions of the Euler equations

$$\frac{\partial \mathbf{u}}{\partial t} = \mathbf{u} \times \omega - \nabla P, \qquad \nabla \cdot \mathbf{u} = 0 \tag{1}$$

where $\omega = \text{curl } \mathbf{u}$, $P = p/\rho + (1/2)\mathbf{u}^2$. By solving the Poisson equation $\nabla^2 P = \nabla \cdot (\mathbf{u} \times \omega)$ for P, and substituting back in (1), this may equally be written

$$\frac{\partial \mathbf{u}}{\partial t} = (\mathbf{u} \times \omega)_S \tag{2}$$

where the suffix S represents *solenoidal projection.* Just as Beltrami flows satisfy the condition $\mathbf{u} \times \omega \equiv 0$, so Euler flows satisfy the weaker condition $(\mathbf{u} \times \omega)_S \equiv 0$.

For a turbulent flow, we may define a *Beltramisation parameter*

$$\beta = \langle B \rangle / \langle B \rangle_G$$

where $B = (\mathbf{u} \times \omega)^2$, and $\langle B \rangle_G$ is the value of $\langle B \rangle$ for a Gaussian field $\mathbf{u}_G(\mathbf{x})$ with the same energy spectrum as $\mathbf{u}$. Thus $\beta = 0$ if (and only if) the field is a Beltrami field. Similarly (Kraichnan & Panda, 1988) we may define an *Eulerisation parameter*

$$\varepsilon = \langle E \rangle / \langle E \rangle_G$$

where $E = (\mathbf{u} \times \omega)_G^2$, so that $\varepsilon = 0$ if (and only if) the flow is an Euler flow. The value of ε, obtained from direct numerical simulation, then gives a measure of how near to an Euler flow the turbulent flow actually is (on average) in the function space. Kraichnan & Panda found that ε decreased from unity (with a Gaussian initial condition) to ≈ 0.57 in decaying homogeneous turbulence, suggesting a significant tendency towards *Eulerisation.*

The simplest example of an Euler flow is a vortex ring which propagates without change of structure in a fluid at rest at infinity. Such a flow is steady in a frame of reference that moves with the ring, and in this frame $(\mathbf{u} \times \omega)_S = 0$. If we consider two vortex rings with different propagation velocities U_1, U_2, then there is no frame of reference in which the composite flow is steady. This immediately presents a difficulty in using $(\mathbf{u} \times \omega)_S$ as the measure of dynamical interaction between the two rings; somehow we should subtract out the intrinsic propagation velocity of each ring:

$$E' = [(\mathbf{u} - \mathbf{\tilde{U}}) \times \omega]_S^2$$

where $\mathbf{\tilde{U}}(\mathbf{x})$ is the local propagation velocity, U_1 for the ring 1, U_2 for ring 2.

This sort of difficulty has been encountered by Legras et al. (1988) in a study of the final stages of decay of 2-D turbulence when the flow consists of a system of well-separated weakly interacting vortex patches, each patch moving in the velocity field induced by the others. Here, Eulerisation is detected through the extent to which points $(\psi, \nabla^2\psi)$ collapse into a single curve - but again ψ must first be adjusted to allow for the transitional motion of each patch in the field of the others.

The fact that ε is substantially less than unity suggests that turbulence does have a tendency to organise itself into units (which may be identifiable with coherent structures) which are long-lived because they approximate to steady Euler flows; energy cascade to small scales and dissipation of energy is then associated with dynamical interaction of these structures. Identification of the typical structures and of the processes of interaction of structures is then a central objective of turbulence theory.

We propose the following questions in relation to this objective:

Question 1: Is ε the best measure of Eulerisation as conceived above? How should the intrinsic propagation velocity of vortex structures be accommodated? How is the random function $E(\mathbf{x},t)$ correlated with other turbulent variables, e.g. vorticity, dissipation, pressure,...? (Shtilman, 1991). What is the spatial distribution of $E(\mathbf{x},t)$? Does is exhibit intermittent bursts in regions where interaction (vortex reconnection?) occurs?

Question 2: How does $E(\mathbf{x},t)$ or an appropriately adjusted form of E vary in elemental interactions of vortex rings, of the type identified by Aref & Zawadski (this volume)? Does the same *signature* of elemental vortex interactions occur in turbulent flow as revealed by DNS? Does a similar signature appear in the bursts that appear in the transition to turbulence in boundary layers and related flows (Gaster, this volume)?

Question 3: How sensitive are the statistics of E(**x**,t) to the inclusion of other influences, e.g. (i) buoyancy forces in Boussinesq turbulence, (ii) Coriolis forces in turbulence in rotating fluids, (iii) Lorentz forces in conducting fluids in an applied magnetic field, (iv) the effect of polymer additives: reduction of nonlinearity may have an important bearing on the problem of drag reduction in turbulent boundary layers and pipe flows.

References:

Kraichnan, R. H. & Panda, R. (1988), Phys. Fluids, **31**, 2395 - 2397.
Legras, B., Santangelo, P. & Benzi, R. (1988), Europhys. Lett. **5**, 37 - 42
Shtilman, L. (1991), On the solenoidality of the Lamb vector (preprint)

Group IV **Prospects of statistical approaches to the turbulence problem - how statistical methods, e.g. the theory of random fields, should be used in view of recent developments regarding the structure of turbulence, dynamical chaos, strange attractors, fractals**

Participants B. Castaing, Y. Gagne, J. Herring, J. Jimenez, N. Malik, E. Novikov (Chairman), G. Pedrizzetti, B. Rys, M. Vergassola, A. Yaglom, G. Zaslavsky

First, about the applications of the theory of random fields. One of the issues which we found interesting is the problem of non-universality of the local structure of turbulence. It is connected with the phenomena of intermittency, described in terms of break-down coefficients and multifractals. Special attention should be paid to the logarithmically - periodic modulation (lacunarity) and to a physically distinguished scale factor (see E. Novikov, 1990, Phys. Fluids, **A2** (5), 814). This scale factor leads to a generalisation of the concept of similarity and potentially to a deeper understanding of the internal mechanism of turbulent flows.

In connection with this, it will be very interesting to look at various non-Gaussian intermittent random fields, which can be compatible with the Navier-Stokes equations. This can lead to physically reasonable turbulence closures, which can be used in many ways, particularly in large-eddy simulation. This direction of research is important, not only for turbulence modelling, but also for a variety of physical systems with strong interaction.

We also discussed the concept of conditional averaging of local characteristics of motion with self-amplification (vorticity in 3-D and vorticity gradient in 2-D). By using this approach we can, to a certain extent, separate internal mechanism of strong interaction in turbulence from the effects of boundary and initial conditions. Then, we can move to a more general program, by using a small parameter $R^{-1/2}$ to include the effects of boundary and initial conditions. Cooperation with laboratory and field measurement, as well as with numerical simulations, is very important in this program. The concept of conditional averaging also opens a new perspective for sub-grid modelling in large-eddy simulations.

Then, of course, there is a problem of coherent structures, which can potentially be interpreted in terms of collective variables interacting with the sea of "equilibrium" turbulence. There are links between sub-grid modelling, coherent structures and the mathematical theory of inertial manifolds (or approximate inertial manifolds). All three are connected with the reduction of the number of effective degrees of freedom. Detailed investigation of these links can be very fruitful.

Finally, we discussed the imbedding of the theory of turbulence into the theory of Markov processes with dependent increments, consistent with the Navier-Stokes equations. This approach leads to a simple system of ordinary differential equations, which simultaneously gives Lagrangian and Eulerian descriptions of turbulence. (E. Novikov, 1989, Phys. of Fluids, **A1** (2), 326; A Rapid Communication, 1992, Phys. Rev., in press). The effects of stratification and topography can be included in this scheme, which is useful in particular for the environmental problems. At the same time, this simple scheme is consistent with the Navier-Stokes equation, at least up to a moment of third order, and this consistency can be increased by incorporating high order statistics. These are actually all the problems we discussed.

Discussion

Yaglom: What type of structures do you have in mind?

Novikov: The type which was observed in real experiment, in numerical experiment, this kind of vortex coherent structures. Maybe they have an interpretation in terms of collective variables. It is a nontrivial problem because they arise from this sea of small-scale structures, they live for some time and they disappear again into the sea. They cannot be completely separated from turbulent flow, from the small-scale motion, but at the same time, they have some kind of independent existence. So, direct observations, numerical observations can help us to understnad a little bit more about how they appear and how they dissolve, they can help us to construct a theory.

Kida: May I pose a question on the chaos problem. Chaos theory is very successful in the state of low dimensional degrees of freedom, but turbulence has enormous degrees of freedom, and I wonder what is the transition from the low-dimensional chaotic state to fully developed turbulence, and is there any view about the transition?

Aref: That is one of the big questions. I am happy that you said that the state of the chaos theory is now such that it can only deal with low-dimensional systems. One can hope that that state will change. I think that the other question is, do we really know that turbulent flow at every instant in time requires everywhere in space very many degrees of freedom, or is there some notion that maybe the turbulent flow can be broken down into subunits or sets that have a better description? I think that it is interesting to see from this point of view people working in such complicated areas as boundary layers, trying to identify structures within a few wall units. I mean, initially, one might think the boundary is some terribly large uncorrelated mess, but as we know, we hope to find smaller active units that can be used as building blocks. So one might hope, for example, that if there are smaller units in other flows, these smaller units could be addressed using ideas from chaos theory, and then the coupling of these smaller units can also be addressed using ideas of dynamical systems and somehow making a hierarchy of units.

Frisch: One still has to take into account the fact that the simplest structures that we have been capturing so far, like vortices or these vortex tubes etc., are usually accompanied by phenomena which require an infinite dimensional description. For example, when we have interaction of vortices, we observe such things as filamentation, which is very well documented. The filaments which are needed are typical - have a rather complex shape; they can hardly be described as a discrete object. The vortex worms, the spaghettis, are not just straight, they are strings, and strings are infinite dimensional objects. So it is not the same sort of infinity, it's a lower infinity, but still, discrete description seems perhaps an illusion, we may have to work with continous and thus infinite dimensional entities.

Aref: The "entity" might be a thing like the non-linear Schrödinger equation or something like that. I think that it is interesting that we have a very reasonable description of a single filament in terms of an integrable dynamical system, which is the non-linear Schrödinger equation. So maybe one should be looking at some set of coupled non-linear Schrödinger equations.

Frisch: I just want to avoid a misunderstanding. It's not that I am taking issue with the, say, optimistic view continuously expressed by Hassan Aref, that we may reduce these things to an object that can be described in a contracted fashion, but even after the reduction I want to warn not just Hassan and Javier, but other people who might misunderstand, that what remains is an infinite dimensional system.

Zaslavsky: First of all, we have to ask something, and the problem is that we don't know what to ask for an infinite-degrees-of-freedom system. We cannot decide on any kind of proposition in this conext. But one of the questions could be to clarify what spatial-temporal chaos is; which the main problem of turbulence is. We have to answer how chaos works for large Reynolds numbers, for small Reynolds numbers and so on. We have to go step by step. For a small number of degrees of freedom it is possible to study the problem and it is almost clear how to study it. The next step should be to understand how the spatial disorder can get smoothed with time. There is some hope that if there are such quasi substructures, maybe this case can be understood some way. We have to look at these data which exist in California. Maybe that is one of the ways.

Castaing: I don't understand what is ment by describing and reducing the number of degrees of freedom. For example, if one makes a measurement for a given Reynolds number and one computes different moments of various quantities, generally, everybody would expect that making the same with another experiment with the same Reynolds number would give exactly the same statistics for the moments. It's not totally impossible to reduce the description to a small-number of parameter. It depends on what we want to do. For example, what can I do where a fully turbulent state exists, what is homogeneous turbulence, what is inhomogeneous turbulence? That is, I think, what we will have to define.

Moffatt: Between dynamical systems and turbulence, infinite dimensional systems come up again and again. If the dynamical systems approach is ever to be useful, the standard approaches in dynamical systems must be applicable. The first standard approach is to look for the fixed points of the dynamical system. Secondly, to consider the stability of these fixed points and then to go to more sophisticated things to look for periodic orbits and to look at the stability of these and to look at the common homoclinic connections between unstable fixed points etc. Now, we take the very first step in that program for turbulence. What are the fixed points? These are, as I mentioned, the steady solutions of the Euler equations. I looked for the whole family of steady solutions of the Euler equations by techniques I don't want to go into because they would take too long, but the conclusion is this: you give me a flow with arbitrary complex streamline topology and I maintain that there is a steady solution of the Euler equations. In other words, a fixed point at the time we are talking about with that same streamline topology with the added rider that that flow, the Euler flow, has vortex sheets imbedded in it.

Frisch: I think it's a very important remark, which particularly emphasises the huge difference between 2-D and 3-D. In 2-D we get these vortices that are well documented and that can be described to some extent by a system of point vortices, so we have through these links to find the dimensionality. In 3-D, even with simple objects like these tubes, we have interacting strings, which is infinitely more complicated. Those are infinite dimensional Hamiltonian systems.

Moffatt: I think they nearly all have vortex sheets embedded in them somewhere, and they are nearly all unstable. And one example, actually, is the Beltrami flow that George Zaslavsky talked about. I believe all the tube flows with a single length scale are all unstable in regard to the solutions of the Euler equations. That doesn't matter, that makes them more interesting.

Aref: Uriel simplified things a bit too much. There are "zoos" known in 2-D too. If you look at, say, vortices of finite size, there is enormous degeneracy. So, if you said, for example, you were not willing to entertain δ-function singularities or regions of constant vorticity, then the "zoo" there is also very large.
I would also like to reply to Castaing's remark: I tend to distinguish between what the degrees of freedom of the fluid are and things like the statistics, e.g. an energy spectrum, which I see as diagnostics applied to the degrees of freedom. Sometimes this gets a little bit confused in turbulence because one kind of very simple turbulence model is to write a differential equation for the energy spectrum, and this was a game that was played already 30 years ago. It may be the ultimate reduction down to a very small system of differential equation. I think what we are talking about here is not that extreme a reduction. Closures, I think, may be viewed as an extreme reduction down onto a few degrees of freedom. I think we are trying to retain more honest dynamical variables in some sense that reflect the spatial structure rather than a bunch of interacting Fourier modes, because we think that these variables provide a more genuine description of the actual process.

Frisch: This has parallels with the description of social-economical events where one could have a very dramatic reduction. One could describe things like the dynamics of wealth, i.e. the distribution of the number of people who have a fortune between x and dx and then write a couple of differential equations for that, but it is probably better to take into account a few spatial structures like countries and borders. It is never clear how much one can get away with a contracted description. It all depends on what one wants to achieve. In the realm of human events one is rather modest. One does not ask for the ultimate theory. But in turbulence we are still looking for the ultimate theory and that may be again asking for too much.

Summary

W. C. Reynolds and U. Frisch

Summary of the Symposium

William C. Reynolds

I would like to share with you some things that I heard at this meeting, and I heard a great deal. Maybe you can correct me if you did not hear them, that might be useful. We started off talking about intuition and I found this to be a delightful statement and it really sets a tone for the whole meeting, and I think there has been a lot of progress on that. What surprised me most was that I was unaware of the extent of the data and the difference of opinions with respect to what goes on in the cascade-process with all the small scales, etc. I didn't realize how widely shared some of these speculations were. The subject which I noted to be of considerable interest was the difference in views on spectral representations. As I understood what was said earlier in the meeting was that the concept of infinite Reynolds number as a limit may not anymore be a meaningful concept, but the viscosity may act at all scales. I think that is what this graph means, which we saw in a couple of the early papers. Now along comes Keith Moffat with a reminder that spiral structures can also produce something which looks pretty similar to the fractal structure. Does it take a multi-fractal set of scales, a wide range of scales to produce turbulence or can you do it with a single structure and a single scale? These are the different set of views.

I think we had an important morning for the experimentalists. I think the debate is now ranging between the east coast and the west coast of the United States about fractals as found in scalar fields. On possible singularities in the Euler equation, I like the comment which came from the experimentalists with respect to that. Now, with the regards to simulations and experiments I think the work we saw comparing experiments and simulations really is very important. What is done, really, is two things: The validity of simulations for a certain kind of information is established. But more than that, it has established the simulations as standard for comparison of experiments, at least in the wall-region of the turbulent boundary-layer. That's a very important thing to realize that experimentalists are now comparing the simulations with experiments. Talking about experiments, I want to mention some of the techniques that were not discussed at the meeting and a number of techniques, which are coming out and offer some real potential to understand the turbulence. We heard about the particle imaging velocimetry technique, which is going to be very good for very slow subsonic velocity fields. There are techniques, called the planar laser view fluorescence-technique, which was developed principally by Ron Hanson at Stanford. You will be surprised to learn that in addition to measurements of velocities in a plane at an instant in transsonic and supersonic flows you can get temperature fields, pressure fields, density fields and species concentration fields by these techniques, and you get them basically at whatever resolution you can image with a visualisation system. I think these techniques, if applied thoughtfully, can provide a lot of new information on turbulence, especially in hostile situations. Miles' Relief Technique was mentioned; it is an interesting one. Also Henslink at Stanford has a technique, in which he writes holograms in the fluid and reads out the rotation rate from the hologram with an interrogation beam, and this gives a direct measurement of vorticity, or strain rate, without having to do a differentiation. So I think that's going to be a powerful technique for a number of circumstances. I want to return to Mike Gaster's talk, which I thought was really good. I think he had a boundary-layer which only had an inviscid instability, growing slowly. There is a thick layer of some sort which has a disturbance on it which is amplified at the scale of the thickness of the shear-layer, which produces smaller shear layers, which then undergo the same basic processes on their own scales, getting smaller and so forth down the line, and of course the time region gets shorter. I think that is an interesting picture. The problem is that it is a representative picture of at least some of the important processes in turbulence. We saw something similar in Wyganski's experiment. I'll come back to that in a moment. Another thing that we did is to present some compelling evidence that the non-linear behaviour of wave-trains is really that of packets. I came up with a simple way of explaining this to my students, which I will share with you. If you imagine that you have a set of exactly equallly spaced vortices of alterning opposite signs we know. It's only that the slight irregularity in space based on some change which takes place as each pair also wants to know exactly where it was, that you have essentially a packet when you have the three vortices like that you very quickly find the induction which will cause some change to take place. This is a nice way to visualise why the wave-train is slow to respond non-linearly and as opposed to the vast response of packets, statistically perhaps. Let me return to the wall- set experiment just for a moment, which was discussed by Wygnanski, because he did exite the wave-train. This raises some questions: would it behave differently if it were exited by a wave packet? Is there an optimal packet for control? This is an interesting question raised by these two experiments. What he noted is that the shear stress was reduced in this flow when he added the exitation. There is a simple explanation that you can give for this: In the shear-layer the production caused by the coherent motions, which have perhaps some ...... about them, suddenly will be positive, and there is a transfer of the main energy form the main (mean)-field to the wave-field, so the wave-train will be positive, but the mean velocity-gradient in the point where all the action is, at the inflectial point, is negative, and consequently the B-part will be positive if the coherent motion is positive. That joined the negative UB(?) of the and that was impressed on the wall so that the wall sees essentially a reduction of its shear stress. (us). I think this kind of thinking can probably be used to explain other experiments of this general type. We heard a little bit about fractals(?) and coherent structures and simulations and soft worms and soft hairpins and all kind of things of that sort but I think we were hearing a good message at one point. Well, I think I want to say again that I learned a great deal at this meeting. One of the things that I learned was that there is an awful lot known about turbulence that I myself do not yet understand, but I was pleased to learn that I'm not alone.

Uriel Frisch

I will start with something a bit technical. There is a great friend of all of ours, who unfortunately wasn't able to attend, but I'm sure is eavesdropping on us and will enjoy this. One word which has hardly been used in this meeting and which is very dear to Arkady is "helicity". So, here I want to advertise a piece of work which is not very well-known yet, due to Kuz'min from Novosibirsk. Kuz'min's (1983) work was published in a western journal not readily available in the Soviet Union (G. A. Kuz'min, 1983, Phys. Lett. 96A, 88-90) and redicsovered independently by V. Oseledets (V. I. Oseledets, 1989, Moskov. Matemat. Obshch. 44, No. 3 (267), 169-170). I was made aware of Kuz'min's contribution during the Monte Verità meeting by E. Novikov. Kuz'min/Oseledets found that helicity conservation can actually be slightly modified into a pointwise conservation law just like vorticity in two dimensions. It is a very simple thing. The incompressible Navier-Stokes equations in any dimension can be re-written as follows:

$$\partial_t \underline{\gamma} + \underline{v} \cdot \nabla \underline{\gamma} = -\underline{\gamma} \cdot (\nabla \underline{v})^T + \nu \nabla^2 \underline{\gamma}$$
$$\underline{v} = \underline{\gamma} + \nabla \varphi \, ; \qquad \nabla \cdot \underline{v} = 0.$$

In this strange-looking equation, T stands for 'transpose', and you can easily check that the equation for the $\underline{\gamma}$ field is exactly what you would obtain if you were to write the equation for a passively advected scalar at Schmidt number one and then consider the equation as its gradient. But the $\underline{\gamma}$ field is not a gradient here. It is easily checked that the above equations are equivalent to the Navier Stokes equations. Now, consider the well-known vorticity equation in three dimensions:

$$\partial + \underline{\omega} + \underline{v} \cdot \nabla \underline{\omega} = \underline{\omega} \cdot \nabla \underline{v} + \nu \nabla^2 \underline{\omega}$$

You see that the left-hand side has the same structure as above, while in the right-hand side you have a plus sign and no transpose. It immediately follows that the Lagrangian derivative of the dot product $\underline{\gamma} \cdot \underline{\omega}$ reduces to viscous contributions. This means that $\underline{\gamma} \cdot \underline{\omega}$ is conserved along particle trajectories at zero viscosity. Now, $\underline{\gamma} \cdot \underline{\omega}$ is actually related to what we usually call 'helicity'. That is easily checked by considering the integral over the whole domain of $\underline{\gamma} \cdot \underline{\omega}$. Because $\underline{\gamma}$ and $\underline{v}$ differ only by a gradient, it immediately follows that, in the absence of boundaries, the integrals of $\underline{\gamma} \cdot \underline{\omega}$ and of $\underline{v} \cdot \underline{\omega}$ will be equal and the integral of $\underline{v} \cdot \underline{\omega}$ is the total helicity of the flow.

Note that helicity conservation in the presence of boundaries is a problem because if you take an ideal fluid, the condition of the vanishing of the normal component of the velocity at the boundary is not enough to cancel the boundary term obtained in trying to prove the conservation of helicity. Ordinary helicity in an ideal fluid with the usual boundary condition, it is not a conserved quantity, while $\int \underline{\gamma} \cdot \underline{\omega} \alpha^3 r$ is conserved. This is a trivial consequence of the fact that $\underline{\gamma} \cdot \underline{\omega}$ is pointwise conserved, so that boundaries do not interfere with conservation. Actually, the integral of any power of $\underline{\gamma} \cdot \underline{\omega}$ is likewise conserved (just as vorticity in two dimensions). I want to stress that the $\underline{\gamma}$-vector is a rather interesting vector because it sort of gets stretched in a way that it tends to become perpendicular to the vorticity. The vorticity gets stretched by the strain rate, and if we had a completely wrong, naive dynamics, where we could prescribe the strain rate and then just look at the vorticity as a passive vector, it would then align with the direction associated with the largest eigenvalue of the strain rate tensor, whereas the $\underline{\gamma}$-vector under the same non-realistic conditions would then align with the eigenvector corresponding to the lowest eigenvalue. This is of course not what is happening to vorticity, but it is very interesting that this $\underline{\gamma}$-vector is being stretched probably as much as the vorticity or the gradient of a passive scalar. Still, $\underline{\gamma} \cdot \underline{\omega}$ remains unaffected as long as viscosity can be ignored. I suggest that tracking of $\underline{\gamma} \cdot \underline{\omega}$ could be useful as an indicator of viscous reconnection.

Turning to other matters I will discuss some items I find of interest.

One question which was mentioned a few times is that there seems to be some change in the vision of what is happening with the Euler equation of the ideal incompressible fluid in three dimensions. Until about ten years ago it used to be said that the commonly accepted view (with the noticeable exception of G. I. Taylor) was that generically the solution of the three-dimensional Euler equation blows up in finite time. Computer simulations were then done by spectral methods, which are quite reliable. However, such calculations, using a uniform grid, cannot do more than a certain dynamical range of scales. Ten years ago it was somewhere between 64 and 256, depending on the flow symmetries. In 1991 it is between 256 and 864. All the spectral calculations done so far indicate that things are rather boring, that in a reliable computation of this sort you find that you get pancakes and that they flatten down exponentially, and there is no evidence at all that things become singular in finite time. During the last six years, I believe, there have been attempts, mostly by Pumir and Siggia, to get out of these very strong constraints of doing calculations on a uniform grid. The numerics are very difficult. At one point some evidence was reported for a finite-time singularity of the Euler- and possibly even of the Navier-Stokes equation. This was around '85, I believe. Then, more refined calculations indicated again that things flatten exponentially. Very recently, in the preprint of Pumir and Siggia (1991), which Keith Moffat very nicely discussed, reasonable evidence, not overwhelming evidence, is presented suggesting that there may after all be a finite time blow-up for 2 1/2 dimensional turbulence. So that is a very interesting question. It doesn't have too many direct applications because in practice whether something is exponential or blows up after finite time, in the presence of viscosity it hardly makes any difference. It is a challenging mathematical problem with little or no direct consequences for turbulence in the sense of the physicist and even less so in the sense of the engineer.

Now I will turn to the second item. There are actually two related items: Depletion and what Pomeau, Young and others called punctuated Hamiltonian dynamics. Depletion, you remember from yesterday, is this tendency for the dynamically important solutions of high Reynolds number flows in both two and three dimensions to develop structures which have the non-linearity which is cancelling to leading order. The punctuated Hamiltonian dynamics has to do with the interaction of such structures. Suppose you have individual animals in which the non-linearity dies, essentially vanishes, like circular vortices, then you have many such things, and their interaction is to a good approximation describable as a Hamiltonian system. Conservative dynamics play a very important role on such scales because essentially dissipation is negligible. By the way, dissipation is negligible most of the time there, precisely because of depletion. When you have strong non-linearities, then excitation (energy or enstrophy) is being transferred to small scales very quickly, essentially in one turnover time, where it is then dissipated, and you cannot neglect dissipation. But when you have depleted non-linearity, the time scale on which things can be dissipated is much longer, so you can safely ignore dissipation, at least for a relatively long stretch of time. But the punctuated story is the revenge. Occasionally, there are interactions of such depleted animals and these interactions can be extraordinarily violent. You have seen vortex merging. There are other things such as filamentation. Such very short and very rare events actually play a considerable role. They are probably even more frequent in three dimensions. The vortex tubes have a tendency to interact, probably more often than isolated vortices in two dimensions - I would say, for topological reasons. If you have two such things, it is easy for them to cross, easier than for two vortices, and when these interactions take place, of course there is a considerable reorganization of topology, and you have new animals being born, and pieces of vorticity can be ejected all over the place (that is more or less what happens).

My third item relates to small-scale anisotropy. At the meeting the statements I made generated quite some heat (as evidenced by the discussion). Here one will find a more careful phrasing of what I meant. I believe that there is a broad concensus that, at high Reynolds numbers, the smallest scales of the inertial range are isotropic. The largest inertial range scales can be contaminated by almost unavoidable anisotropies of the large energy-carrying eddies. I wish to stress that a similar contamination can take place at very small scales, well within the dissipation range. Indeed, there is an argument analogue to Landau's objection to the universality of the Kolmogorov constant, which indicates that very minute fluctuations in the rate of turbulence production can produce gigantic fluctuations in the far dissipation range. This argument may be found in Kraichnan (1967,

Phys. Fluids 10, 2080) or in Frisch and Morf (1981, Phys. Rev. A23, 2673). Such anisotropies, if they exist, may be hard to detect since they may show up only in high order derivatives of the velocity (maybe not so high for the temperature field?).

I will now turn to an aspect of CFD, namely "visualising what and how". There are several problems I would like to mention. First, that there has been some emphasis on possible artefacts in visualisation. One has to be very careful in interpreting three-dimensional visualisations, to some extent also two-dimensional ones, For example, some of the evidence for vortex tubes has to be carefully analysed because depending on where you put your level surfaces, the vortex tubes may look very circular or not circular at all. I mean if the peak vorticity is achieved along a line, and then you look at the places where you are close to the peak, you can expect to find things which are rather nicely round. If you go further out, they may be less so and it is rather important to know if these vortex tubes are depleted. If they have circular cross-section and are reasonably straight, the non-linear interactions are vanishing there. The only interactions which remain are the interactions with external fields, straining field interaction and interactions between different vortices. So it may be necessary to do a more careful job on deciding where to put the iso-vorticity surfaces, and it is mostly a problem of capturing a sizeable fraction of the circulation. One should not do the contouring at a point where only a small fraction of the circulation of these vortices is captured. Another thing is that people who do computations sometimes - for reasons that have to do with the structure of the code - have to do a bit of extra work to extract the pressure. You should always do that extra work. The pressure is really always a very interesting animal in both two and three dimensions. For example, it can give you an early indication of where vortical structures are going to appear. Although it is not easy to measure pressure, it is easy to visualise it because it is a scalar. So one can still learn a lot from that. A more general observation on CFD: It seems to me that to do a good job in developing new schemes and de-bugging them it is probably better to be a theoretician because you can think of non-trivial tests and things like that. But in order to analyse the results of computation, the best thing is to involve an experimentalist, or at least to have very close links to experimentalists. The quality of what you get and the relevance of the questions which are asked seems to be much greater.

Now I will turn to this all-important question of structures, which we touched upon already before. One specific question is if there are structures in addition to those which were already detected many years ago, and if there are vortex filaments, for example. How to incorporate them into a modified phenomenology of the turbulence in boundary layers? For example, what will be the effect on the so-called log law? Is it sufficiently robust that it will survive or has it to be completely revisited?

Finally, I want to stress that there have been some oscillations in the history of the subject, between the emphasis on structures and the emphasis on a statistical description. Probably the oldest documented descriptions of structures is in the famous drawings by Leonardo da

Vinci, when he was studying the rapids in the river Arno, and that was almost exactly five centuries ago. I am not going to summarise what happened between 1500 and the present time. Everybody knows that for example with von Kármán and even more so with Kolmogorov the emphasis was on the statistical description, although it isn't clear if one should say statistical description or scaling arguments. But the scaling arguments were definitely applied to average quantities, that is statistical aspects. Structure was present, in Kolmogorov in the fuzzy form of 'eddies', mostly in footnotes. With the advent of better techniques for looking at turbulent flows, the pendulum has been swinging back to structures, starting in the sixties (large-scale coherent structures). Siggia started reporting vortex tubes in the late seventies (New York Acad. Sci. meeting, Dec. 1979). Advances in computations and even more in visualisation of CFD output have turned vortex tubes into a hot topic. What is next? Merging structures and statistical mechanics. Even Javier Jimenez, who is by training an engineer, stresses the need to develop a statistical mechanics of vortex tubes.

A word of warning concerning the use of statistical mechanics. If you do statistical mechanics on the wrong object, well you may capture something but also miss a lot. For example you can do statistical mechanics on the Fourrier components of the truncated Euler equation. This interesting game was even started by such respectable people as J.M. Burgers and T.D. Lee . One is learning something. It gives, for example, some hints as to the right directions of cascades, as Kraichnan has shown. But there isn't too much mileage in such approaches because you are doing the statistical mechanics on the wrong objects. You have to first find the relevant object before you can attempt seriously to do statistical mechanics. There are many instances in condensed matter physics where one has this picture: one has been making significant progress only after finding the right things on which to apply statistical tools, including the renormalisation group. For example, I will not rule out that renormalization methods will have a lot to say about turbulence once they are applied to the right objects, which has not been the case so far.

Discussion

Lesieur: I have a question on the singularity at finite time of the Euler equation. I think what you said was not totally complete because you have other evidence of singularities, for instance if you rely upon two-point closures. Since you believe that closures give an answer for the issue of isotropy, you must trust them a little bit.

Frisch: But that's not a compelling argument.

Lesieur: Nearly 20 years ago I showed some exact theorems within closures concerning enstrophy blow-up at finite times. Another point is that there is an exact equation for the enstrophy evolution obtained from the Navier-Stokes equation in terms of the skewness factor. (I'm talking of the velocity derivative skewness.) If this skewness doesn't decrease too fast to zero, then it implies that the enstrophy blows up at a finite time. I think all the direct numerical simulations which were done these last few years, and Jack Herring's is among these, show this.

Frisch: I would be very careful in statements about this very delicate question of singularities and there is some good reason for that. For example, in 1978 - 1979 there were studies of two-dimensional MHD turbulence, which was a possible candidate for a singularity because the presence of the magnetic field has the consequence that vorticity is not conserved any more and actually could increase considerably (A. Pouquet, 1978, J. Fluid Mech. 88, 1 and S. A. Orszag and C. M. Tang, 1979, J. Fluid Mech. 90, 129). When closure was applied to this problem, it gave a finite time singularity. However, two-dimensional problems can be studied on the computer at very high resolutions. When that became feasible, it was found that there is no singularity and the reason is very instructive. (U. Frisch, A. Piuguet, P. I. Julem and M. Meneguzz, 1983, J. Méc. Théor. Appl., special issue, Two-dimensional turbulence, pp. 191 - 216). Suppose you have a point of vanishing magnetic field with two magnetic field lines through it (magnetic X point). If you have a situation like this initially, then very quickly this thing flattens down, the angle between the two field lines decreases exponentially, and you end up having ribbon-like structures, very flat structures, the orientation of which are as random as the initial characteristics of such X points. If you have a situation where the box is much larger than the integral scale, then you have lots of these sheets with random orientations. The reason that it is very difficult to get a singularity, now fully understood on the basis of an asymptotic analysis, is that once things have flattened down to these thin ribons, which are just the equivalent of Betchov's pancakes in two dimension, the non-linearity is strongly reduced. The non-linearity is so strongly depleted, that things are then proceeding only exponentially, where dimensional analysis would predict that they would proceed in an explosive way. And because the ribons are oriented at random, it is very easy to see that if you are working within a framework where you only have second order moments, you will never capture this phenomenon. You have to include at least fourth-order moments. Now, I wouldn't make a statement that a fourth-order closure wasn't be able to capture this, but who wants to do that? So, second-order closures are just not able to capture these strong depletions and therefore they systematically over-estimate the strength of the non-linearity. This is a very important issue. Now, as to the the argument of skewness: as you know, there is an exact relation for homogeneous isotropic turbulence, relating the rate of change of enstrophy and the skewness of the velocity derivative, which you can find in any text-book.

Yaglom: The skewness of what?

Frisch: The skewness of the velocity derivative. As a consequence of this exact relation, if you have a homogeneous *isotropic* turbulence, and if the skewness stays bounded from below by a positive quantity at all times, necessarily you have a singularity at finite time. In computer similations of the 3-D Euler equations, the skewness is usually found to grow at first and then to level off somewhat; it has never been observed to come down and become negative. Is it possible from there to infer, with M. Lesieur, that a singularity will take place after a finite time? Yes, if their behaviour persists for longer times (requiring much higher resolutions). Also, it was pointed out by P.L. Sulem (1991, private communication) that individual realizations of the Euler flow being highly anisotropic (pancakes!), the skewness relation does not apply. But then, I have a question to Marcel Lesieur: he is saying that the presence or not of singularities has some actual physical implication for the study of transition.

Lesieur: Well, I don't know exactly; if we believe in singularities, such as we observe going back to Leray's paper where the vorticity should grow like $\frac{1}{t - t^*}$ or the enstrophy like $\frac{1}{(t - t^*)^2}$, then t* depends of course on the amplitude of the initial perturbation. It is related to the enstrophy of the initial state. It seems to me that you always have a small perturbation, so if you have this type of algebraic singularity, then you will never be able to prevent turbulence from developing, on the other hand, if you have something which grows exponentially, maybe you can expect to always delay the appearance of turbulence.

Yaglom: I am grateful to Uriel Frisch for a number of interesting remarks, which make the discussion more lively. However, some of his statements are unclear to me and I am not sure that everything said is correct. I did not hear earlier that there are experimental data demonstrating the strong anisotropy of the smallest-scale fluctuations accompanied by almost ideal isotropy of fluctuations with scales from the inertial range. It was discovered about twenty years ago (see, e.g. the paper by R. W. Stewart, 1969, Radio Sci., Vol. 4, 12) that the skewness of the temperature derivative differs from zero though it must vanish for an isotropic scalar field. This seemed to be strange since it was known that the -5/3 power law is valid with high accuracy for the temperature spectrum and it was therefore thought that the temperature fluctuations with scales from inertial range are isotropic. Later, however, it was also discovered that temperature structure functions of higher orders in the inertial range are also different from zero; so, apparently, the anisotropy of temperature fluctuations in the inertial range of scales is of the same order of magnitude as the anisotropy in the dissipation range. Nevertheless, it is quite interesting to study carefully the structure of the smallest fluctuations in the far dissipation range. My feeling is that up to now there are no reliable data related to this range and therefore all the discussions about the degree of anisotropy are based only on some speculative theoretical considerations. Of course, if the smallest fluctuations are in fact always anisotropic, then this will force us to revise considerably all the present ideas concerning such fluctuations.

Frisch: No, it doesn't force us at all. That's my point. I wasn't referring to inertial range scales, I was referring to far dissipation scales. The revision that you are referrring to would be necessary only if we had in the inertial range anisotropies, say at scales comparable to the Taylor micro-scale. I am selecting that scale because it is roughly in the middle of the inertial range, so that we are the most distant place from edge-of-inertial-range effects. If we had some evidence - which we don't - that at the Taylor-scale the anisotropy does not go to zero when the Reynolds number goes to infinity, then indeed we would have to revise many ideas, about turbulence. It would be great news, but we don't have such evidence. But don't we have some evidence that at dissipative scales things are anisotropic?

Yaglom: I don't know such evidence. Of course, there is a definite influence of the intermittency on small-scale fluctuations and this influence is strong enough, but how does it produce the small-scale anisotropy? Let me also comment on your remark about the relation between the large-scale structures and the statistical description of turbulence. This was in fact just the topic stressed in my talk at the beginning of the symposium. I think that at present there is too little emphasis on the incorporation of organised structures into the statistical theory of turbulence. At present, all the numerical simulations and the modern experimental techniques make possible to discover a lot of coherent structures, but the question of how to incorporate these structures into the theory has not been answered yet. Many beautiful pictures of organised structures can be found in the available literature, but how must we include these findings into statisticl expression? Do the structures affect such fundamental laws as resistance (i.e. friction) laws and heat and mass transfer laws? Make them contributions to the equations for second-order moments? I mentioned in my talk the works by Tony Perry and his collaborators, which seem to me very interesting - though probably not exactly correct - in their approach to the problem. (The data of numerical simulations including those presented at this symposium do not agree well with the picture of a forest of Λ - shaped vortices postulated by Perry, A. E. and Chong, M. S. (J. Fluid Mech., 1982, Vol. 119, 173-217). But these works try to incorporate the coherent structures into statistical models and they signify the beginning of the new approach to the statistical theory, which, as I hope, will be actively developing in the near future.

Novikov: I want to stress that we have to develop some kind of new statistical mechanics. If you look at the history of aerodynamics, it inspired many new mathematical approaches which have not been known before, because a problem of hydrodynamics is usually so non-trivial that the old approaches just give us nothing. So we have to develop quite new concepts, new approaches, and I think that the notion that these turbulence structures are a very important part of turbulent flow have inspired many people, not only experimentalists and theoreticians, but also pure mathematicians, etc., to develop quite new and very powerful statistical mechanics, which include these turbulent structures as an essential part.

Adrian: It seems to me that there is a fairly clear linkage between coherent structures and statistical theories when you use conditional averages of the type that Novikov mentioned for velocity fields. The question is, how does one link conditional averages to what we know in terms of structure functions. It has been shown that the stochastic estimation process does that by transforming the correlation functions into vector fields. For example, stochastic estimation of channel flow yields structures that consist of streaks on the wall and streamwise vortices and hairpin vortices and vortex rings far away from the wall. All that information is contained in the classical correlation function for the channel flow.

Moffatt: I would like to make a brief comment about the theorem of Kuz'min and Oseledets that Frisch referred to. There is the very useful analogy between velocity field and magnetic field and between vorticity and magnetic field in a perfectly conducting fluid. The magnetic field is a frozen-in field and it is a solenoidal field and one can therefore introduce a vector potential A for that field and you have a certain freedom in the choice of the gauge of A. In fact, Elsässer in the late 1940's showed that there was a particular choice of the gauge of A which was such that $\frac{d}{dt}(A \cdot B)$ is equal to zero. So it is not unlike the analogue situation in fluid-mechanics. Now, in fact, Woltjer went on to show that the integral of $A \cdot B$ is a constant inside any material surface, on which $B \cdot n$, the normal component, is equal to zero. Then this is the magnetic helicity invariant. I think it's a very important thing that this invariant is actually also a guage invariant, and if you replace A by A plus, the gradient of any scalar, then the value of the magnetic field is unaltered, its gauge invariant also.

Frisch: Hassan Aref showed us the other day an instance, a bit contrived but still consistent with the equations, where two unlinked vortices managed to link so that there is a change in helicity. (As Keith Moffat showed years ago, the linkage of vortices directly measures the helicity.)
In this context let me recall that in the seventies André and Lesieur used closure to show that you can have a situation with no helicity initially and where helicity is generated by *dissipation*. It's just that helicity, not being a definite positive quantity, doesn't have to be decreasing in time, like energy. So, due to the viscous effect, helicity can actually be generated. But of course such flows are not parity-invariant, they just happen to have a vanishing total helicity. And I suppose that the case considered by Hassan didn't have any centre of symmetry.

Aref: A couple of remarks on this γ-field. I think it's actually rather interesting; I'm not sure, it doesn't implicitly arise in some of the numerical simulations because it's essentially like a vorticity field that doesn't have divergence, and actually what surprises me is that vortex methods at an intermediate stage have something that is advected with stress like a vorticity field and you have to remove the non-divergence of the parts. So I am not sure that this may not be useful in the context of numerical analysis.
The other remark that was made was that when one wants to do statistical mechanics, one should make sure that one doesn't do it on the wrong object. It reminded me of two things: one was a paper written by Onsager in 1949 (L. Onsager, Nuovo Cimento, Suppl. 6, 274) in which he did statistical mechanics of singular vortices and used that argument to argue that large coherent structures would form particularly in two-dimensional flows. I think the year 1949 is quite interesting because similar predictions based on statistical theories using Fourier modes were not made for at least another 15 years.

Frisch: No, Burgers did it earlier, in the twenties (J. M. Burgers, 1929, Verh. K. Akad. Wet. Amsterdam, 32, 643). But I don't think he was aware of all the consequences.

Aref: The other thing which I was reminded of was a conversation I had with Joe Keller a couple of years ago, which didn't have so much to do with doing statistical mechanics on the wrong object, but doing statistical mechanics with the wrong assumptions. Joe Keller's idea was that the statistics of turbulent flow might be obtainable by doing statistical mechanics using ensembles that were more general than the ones we usually use in equilibrium statistical mechanics, i.e. in two-dimensional flows. Instead of using an ensemble in which you put in conservation of energy and enstrophy, he wanted to put in the conservation of energy, enstrophy and energy transport. He claimed that if you did that, you could actually pull out the k^{-3} spectral range. He also tried to do that in three dimensions, but it was not nearly as successful. But he at least suggested that maybe generalizing the idea of how you could do statistical mechanics and thinking about a bit more general ensembles rather than the ones we know from equilibrium statistical mechanics...

Stihlman: Some comments related to helicity: Even in the direct numerical simulation one can support what was said about the work of André Lesieur. If you start from zero initial helicity, then you will have positive or negative helicity while fluctuations of total helicity, as far as I observed, are never large, it is just normal fluctuations of total helicity.
Another thing is related to Keith's comments about the relation between helicity and structures. Of course it sounds reasonable that they are related, but helicity is well-defined on a particular surface and if you want to indentify helicity with any particular structure, you have to define this proper surface, and this in practice would be very difficult. So it would be very difficult to pull the helicity of any particular structure if you want to deal with a well-defined flow.

Dracos: Further questions or comments? We shall close here, and I think with this we can also close this symposium as there is no other activity planned. What I want to do here is to thank the people sitting down there, who really helped to make this symposium possible, especially Drs. Studerus, Gyr, Shen, Hoyer, Virant. I hope they managed to register everything which was said here.

Lesieur: I think we have to also thank Prof. Dracos for organizing this very interesting and stimulating meeting. (Applause)

Frisch: Don't forget Arkady.

ZAMP

Zeitschrift für angewandte Mathematik und Physik
Journal of Applied Mathematics and Physics
Journal de Mathématiques et de Physique appliquées

The ***Journal of Applied Mathematics and Physics (ZAMP)*** publishes papers of high quality in fluid mechanics, mechanics of solids and differential Equations / applied mathematics. The readers of ZAMP will find not only articles in their own special field but also original work in neighbouring domains. This will lead to an exchange of ideas; concepts and methods which have proven to be successful in one field may well be useful to other areas. ZAMP attempts to publish articles reasonably quickly. Longer papers are published in the section "Original Papers", shorter ones may appear under "Brief Reports". Publication of contributions to the "Brief Reports" section is particularly rapid.
The journal includes a "Book Review" section and provides information on activities (such as upcoming symposia, meetings, special courses) which are of interest to its readers.

Selected and forthcoming articles:
H.W. Broer, S.N. Chow and Y. Kim, Groningen: *A normally elliptic Hamiltonian bifurcation*
C.F. Delale, G.H. Schnerr and J. Zierep, Karlsruhe: *The mathematical theory of thermal choking in nozzle flows*
B. Dionne, Waterloo: *Planforms in three dimensions*
V.I.Fabrikant, Montreal: *Continuation of some solutions in potential theory and their applications to elastic contact and crack problems*
Z. Feng, Santa Barbara and S. Wiggins, Passadena: *On the existence of chaos in a class of two-degree-of-freedom, damped, strongly parametrically forced mechanical systems with broken O(2) symmetry*
Ch. Lubich, Würzburg: *Integrating of stiff mechanical systems by Runge-Kutta methods*
D.W. Schwendemann, M.C.A. Kropinski and J.D. Cole, Troy: *On the construction and calculation of optimal nonlifting critical airfoils*

Subscription information:
1993 subscription,
volume 44 (6 issues)
sFr. 728.– / DM 866.– / US$ 498.00
(plus postage & handling)
Single copy:
sFr. 139.– / DM 164.– / US$ 98.00
(plus postage & handling)

Editors:
U. Kirchgraber, Zürich
M. Renardy, Blacksburg
I.L. Ryhming, Lausanne
M. Sayir, Zürich

Assistant Editor:
K. Nipp, Zürich

Editorial Office:
ZAMP, P.O.Box 274
CH-4024 Basel
Switzerland

Co-editors:
F. Bark, Stockholm
F.G. Blottner, Albuquerque
S.R. Bodner, Haifa
F.H. Busse, Bayreuth
G.–Q. Chen, Chicago
T.K. Fannelöp, Zürich
H. Gao, Stanford
J. Glimm, Stony Brook
M. Hiller, Duisburg
T. Maxworthy, Los Angeles
P.A. Monkewitz, Los Angeles
A. Moser, Zürich
P.M. Naghdi, Berkeley
N. Rott, Palo Alto
N.M. Temme, Amsterdam
H. Thomann, Zürich
J. Waldvogel, Zürich

Birkhäuser

Birkhäuser Verlag AG
Basel · Boston · Berlin

You may order through your subscription agency or directly from the publisher:

Birkhäuser Verlag AG, P.O. Box 133, CH-4010 Basel / Switzerland, Fax: ++41 / 61 / 721 76 66

GPSR Compliance

The European Union's (EU) General Product Safety Regulation (GPSR) is a set of rules that requires consumer products to be safe and our obligations to ensure this.

If you have any concerns about our products, you can contact us on ProductSafety@springernature.com

In case Publisher is established outside the EU, the EU authorized representative is:

Springer Nature Customer Service Center GmbH
Europaplatz 3
69115 Heidelberg, Germany

Batch number: 10726517

Printed by Printforce, the Netherlands